FUNDAMENTALS OF MATRIX COMPUTATIONS

FUNDAMENTALS OF MATRIX COMPUTATIONS

Edited by:

Olga Moreira

ARCLER PRESS

www.arclerpress.com

Fundamentals of Matrix Computations

Olga Moreira

Arcler Press
2010 Winston Park Drive,
2nd Floor
Oakville, ON L6H 5R7
Canada
www.arclerpress.com
Tel: 001-289-291-7705
001-905-616-2116
Fax: 001-289-291-7601
Email: orders@arclereducation.com

ISBN: 978-1-77407-144-1 (Hardcover)

DECLARATION

Some content or chapters in this book are open access copyright free published research work, which is published under Creative Commons License and are indicated with the citation. We are thankful to the publishers and authors of the content and chapters as without them this book wouldn't have been possible.

ABOUT THE EDITOR

Olga Moreira obtained her Ph.D. in Astrophysics from the University of Liege (Belgium) in 2010, her BSc. in Physics and Applied Mathematics from the University of Porto (Portugal). Her post-graduate travels and international collaborations with the European Space Agency (ESA) and European Southern Observatory (ESO) led to great personal and professional growth as a scientist. Currently, she is working as an independent researcher, technical writer, and editor in the fields of Mathematics, Physics, Astronomy and Astrophysics.

TABLE OF CONTENTS

LIST OF CONTRIBUTORS

Dan-Daniel Erdmann-Pham
Mathematics Department, Jacobs University, Bremen, 28759, Germany

Aviv Gibali
Mathematics Department, ORT Braude College, Karmiel, 21982, Israel

Karl-Heinz Küfer
Optimization Department, Fraunhofer - ITWM, Kaiserslautern, 67663, Germany

Philipp Süss
Optimization Department, Fraunhofer - ITWM, Kaiserslautern, 67663, Germany

Duanmei Zhou
College of Mathematics and Computer Science, Gannan Normal University, 341000 Ganzhou, People's Republic of China

Guoliang Chen
Department of Mathematics, East China Normal University, 200241 Shanghai, People's Republic of China

Guoxing Wu
Department of Mathematics, Northeast Forestry University, 150040 Harbin, People's Republic of China

Xiaoyan Chen
Library, Gannan Normal University, 341000 Ganzhou, People's Republic of China

A. H. Refahi Sheikhani
Department of Applied Mathematics, Faculty of Mathematical Sciences, Lahijan Branch, Islamic Azad University, Lahijan, Iran

S. Kordrostami
Department of Applied Mathematics, Faculty of Mathematical Sciences, Lahijan Branch, Islamic Azad University, Lahijan, Iran

Muhammad O Bin Saeed
Electrical Engineering Department, King Fahd University of Petroleum & Minerals, Dhahran 31261, Saudi Arabia

Azzedine Zerguine
Electrical Engineering Department, King Fahd University of Petroleum & Minerals, Dhahran 31261, Saudi Arabia

Salam A Zummo
Electrical Engineering Department, King Fahd University of Petroleum & Minerals, Dhahran 31261, Saudi Arabia

Mahdi Tohidian
Department of Electrical Engineering, Amirkabir University of Technology, Tehran, Iran

Hamidreza Amindavar
Department of Electrical Engineering, Amirkabir University of Technology, Tehran, Iran

Ali M Reza
Electrical Engineering and Computer Science, University of Wisconsin-Milwaukee, Milwaukee, WI 53201–0784, USA.

Lu Wang
INSERM, UMR 1099, F-35000 Rennes, France
LTSI, Université de Rennes 1, Bât. 22, Campus de Beaulieu, F-35000 Rennes, France
Centre de Recherche en Information Biomédicale Sino-Français (CRIBs), F-35000 Rennes, France

Laurent Albera
INSERM, UMR 1099, F-35000 Rennes, France
LTSI, Université de Rennes 1, Bât. 22, Campus de Beaulieu, F-35000 Rennes, France

Centre de Recherche en Information Biomédicale Sino-Français (CRIBs), F-35000 Rennes, France
Centre INRIA Rennes-Bretagne Atlantique, Bât. 12G, Campus de Beaulieu, F-35042 Rennes, France

Amar Kachenoura
INSERM, UMR 1099, F-35000 Rennes, France
LTSI, Université de Rennes 1, Bât. 22, Campus de Beaulieu, F-35000 Rennes, France
Centre de Recherche en Information Biomédicale Sino-Français (CRIBs), F-35000 Rennes, France

Huazhong Shu
Laboratory of Image Science and Technology, School of Computer Science and Engineering, Southeast University, Qun Xian Building, No. 2 Sipailou, 210096 Nanjing, China
Centre de Recherche en Information Biomédicale Sino-Français (CRIBs), F-35000 Rennes, France

Lotfi Senhadji
INSERM, UMR 1099, F-35000 Rennes, France
LTSI, Université de Rennes 1, Bât. 22, Campus de Beaulieu, F-35000 Rennes, France
Centre de Recherche en Information Biomédicale Sino-Français (CRIBs), F-35000 Rennes, France

Hong Sun
School of Electronic Information, Wuhan University, Luojia Hill, Wuhan 430072, China
Signal and Image Processing Department, Telecom ParisTech, 46 rue Barrault, 75013 Paris, France

Cheng-wei Sang
School of Electronic Information, Wuhan University, Luojia Hill, Wuhan 430072, China

Didier Le Ruyet
CEDRIC Laboratory, CNAM, 292 rue Saint Martin, 75003 Paris, France

Jicheng Li
School of Mathematics and Statistics, Xi'an Jiaotong University, No. 28, Xianning West Road, Xi'an, 710049, China

Zisheng Liu
School of Mathematics and Statistics, Xi'an Jiaotong University, No. 28, Xianning West Road, Xi'an, 710049, China
School of Statistics, Henan University of Economics and Law, No. 180, Jinshui East Road, Zhengzhou, 450046, China

Guo Li
College of Mathematics and Statistics, Shenzhen University, No. 3688, Nanhai Ave, Shenzhen, 518060, China

Eduardo L. O. Batista
LINSE—Circuits and Signal Processing Laboratory, Department of Electrical and Electronics Engineering, Federal University of Santa Catarina, Campus Universitário Trindade, Florianópolis, Brazil

Rui Seara
LINSE—Circuits and Signal Processing Laboratory, Department of Electrical and Electronics Engineering, Federal University of Santa Catarina, Campus Universitário Trindade, Florianópolis, Brazil

Ruiping Wen
Key Laboratory of Engineering & Computing Science, Shanxi Provincial Department of Education/Department of Mathematics, Taiyuan Normal University, Jinzhong, P.R. China

Shuzhen Li
Department of Mathematics, Taiyuan Normal University, Jinzhong, P.R. China

Yonghong Duan
Department of Mathematics, Taiyuan University, Taiyuan, P.R. China

Grigorios Tsagkatakis
Institute of Computer Science, Foundation for Research & Technology - Hellas (FORTH), Crete, Greece

Baltasar Beferull-Lozano
Lab Intelligent Signal Processing & Wireless Networks (WISENET), Department of Information and Communication Technologies, University of Agder, Grimstad, Norway

Panagiotis Tsakalides
Institute of Computer Science, Foundation for Research & Technology - Hellas (FORTH), Crete, Greece
Department of Computer Science, University of Crete, Crete, Greece

Liejun Xie
Department of Mathematics, Ningbo University, Fenghua Road 818, Jiangbei District, Ningbo City 315211, Zhejiang Province, People's Republic of China

Cailian Zhou
Department of Mathematics, Ningbo University, Fenghua Road 818, Jiangbei District, Ningbo City 315211, Zhejiang Province, People's Republic of China

Song Xu
Department of Mathematics, Ningbo University, Fenghua Road 818, Jiangbei District, Ningbo City 315211, Zhejiang Province, People's Republic of China

LIST OF ABBREVIATIONS

ATC	Adapt-Then-Combine
ADM	Adomian decomposition method
ACDC	Alternating columns and diagonal center
ADMM	Alternating Directions Method of Multipliers
ALS	Alternating least squares
ALM	Augmented Lagrange multiplier
BSS	Blind source separation
BSM	B-spline method
CS	Compressed Sensing
CFP	Convex Feasibility Problem
CSM	Cubic spline method
DTM	Differential transform method
DRC	Diffusion blind block recursive Cholesky
DFT	Discrete Fourier transform
EVD	Eigenvalue decomposition
ELS	Enhanced line search
EM	Expectation Maximization
FFDIAG	Fast Frobenius diagonalization
FDM	Finite difference method
FIR	Finite impulse response
FFDM	Fourth order finite difference method
GSVD	Generalized singular-value decomposition
HO	Higher order
HOSVD	singular value decomposition
IDTM	Improved differential transform method
ICA	Independent component analysis

INDSCAL	Individual differences in scaling
IoT	Internet-of-Things
LMS	Least-mean-square
MRS	Magnetic resonance spectroscopy
MC	Matrix Completion
MALM	Modified Lagrange multiplier
MIMO	Multiple-input multiple-output
MSSA	Multivariate Singular Spectrum Analysis
NN-COMP	Nonnegative compression algorithm
NMF	Nonnegative matrix factorization
NTF	Nonnegative tensor factorization
NPCSM	Non-polynomial cubic spline method
PSNR	Peak signal-to-noise ratio
PCA	Principal Component analysis
PPB	Probabilistic patch-based
RPCA	Robust principal component analysis
SSALM	Semi-smoothing augmented Lagrange multiplier
SEM	Series expansion technique
SSMs	Signal subspace methods
SNR	Signal-to-noise ratios
SBVPs	Singular boundary value problems
SSA	Singular Spectrum Analysis
SS-MC	Singular Spectrum Matrix Completion
SVD	Singular value decomposition
SVH	Singular Value Homogenization
SSIM	Structural similarity index metric
PCA	Principal component analysis
PPB	Probabilistic patch-based
RC	Recursive Cholesky
RPCA	Robust principal component analysis
SSALM	Semi-smoothing augmented Lagrange multiplier
SISO	Single-input-single-output

SBVPs	Singular boundary value problems
SSA	Singular Spectrum Analysis
SS-MC	Singular Spectrum Matrix Completion
SVD	Singular value decomposition
SVH	Singular Value Homogenization
SALM	Smoothing augmented Lagrangemultiplier
SSIM	Structural similarity index metric
TSM	Taylor series method
TMC	Toeplitz matrix completion
VFFRS	Variable forgetting factor RS
VFF	Variable forgetting factor
VIM	Variational iteration method
WSN	Wireless sensor network

PREFACE

Matrix Computations research is mainly concerned with developing of fast processing and finite precision algorithms that use properties of vectors and matrices to automatically solving problems of continuous mathematics. Algorithms based on matrix decomposition, for instance, are often used for solving linear systems of equations, locating eigenvalues, performing least squares optimization, modeling differential equations. Matrix computations metholodies are used to solve a wide range of engineering and computational science problems such as signal processing, telecommunication, fluid dynamics, materials science simulations, data mining, bioinformatics.

The following specific numerical methods and matrix-based algorithms are included in this book:

- a singular value homogenization method for solving convex optimization (Chapter 1) ;
- perturbation bounds for eigenvalues of diagonalizable matrices and for singular values of square matrices (Chapter 2);
- iterative methods for computing generalized singular and corresponding vectors of large sparse matrices (Chapter 3);
- two algorithms based on singular value decomposition and Cholesky factorization for blind signal estimation (Chapter 4).
- Discrete Fourier Transform methods for singular value and eigenvalue decomposition of polynomial matrices (Chapter 5);
- LU and QR matrix factorizations for solving the canonical polyadic decomposition problem of semi-nonnegative semi-symmetric matrices (Chapter 6);
- the sparse signal subspace decomposition (3SD) method that can be used for feature extraction, solving inverse problems, or machine learning (Chapter 7).
- a singular value thresholding algorithm for low-rank matrix approximation problem (Chapter 8);
- a reduced-rank method that uses the singular value decomposition

for obtaining reduced-complexity implementations of Volterra filters (Chapter 9).

- the semi-smoothing augmented Lagrange multiplier (SSALM) algorithm for completing a low-rank Toeplitz matrix (Chapter 10);
- a singular spectrum matrix completion (SS-MC) algorithm for simultaneous the recovery of low-rank matrices from a minimal set of measurements and the prediction of future behavior in the absence of complete measurement sets (Chapter 11);
- an improved differential transform method that can solve singular boundary value problems based on the decomposition of Adomian polynomial matrices (Chapter 12);

This edited book is directed towards the numerical linear algebra community, including computational scientists, engineers or anyone whose research work requires the solution to a matrix problem.

Chapter 1

Singular Value Homogenization: a Simple Preconditioning Technique for Linearly Constrained Optimization and its Potential Applications in Medical Therapy

Dan-Daniel Erdmann-Pham[1] , Aviv Gibali[2], Karl-Heinz Küfer[3] and Philipp Süss[3]

[1]Mathematics Department, Jacobs University, Bremen, 28759, Germany

[2]Mathematics Department, ORT Braude College, Karmiel, 21982, Israel

[3]Optimization Department, Fraunhofer - ITWM, Kaiserslautern, 67663, Germany

ABSTRACT

A wealth of problems occurring naturally in the applied sciences can be reformulated as optimization tasks whose argument is constrained to the solution set of a system of linear equations. Solving these efficiently

Citation (APA): Erdmann-Pham, D. D., Gibali, A., Küfer, K. H., & Süss, P. (2016). Singular Value Homogenization: a simple preconditioning technique for linearly constrained optimization and its potential applications in medical therapy. Journal of Mathematics in Industry, 6(1), 1. (11 pages), DOI: https://doi.org/10.1186/s13362-016-0017-5

typically requires computation of feasible descent directions and proper step sizes-the quality of which depends largely on conditioning of the linear equality constraint. In this paper we present a way of transforming such ill-conditioned problems into easily solvable, equivalent formulations by means of directly changing the singular values of the system's associated matrix. This transformation allows us to solve problems for which corresponding routines in the LAPACK library as well as widely used projection methods converged either very slowly or not at all.

Keywords: singular value decomposition; ill-conditioned matrix; projection methods; linear least squares; spectral regularization

INTRODUCTION

In many experimental settings the information[a] $z \in \mathbb{R}^n$ to be processed and analyzed computationally is obtained through measuring some real world data $x \in \mathbb{R}^m$. The action of performing such measurement oftentimes introduces distortions or errors in the real data which, given that the distortion $A : \mathbb{R}^m \to \mathbb{R}^n$ is known, may be inverted to recover the original data. A particularly common case (e.g. in image processing, dose computationb or convolution and deconvolution processes in general [1,2]) occurs when this relation A between measurements and data is in fact linear or easily linearizable, i.e. if $A \in \mathbb{R}^{m \times n}$.

It is thus natural to consider the following optimization problem

$$\min_{x \in \mathbb{R}^m} f(Ax), \quad (1.1)$$

where $f : \mathbb{R}^n \to \mathbb{R}$ is a continuously differentiable function and A is a real $m \times n$ matrix. Typical (first order) approaches for solving (1.1) involve estimates of the gradient, see for example the classical works of Levitin and Polyak [3], Goldstein and Tretyakov [4] and more recent and related results [5,6]. Hence there is the need to evaluate the term

$$\nabla_x f(Ax) = A^T \cdot \nabla_z f(z), \quad (1.2)$$

where $z=Ax$. In the case of ill-conditioned A, (1.2) gives only little information and hence long run-times ensue, see also [7,8].

The purpose of this paper is introduce a new preconditioning process through altering the singular value spectrum of A and then transforming (1.1) into a more benign problem. Our proposed algorithmic scheme can be used as a preconditioning process in many optimization procedures; but due to their simplicity and nice geometrical interpretation we focus here on*Projection Methods*. For related work using preconditioning in optimization with applications see [9,10] and the many references therein.

The paper is organized as follows. In Section2we present some preliminaries and definitions that will be needed in the sequel. Later, in Section3the new Singular Value Homogenization (SVH) transformation is presented and analyzed. In Section4we present numerical experiments to linear least squares and dose deposition computation in IMRT; these results are conducted and compared with LAPACK solvers and projection methods. Finally we summarize our findings and put them into larger context in Section5.

PRELIMINARIES

In our terminology we shall always adhere to the subsequent definitions. We denote by $C^1(\mathbb{R}^m)$ the set of all continuously differentiable functions $f:\mathbb{R}^m \to \mathbb{R}$.

Definition 2.1

Let $a \in \mathbb{R}^n$, $a \neq 0$ and $\beta \in \mathbb{R}$, then $H_-(\alpha,\beta)$ is called ahalf-space, and it is defined as

$$H_-(\alpha,\beta) := \{z \in \mathbb{R}^n \mid \langle a,z\rangle \leq \beta\}. \qquad (2.1)$$

When there is equality in (2.1) then it is called ahyper-planeand it is denoted by$H(\alpha,\beta)$.

Definition 2.2

Let$C$be non-empty, closed and convex subset of$\mathbb{R}^n$. For any point$x \in \mathbb{R}^n$, there exists a point$P_C(x)$in$C$that is the unique point in$C$closest to$x$, in the sense of the Euclidean norm; that is,

$$\|x - P_C(x)\| \leq \|x-y\| \quad \text{for all } y \in C. \qquad (2.2)$$

The mapping$P_C:\mathbb{R}^n \to C$is called theorthogonalormetric projectionof $\mathbb{R}^n$onto$C$. The metric projection$P_C$is characterized [11], Section3, by the following two properties:

$$P_C(x) \in C \qquad (2.3)$$

and

$$\langle x - P_C(x), P_C(x) - y\rangle \geq 0 \quad \text{for all } x \in \mathbb{R}^n, y \in C, \tag{2.4}$$

where equality in (2.4) is reached, if$C$is a hyper-plane.

A simple example when the projection has a close formula is the following.

Example 2.3

The*orthogonalprojection of a point*$x \in \mathbb{R}^n$ onto $H_{-}(\alpha, \beta)$is defined as

$$P_{H_{-}(\alpha,\beta)}(x) := \begin{cases} x - \frac{\langle a,z\rangle - \beta}{\|a\|^2} a & \text{if } \langle a, x\rangle > \beta, \\ x & \text{if } \langle a, x\rangle \leq \beta. \end{cases} \tag{2.5}$$

Projection Methods

Projection methods (see, e.g., [12,13,14]) were first used to solve systems of linear equations in Euclidean spaces in the 1930s and were subsequently extended to systems of linear inequalities. The basic step in these early algorithms consists of a projection onto a hyper-plane or a half-space. Modern projection methods are more sophisticated and they can solve the general Convex Feasibility Problem (CFP) in a Hilbert space, see, e.g., [15].

In general, projection methods are iterative algorithms that use projections onto sets while relying on the general principle that when a family of (usually closed and convex) sets is present, then projections onto the given individual sets are easier to perform than projections onto other sets (intersections, image sets under some transformation, etc.) that are derived from the given individual sets. These methods have a nice geometrical interpretation, moreover their main advantage is low computational effort and stability. This is the major reason they are so successful in real-world applications, see [16,17].

As two prominent classical examples of projection methods, we avail the Kaczmarz [18] and Cimmino [19] algorithms for solving linear systems of the form$Ax=b$as above. Denote by$a^i$the$i$th row of$A$. In our presentation of these algorithms here, they are restricted to exact projection onto the corresponding hyper-plane while in general relaxation is also permitted.

Algorithm 2.4

(Kaczmarz method)

Step 0::

Letx^0*be arbitrary initial point in*$\mathbb{R}^n$*,and set*$k=0$.

Step 1::

Given the current iterate x^k*, compute the next iterate by*

$$x^{k+1} = P_{H(a^i,b_i)}\left(x^k\right) := x^k + \frac{b_i - \langle a^i, x^k \rangle}{\|a^i\|^2} a^i, \tag{2.6}$$

where $i = k \bmod m + 1$.

Step 2::

Set $k \leftarrow (k+1)$ *and return to Step* 1.

Algorithm 2.5

(Cimmino method)

Step 0::

Let x^0 *be arbitrary initial point in* $\mathbb{R}^n$*, and set* $k=0$.

Step 1::

Given the current iterate x^k*, compute the next iterate by*

$$x^{k+1} := \frac{1}{n} \sum_{i=1}^{n} \left(x^k + 2\frac{b_i - \langle a^i, x^k \rangle}{\|a^i\|^2} a^i \right). \tag{2.7}$$

Step 2::

Set $k \leftarrow (k+1)$ *and return to Step* 1.

Moreover, in order to develop the process by which we improve a matrix's condition, understanding of the following concepts is essential.

Definition 2.6

Let *A* be an $m \times n$ real (complex) matrix of rank *r*. The singular value decomposition of *A* is a factorization of the form $A = U\Sigma V^*$ where *U* is an $m \times m$ real or complex unitary matrix, Σ is an $m \times n$ rectangular diagonal matrix with non-negative real numbers on the diagonal, and V^* is an $n \times n$ real or complex unitary matrix. The diagonal entries σ_i of Σ, for which holds $\sigma_1 \geq \sigma_2 \geq \cdots \geq \sigma_r > 0 = \sigma_{r+1} = \cdots = \sigma_n$, are known as the singular values of *A*. The *m* columns $\langle u_1, \ldots, u_m \rangle$ of *U* and the *n* columns $\langle v_1, \ldots, v_n \rangle$ of *V* are called the left-singular vectors and right-singular vectors of *A*, respectively.

Definition 2.7

The condition number $\kappa(A)$ of an $m \times n$ matrix *A* is given by

$$\kappa(A) = \frac{\sigma_1}{\sigma_r} \quad (2.8)$$

and is a measure of its degeneracy. We speak of A being well-conditioned if $\kappa(A) \approx 1$ and the more ill-conditioned the farther away $\kappa(A)$ is from unity.

SINGULAR VALUE HOMOGENIZATION

The ill-conditioning of a linear inverse problem $Ax=z$ is directly seen in the singular value decomposition (SVD) $A=U\Sigma V^T$ of its associated matrix, namely as the ratio of $\sigma_{max}/\sigma_{min}$. Changes in the data along the associated first and last right singular vectors (or more generally along any two right singular vectors whose ratio of corresponding singular values is large) are only reflected in measurement changes along the major left singular vector-which poses challenges in achieving sufficient accuracy with respect to the minor singular vectors.[3]

A new geometrical interpretation of the above can be described in the language of projection methods. This conflicting behavior along singular vectors corresponds to projections onto hyper-planes whose normal vectors are to a high degree identically aligned, i.e. for any two such normal vectors $n_1, n_2 \in \mathbb{R}^n$ their dot product is close to unity. A toy example for $A \in \mathbb{R}^{3\times 2}$ that will be used for visualization is provided on the left in Figure1.

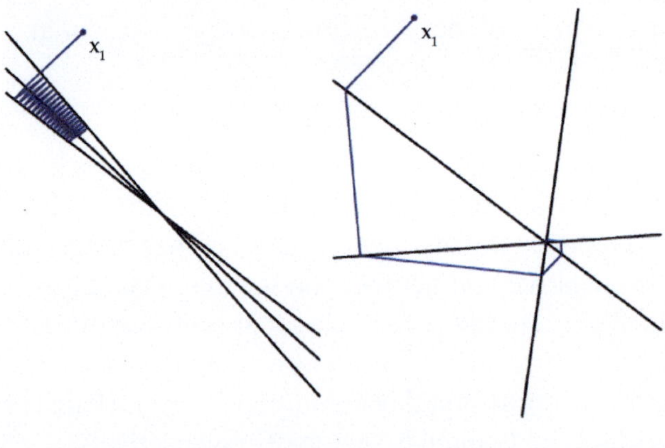

Figure1: Illustration of ill-conditioning (left) as a challenge to linear comparing to the preconditioning step (SVH) described herein (right).

Such high degree of alignment poses challenges to classical projection methods since the progress made in each iteration is clearly humble. A much more favorable situation applies when the normal vectors' directions are spread close to evenly over the unit circle so as to lower the conditioning of the problem. The system depicted on the right in Figure1is obtained from the previous ill-conditioned one through the easily invertible Singular Value Homogenization (SVH) transformation (described below) and visibly features such better condition. Also plotted is the progress made by the classical Kaczmarz projection method which confirms the improved run-time (left: first 50 iterations without convergence, right: convergence after seven steps).

The Transformation

To achieve better condition number $\kappa(A)$ of A we directly manipulate its SVD through introducing the SVH matrix $\Gamma = \mathrm{diag}(\gamma_1, \gamma_2, \ldots, \gamma_n) \in \mathbb{R}^{n\times n}$ $(\gamma_i \neq 0)$ to multiply the singular values $(\sigma_1, \sigma_2, \ldots, \sigma_r)$, where $r \leq \min\{n,m\}$ is the rank of A:

$$\tilde{A} = U\Sigma\Gamma V^T. \tag{3.1}$$

By proper choice of $\gamma_1, \ldots, \gamma_r$, the singular values $\tilde{\sigma}_1, \ldots, \tilde{\sigma}_r$ of $\tilde{A}$ can be set to any arbitrary values. In particular, they may be chosen such $\kappa(\tilde{A}) = 1$. Consequently, solving the transformed problem

$$\tilde{A}x = z \tag{3.2}$$

iteratively does not pose difficulties to most (projection) solvers. Assume (3.2) admits a solution $\tilde{x}_0$, the question then is whether we can recover (easily) a solution x_0 satisfying

$$Ax_0 = z \tag{3.3}$$

that is, the original linear subproblem.

Since Γ leaves the range of $A \subset \mathbb{R}^n$, invariant, solutions to (3.2) exist if and only if (3.3) admits such. Moreover, setting

$$x_0 = V\Gamma V^T \tilde{x}_0 \tag{3.4}$$

a solution to (3.3) is obtained:

$$Ax_0 = \left(U\Sigma V^T\right)\left(V\Gamma V^T\right)\tilde{x}_0 = \left(U\Sigma\Gamma V^T\right)\tilde{x}_0 = \tilde{A}\tilde{x}_0 = z. \tag{3.5}$$

Thus by a scaling of the components of $\tilde{x}_0$ in the coordinate system of A's right singular vectors, which computationally does not pose any difficulties, we can solve the original problem (3.3) by working out the solution to the simpler formulation (3.2).

Example 3.1

For geometric intuition, the assignment in (3.4) can be rewritten as

$$x_0 = \left(V\left[(\Gamma - I) + I\right]V^T\right)\tilde{x}_0$$

$$= \tilde{x}_0 + \sum_{i=1}^{n} \tilde{\alpha}_i(\gamma_i - 1) \cdot v_i, \tag{3.6}$$

where $\tilde{\alpha}_i$ are the V-coordinates of $\tilde{x}_0$ and v_i the right singular vectors. Equation (3.6) illustrates that the Γ-transformation is in fact a translation of the solution set along these right singular vectors, proportional to the choice of γ_i. The toy example from Figure1 is used to demonstrate this effect in Figure2.

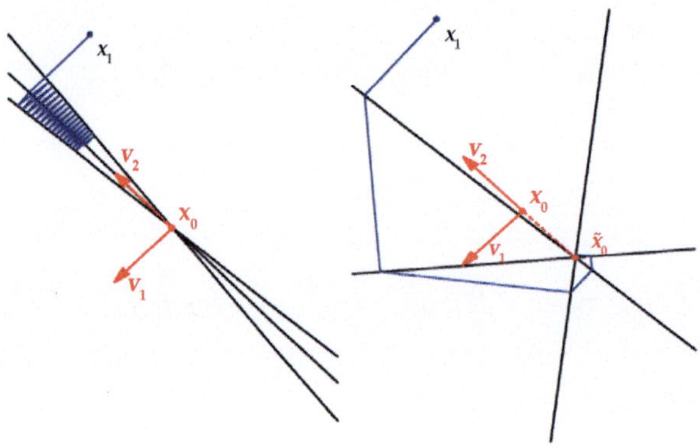

Figure2: Reconstruction of x_0 (= x_{opt}).

Here

$$A = \begin{pmatrix} 1 & 0.8 \\ 1 & 1 \\ 1 & 1.2 \end{pmatrix}, \quad z = A \cdot \begin{pmatrix} 100 \\ 100 \end{pmatrix}, \quad x_0 = \begin{pmatrix} 100 \\ 100 \end{pmatrix} \tag{3.7}$$

with

$$\sigma_1 = 2.46, \qquad \sigma_2 = 0.20, \qquad \kappa(A) = 12.33 \tag{3.8}$$

and right-singular vectors visualized in red.

Applying the Γ-transformation with $\gamma_1 = \sigma_1$ and $\gamma_2 = \sigma_1/\sigma_2$ the transformed inverse problem $\tilde{A}x = z$ is optimally conditioned with $\kappa(\tilde{A}) = 1$ and hence easily solvable with solution

$$\tilde{x}_0 = (100.6, 99.4)^T \tag{3.9}$$

which is exactly the translation expected from (3.6).

Main Result

The application of this preconditioning process to optimization problems with linear subproblems as in (1.1) is the natural next step.

Theorem 3.2

Given a convex function $f \in C^1(\mathbb{R}^n)$*, then the minimization problem*

$$\min_{x \in \mathbb{R}^m} f(Ax) \tag{3.10}$$

has solution

$$x_0 = V \Gamma V^T \tilde{x}_0, \tag{3.11}$$

where Γ *is a diagonal matrix with non-zero diagonal elements and* $\tilde{x}_0$ *solves*

$$\min_{\tilde{x} \in \mathbb{R}^m} f(\tilde{A}\tilde{x}) \tag{3.12}$$

with $\tilde{A}$ *as in* (3.1).

Proof

For the minimizer x_0, we have

$$\nabla_x f(Ax_0) = A^T \cdot \nabla_z f(z)|_{Ax_0} = 0 \tag{3.13}$$

and hence

$$\nabla_z f(z)|_{Ax_0} \in \ker(A^T). \tag{3.14}$$

The statement of the theorem is thus equivalent to showing

$$\nabla_z f(z)|_{\tilde{A}\tilde{x}_0} \in \ker(\tilde{A}^T) \quad \Longrightarrow \quad \nabla_z f(z)|_{Ax_0} \in \ker(A^T) \tag{3.15}$$

which follows from our previous observation that Γ-transformations leave kernel and range of A invariant together with (3.5).

The Algorithmic Scheme

The results of the previous two sections are straightforward to encode into a program usable for actual computation. What follows is a pseudo-code of the general scheme.

Algorithm 3.3

(Singular Value Homogenization)

Step 0::

Let f and A be given as in (1.1).

Step 1::

Compute the SVD of $A=U\Sigma V^T$ *and choose* $\Gamma=\mathrm{diag}(\gamma_1,\ldots,\gamma_m)$ *such that*

$$\kappa\left(\tilde{A} = U\Sigma\Gamma V^T\right) \approx 1. \quad (3.16)$$

Step 2::

Apply any optimization procedure to solve (3.12) *and obtain a solution* x~0x~0.

Step 3::

Reconstruct the original solution x_0 *of* (3.10) *via*

$$x_0 = V\Gamma V^T \tilde{x}_0. \quad (3.17)$$

The optimal choices of Γ in Step 1 and the concrete solver to find $\tilde{x}_0$ in Step 2 are likely problem specific and are as of now left as user parameters. A parameter exploration to find all-purpose configurations is included in the next section.

Furthermore, due to the near-optimal conditioning in Step 1 the time complexity of Algorithm 3.3 is $\mathcal{O}(\min\{mn^2, m^2n\})$ since it is dominated by the SVD of A.

This does not necessarily prohibit from solving large linear systems as in many cases (e.g. in IMRT [20]) either the spectral gap of A is big or large and small singular values cluster together-which allows for reliable k-SVD schemes that can be computed in O(mnlogk) time.

NUMERICAL EXPERIMENTS

All testing was done in both Matlab and Mathematica with negligible performance differences between the two (as both implement the same set of standard minimization algorithms).

Linear Feasibility and Linear Least Squares

The first series of experiments concerns the simplest and most often encountered formulation of (1.1) with

$$f(Ax) = \|Ax - b\|_2^2 \quad (4.1)$$

which corresponds to solving a linear system of equations exactly if a solution exist or in the least squares sense if it has empty intersection (here

$b \in \mathbb{R}^n$ is fixed).

As projection methods in general, and the Kaczmarz and Cimmino algorithms in particular, are known to perform well in such settings, we chose to compare execution of Algorithm3.3to these two for benchmarking. Moreover, to isolate the effects of the Γ-transformation most visibly, these two algorithms are used as subroutines in Step 2 as well.

Performance was measured on a set $\{A_i \in \mathbb{R}^{100\times 3}\}$of 3,000 randomly generated matrices with$\kappa(A_i) \in [1,105]$and data$x_0=(1,1,1)^T$that the respective algorithms were run on. The convergence threshold in all cases was set to $10^{-3}$and Γ chosen[4]such that$\Sigma\Gamma=\sigma_2 \cdot I$. The results are depicted in Figure3and Figure4(the presence of two graphs for Algorithm3.3indicate whether Kaczmarz (green) or Cimmino (red) was used as subroutine).

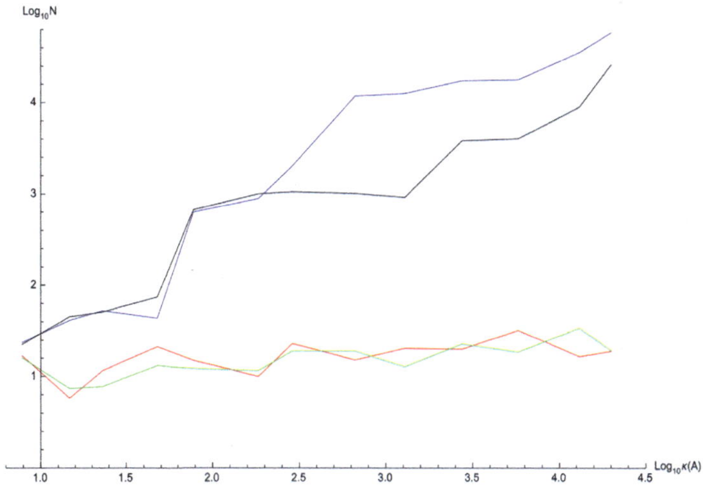

Figure3: Comparison with respect to number of iterations as a function of condition number between Kaczmarz, Cimmino and SVH with Kaczmarz and SVH with Cimmino.Stopping criteria is $\|x_{\text{opt}} - x\| \leq 10^{-3}$, where $x_{\text{opt}} = (1,1,1)^T$.

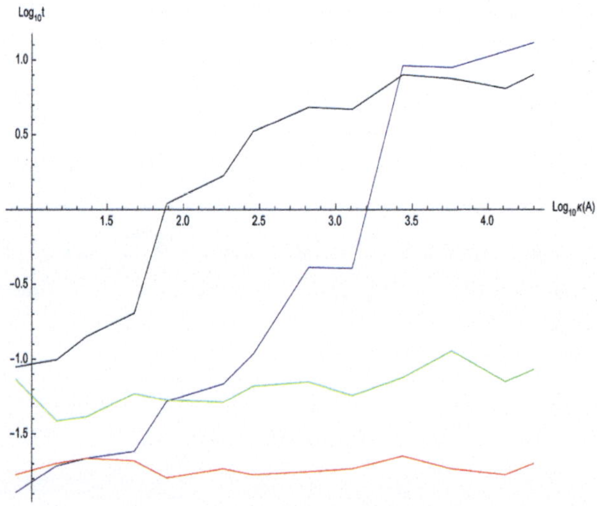

Figure4: Time needed: Algorithm3.3(red, green) and Kaczmarz (blue) and Cimmino (black) methods.

As expected from the results obtained in the preceding sections, both projection solvers scale poorly (indeed exponentially) with the condition number of*A*while Algorithm3.3retains constant time (≈0.02s and[5]≈0.06s respectively) and numbers of iteration (≈10) necessary.

In addition, reducing the accuracy threshold ($<10^{-4}$) or constructing matrices of extreme condition ($\kappa(A)\geq 10^6$ that result in failure to converge of Cimmino, Kaczmarz and LAPACK solvers native to Matlab and Mathematica does not impair the performance of Algorithm3.3. That is, through appropriate Γ-transformation we were able to solve very ill-conditioned linear problems for the first time to 10^{-5}accuracy within seconds.

LpLpPenalties and One-sidedLpLpPenalties

In the biomedical field of cancer treatment planning problems of the kind (1.1) occur often in calculating the optimal dose deposition in patient tissue. A typical formulation involves the linearized convolution*A*of radiation*x*into dose*d*and a reference dose $r \in \mathbb{R}^n$which is to be achieved underL^ppenalties$\|Ax-r\|_p$or their one-sided variations$\|\max\{0,Ax-r\}\|_p$and$\|\min\{0,Ax-r\}\|_p$.

We examined five[6]cases$\{A_i,r_i\}$that were collected from patient data under the penalties

$$f_1(d) = \|d - r\|_2, \tag{4.2}$$
$$f_2(d) = \|d - r\|_8, \tag{4.3}$$
$$f_3(d) = \|\max(0, d - r)\|_2, \tag{4.4}$$
$$f_4(d) = \|\min(0, d - r)\|_2, \tag{4.5}$$
$$f_5(d) = f_1(d \cdot s_1) + f_2(d \cdot s_2) + f_3(d \cdot s_3) + f_4(d \cdot s_4), \tag{4.6}$$

where $s_i \in \{0,1\}^n$ with $\sum s_i = \mathbf{1}_{\mathbb{R}^n}$ is a partition of the unit vector accounting for varied sensitivity of distinct body tissue to radiation.

The performance of Algorithm3.3in comparison to native Matlab and Mathematica methods is given in Table1. $t_i^{\bullet}$ ·is the time in seconds until convergence that Mathematica's NMinimize and Matlab's fminunc routines require on average whereas t_i° seconds are needed for Algorithm3.3to converge. In the case of neither Mathematica nor Matlab finding a solution (*nC*for*not converging*), the accuracy of Algorithm3.3's output$x_0$is tested through the parameterμ. This is done by randomly sampling a neighborhood of$x_0$and counting instances that improve the objective. These hits are then sampled similarly until no further such points can detected.$\mu$is the total number of neighborhoods so checked. In all cases, the improvement in*f*remained below 10^{-4}.

Table1: Comparison for nonlinear objective function

	A1A1	A2A2	A3A3	A4A4	A5A5
dim	504×250	336×192	408×128	457×206	500×82
κ	6×10^{16}	2×10^{16}	2×10^{18}	6×10^{12}	9×10^{9}
$t_i^{\bullet}$	nC	nC	nC	nC	2,381
t_i°	44	123	82	85	2
$t_2^{\bullet}$	nC	nC	nC	nC	2,445
t_2°	44	117	102	90	3
$t_3^{\bullet}$	nC	nC	nC	nC	2,579
t_3°	48	117	98	92	5

$t_4^\bullet$	nC	nC	nC	nC	2,502
t_4°	40	108	106	79	3
$t_5^\bullet$	nC	nC	nC	nC	2,524
t_5°	42	117	105	87	3
μ	2	8	6	3	0

The results are parallel to what could be seen in the linear feasibility formulation and encourage further exploration.

CONCLUSION

We were able to reduce the time needed to solve a general convex optimization problem with linear subproblem for modestly sized matrices. The performance of the proposed algorithm was compared to classical LAPACK and projection methods which showed an improvement in run-times by a factor of up to 1,190. Additionally, in many cases where LAPACK and projection solvers failed to converge, the singular value homogenization found 10^{-4}accurate solutions. These results are promising and encourage further exploration of SVH. Especially its application to structured large matrices and constrained optimization as well as in-depth parameter explorations may well turn out to be worthwhile.

Footnotes

[1]We choseRRonly as it is more pertinent to most practical applications, the extension of all results to $\mathbb{C}$ is straightforward.

[2]Which is particularly important in intensity modulated radiation therapy IMRT from which later numerical experiments will be drawn.

[3]*Major*and*minor*here refer to the size of the singular values associated with a singular vector.

[4]Experimental evidence suggests that in this setting of randomized matrices such homogenization to one singular value represents the most reasonable choice; different Γ display similar behavior with overall longer run-times.

[5]This time difference is due to the higher overhead required for the block projections of the Cimmino algorithm.

[6]The dose calculations are of cancerous tissue in the brain, the neck region and the prostate.

ACKNOWLEDGEMENTS

This work was supported by the Fraunhofer Institute for Industrial Mathematics-ITWM.

AUTHORS' CONTRIBUTIONS

D.-D. Erdmann-Pham, A. Gibali and P. Süss contributed equally to the writing of this paper. K.-H. Küfer, the head of the Optimization Department in Fraunhofer- ITWM provide technical and general support. All authors read and approved the final manuscript.

REFERENCES

1. Sadek RA. SVD based image processing applications: state of the art, contributions and research challenges. Int J Adv Comput Sci Appl. 2012;3:26-34.
2. Rajwade A, Rangarajan A, Banerjee A. Image denoising using the higher order singular value decomposition. IEEE Trans Pattern Anal Mach Intell. 2013;35:849-61.
3. Levitin ES, Polyak BT. Constrained minimization problems. USSR Comput Math Math Phys. 1966;6:1-50.
4. Golshtein EG, Tretyakov NV. Modified Lagrangians and monotone maps in optimization. New York: Wiley; 1996.
5. Erhel J, Guyomarc'h F, Saad Y. Least-squares polynomial filters for ill-conditioned linear system. Thème 4 - Simulation et optimisation de systèmes complexes. Projet ALADIN, Rapport de recherche N∘4175, Mai 2001, 28 pp.
6. Meng X, Saunders MA, Mahoney MW. LSRN: a parallel iterative solver for strongly over- or underdetermined systems. SIAM J Sci Comput. 2014;36:95-118.
7. Engl HW, Hanke M, Neubauer A. Regularization of inverse problems. Dordrecht: Kluwer Academic; 1996.
8. Vogel CR. Computational methods for inverse problems. Philadelphia: Society for Industrial and Applied ematics; 2002.
9. Orkisz J, Pazdanowski M. On a new feasible directions solution approach in constrained optimization. In: Onate E, Periaux J, Samuelsson A, editors. The finite element method in the 1990's. Berlin: Springer; 1991. p. 621-32.
10. Pazdanowski M. SVD as a preconditioner in nonlinear optimization. Comput Assist Mech Eng Sci. 2014;21:141-50.
11. Goebel K, Reich S. Uniform convexity, hyperbolic geometry, and nonexpansive mappings. New York: Marcel Dekker; 1984.
12. Censor Y, Zenios SA. Parallel optimization: theory, algorithms, and applications. New York: Oxford University Press; 1997.
13. Galántai A. Projectors and projection methods. Dordrecht: Kluwer Academic; 2004.
14. Escalante R, Raydan M. Alternating projection methods. Philadelphia: Society for Industrial and Applied ematics; 2011.

15. Bauschke HH, Combettes PL. Convex analysis and monotone operator theory in Hilbert spaces. Berlin: Springer; 2011.
16. Censor Y, Chen W, Combettes PL, Davidi R, Herman GT. On the effectiveness of projection methods for convex feasibility problems with linear inequality constraints. Comput Optim Appl. 2012;51:1065-88.
17. Bauschke HH, Koch VR. Projection methods: Swiss Army knives for solving feasibility and best approximation problems with halfspaces. Contemp . 2015;636:1-40.
18. Kaczmarz S. Angenäherte Auflösung von Systemen linearer Gleichungen. Bull Int Acad Pol Sci Let. 1937;35:355-7.
19. Cimmino G. Calcolo approssimato per le soluzioni dei sistemi di equazioni lineari. Ric Sci, Ser II. 1938;9:326-33.
20. Webb S. Intensity-modulated radiation therapy. Boca Raton: CRC Press; 2001.

Chapter 2

Perturbation Bounds for Eigenvalues of Diagonalizable Matrices and Singular Values

Duanmei Zhou[1], Guoliang Chen[2] , Guoxing Wu[3] and Xiaoyan Chen[4]

[1]College of Mathematics and Computer Science, Gannan Normal University, 341000 Ganzhou, People's Republic of China

[2]Department of Mathematics, East China Normal University, 200241 Shanghai, People's Republic of China

[3]Department of Mathematics, Northeast Forestry University, 150040 Harbin, People's Republic of China

[4]Library, Gannan Normal University, 341000 Ganzhou, People's Republic of China

ABSTRACT

Perturbation bounds for eigenvalues of diagonalizable matrices are derived. Perturbation bounds for singular values of arbitrary matrices are also given. We generalize some existing results.

Citation (APA): Zhou, D., Chen, G., Wu, G., & Chen, X. (2016). Perturbation bounds for eigenvalues of diagonalizable matrices and singular values. Journal of Inequalities and Applications, 2016(1), 113. (11 pages), DOI: https://doi.org/10.1186/s13660-016-1053-9

Keywords: perturbation bound, eigenvalue, diagonalizable matrix, singular value

INTRODUCTION

Many problems in science and engineering lead to eigenvalue and singular value problems for matrices. Perturbation bounds of eigenvalues and singular values play an important role in matrix computations. Let S_n be the set of all $n!$ permutations of $\{1,2,\ldots,n\}$. If $x=(x_1,x_2,\ldots,x_n)$ and $\pi\in S_n$, then the vector x_π is defined as $(x_{\pi(1)},x_{\pi(2)},\ldots,x_{\pi(n)})$. A square matrix is called doubly stochastic if its elements are real nonnegative numbers and if the sum of the elements in each row and in each column is equal to 1. Let $\mathbb{C}^{n\times n}$ be the set of $n\times n$ complex matrices. Let $A=(a_{ij})\in \mathbb{C}^{n\times n}$, we use the notation (see [1,2])

$$\|A\|_p = \left(\sum_{i,j=1}^{n} |a_{ij}|^p\right)^{\frac{1}{p}} \quad \text{for } p \geq 0, \tag{1.1}$$

$$\|A\|_{q,p} = \left(\sum_{j=1}^{n}\left(\sum_{k=1}^{n} |a_{kj}|^q\right)^{\frac{p}{q}}\right)^{\frac{1}{p}} \quad \text{for } p > 0, q > 0, \frac{1}{p} + \frac{1}{q} = 1. \tag{1.2}$$

Let $T\in\mathbb{C}^{n\times n}$ and assume that

$$\Lambda^{(k)} = \operatorname{diag}\left(\lambda_1^{(k)},\lambda_2^{(k)},\ldots,\lambda_n^{(k)}\right) \in \mathbb{C}^{n\times n}, \quad k = 1,\ldots,4,$$

are diagonal matrices. In [3], the following classical result is given:

$$\|\Lambda^{(1)}T\Lambda^{(2)} - \Lambda^{(3)}T\Lambda^{(4)}\|_2^2 \geq s_n^2(T)\sum_{i=1}^{n}|\lambda_i^{(1)}\lambda_{\pi(i)}^{(2)} - \lambda_i^{(3)}\lambda_{\pi(i)}^{(4)}|^2 \tag{1.3}$$

for some $\pi\in S_n$, where $s_n(T)$ is the smallest singular value of T. The inequality has many applications in bounding the (relative) perturbation for eigenvalues and singular values, such as [4,5,6] and the references therein. We generalize (1.3) in Section 2.

Let $\lambda(A)$ denote the spectrum of matrix A. In 1970, Ikramov [7] defined the 'Hölder distance $d_p(\lambda(A),\lambda(B))$ between the spectra' of the matrices A and B, which have the eigenvalues $\lambda_1,\lambda_2,\ldots,\lambda_n$ and $\mu_1,\mu_2,\ldots,\mu_n$, respectively, by the equation:

$$d_p(\lambda(A),\lambda(B)) = \min_{\pi\in S_n}\left(\sum_{i=1}^{n}|\lambda_i - \mu_{\pi(i)}|^p\right)^{\frac{1}{p}}. \tag{1.4}$$

If A and B are Hermitian matrices and $1\leq p<2$, [7] obtained

$$d_p(\lambda(A),\lambda(B)) \leq \|A - B\|_p, \tag{1.5}$$

which partially generalizes the Hoffman-Wielandt theorem [8]. However, for normal matrices (1.5) can no longer be valid. The purpose of this paper is to obtain several inequalities similar to (1.5) for diagonalizable matrices. We exhibit some upper bounds and lower bounds for $d_p(\lambda(A),\lambda(B))$ of diagonalizable matrices A and B in Section2.

Majorization is one of the most powerful techniques for deriving inequalities. We use majorization to get some perturbation bounds for singular values. For simplicity of the notations, in most cases in this paper the vectors in $\mathbb{R}^n$ are regarded as row vectors, but when they are multiplied by matrices we regard them as column vectors. Given a real vector $x=(x_1,x_2,\dots,x_n)\in \mathbb{R}^n$, we rearrange its components as $x_{[1]}\geq x_{[2]}\geq\cdots\geq x_{[n]}$.

Definition 1.1

([9], p.14)

For $x = (x_1, x_2, \dots, x_n), y = (y_1, y_2, \dots, y_n) \in \mathbb{R}^n$, if

$$\sum_{i=1}^{k} x_{[i]} \leq \sum_{i=1}^{k} y_{[i]}, \quad k = 1, 2, \dots, n,$$

then we say that x is weakly majorized by y and denote $x \prec_w y$. If $x \prec_w y$ and $\sum_{i=1}^{n} x_i = \sum_{i=1}^{n} y_i$, then we say that x is majorized by y and denote $x\prec y$.

Let $s_1 \geq s_2 \geq \cdots \geq s_n$ and $\delta_1 \geq \delta_2 \geq \cdots \geq \delta_n$ be the singular values of the complex matrices $A = (a_{ij}) \in \mathbb{C}^{n\times n}$ and $B = (b_{ij}) \in \mathbb{C}^{n\times n}$, respectively. In [10], p.215, and [11], p.199, the following classical result is given:

$$\sum_{i=1}^{n} |s_i - \delta_i|^2 \leq \sum_{i,j=1}^{n} |a_{ij} - b_{ij}|^2. \tag{1.6}$$

We generalize the inequality (1.6) in Section3.

PERTURBATION BOUNDS FOR EIGENVALUES OF DIAGONALIZABLE MATRICES

Let $A\circ B$ denote the Hadamard product of matrices A and B. $\|A\|$ denotes the spectral norm of matrix A. A^T denotes the transpose of matrix A. For two n-square real matrices A,B, we write $A\leq_e B$ to mean that $B-A$ is (entrywise) nonnegative. For $A = (a_{ij}) \in \mathbb{C}^{n\times n}$ and a real number $t>0$, we denote $A^{|\circ|t} \equiv (|a_{ij}|^t) \in \mathbb{C}^{n\times n}$. Let $s_n(A)$ and $s_1(A)$ be the smallest and the largest singular values ofA, respectively. The following entrywise inequalities involve the smallest and the largest singular values.

Lemma 2.1

([9], p.52)

Let $A \in \mathbb{C}^{n\times n}$ and letp,qbe real numbers with0<p≤2andq≥2.Then there exist two doubly stochastic matricesB,C∈$\mathbb{C}^{n\times n}$ such that

$$s_n(A)^p B \leq_e A^{|\circ|p} \tag{2.1}$$

and

$$A^{|\circ|q} \leq_e s_1(A)^q C. \tag{2.2}$$

Theorem 2.2

LetT∈$\mathbb{C}^{n\times n}$ and letp,qbe real numbers with0<p≤2andq≥2.Assume that $\Lambda^{(k)} = \mathrm{diag}(\lambda_1^{(k)}, \lambda_2^{(k)}, \ldots, \lambda_n^{(k)}) \in \mathbb{C}^{n\times n}, k = 1, \ldots, 4$,are diagonal matrices.Then there are permutations$\pi$andνofS_n such that

$$\|\Lambda^{(1)}T\Lambda^{(2)} - \Lambda^{(3)}T\Lambda^{(4)}\|_p^p \geq s_n^p(T)\sum_{i=1}^{n}|\lambda_i^{(1)}\lambda_{\pi(i)}^{(2)} - \lambda_i^{(3)}\lambda_{\pi(i)}^{(4)}|^p \tag{2.3}$$

and

$$\|\Lambda^{(1)}T\Lambda^{(2)} - \Lambda^{(3)}T\Lambda^{(4)}\|_q^q \leq s_1^q(T)\sum_{i=1}^{n}|\lambda_i^{(1)}\lambda_{\nu(i)}^{(2)} - \lambda_i^{(3)}\lambda_{\nu(i)}^{(4)}|^q. \tag{2.4}$$

<u>Proof</u>

Set $T = (t_{ij}) \in \mathbb{C}^{n\times n}$. Then

$$\begin{aligned}\|\Lambda^{(1)}T\Lambda^{(2)} - \Lambda^{(3)}T\Lambda^{(4)}\|_p^p &= \sum_{i,j=1}^{n}|t_{ij}|^p|\lambda_i^{(1)}\lambda_j^{(2)} - \lambda_i^{(3)}\lambda_j^{(4)}|^p \\ &= e^T(T^{|\circ|p} \circ M)e,\end{aligned} \tag{2.5}$$

where $M = (|\lambda_i^{(1)}\lambda_j^{(2)} - \lambda_i^{(3)}\lambda_j^{(4)}|^p) \in \mathbb{C}^{n\times n}, e = (1,1,\ldots,1)^T \in \mathbb{C}^n$. Applying inequality (2.1), we have

$$T^{|\circ|p} \geq s_n^p(T)B,$$

whereB=(b_{ij})is a doubly stochastic matrix. Then

$$\|\Lambda^{(1)}T\Lambda^{(2)} - \Lambda^{(3)}T\Lambda^{(4)}\|_p^p \geq e^T(s_n^p(T)B \circ M)e = s_n^p(T)e^T(B \circ M)e.$$

SinceBis doubly stochastic, by Birkhoff's theorem ([12,13], p.527)Bis a convex combination of permutation matrices:

$B = \sum_{i=1}^{n!} \tau_i P_i, \quad \tau_i \geq 0, \sum_{i=1}^{n!} \tau_i = 1, P_i$ are permutation matrices.

Suppose $e^T(B \circ P_k)e = \min\{e^T(B \circ P_i)e \mid 1 \leq i \leq n!\}$ and P_k corresponds to $\pi \in S_n$. Then

$$\begin{aligned}\|\Lambda^{(1)}T\Lambda^{(2)} - \Lambda^{(3)}T\Lambda^{(4)}\|_p^p &\geq s_n^p(T)e^T(B \circ M)e \\ &= s_n^p(T)e^T\left(\sum_{i=1}^{n!} \tau_i P_i \circ M\right)e \\ &\geq s_n^p(T)\sum_{i=1}^{n!} \tau_i e^T(P_k \circ M)e \\ &= s_n^p(T)e^T(P_k \circ M)e \\ &= s_n^p(T)\sum_{i=1}^{n} |\lambda_i^{(1)}\lambda_{\pi(i)}^{(2)} - \lambda_i^{(3)}\lambda_{\pi(i)}^{(4)}|^p.\end{aligned}$$

Proving (2.3).

Use (2.2), (2.5), and the Birkhoff theorem, we can deduce the inequality (2.4).

Remark 2.3

If we take $p=2$, we get Theorem3.2 in [3]. So, the bound in inequality (2.3) generalizes the bound of Theorem3.2 in [3].

Next, we apply Theorem2.2 to get some perturbation bounds for the eigenvalues of diagonalizable matrices. Let $A = (a_{ij}) \in \mathbb{C}^{n\times n}$ and $B = (b_{ij}) \in \mathbb{C}^{n\times n}$. When $p > 0, q > 0, \frac{1}{p} + \frac{1}{q} = 1$, then (see [2])

$$\|AB\|_p \leq \|A\|_p \|B\|_{q,p}, \tag{2.6}$$

$$\|AB\|_p \leq \left\|A^T\right\|_{q,p} \|B\|_p. \tag{2.7}$$

If B is nonsingular, then we have

$$\frac{\|A\|_p}{\|B^{-1}\|_{q,p}} \leq \|AB\|_p. \tag{2.8}$$

If A is nonsingular, then we have

$$\frac{\|B\|_p}{\|(A^{-1})^T\|_{q,p}} \leq \|AB\|_p. \tag{2.9}$$

For normal matrices the statement of Theorem3 in [7] (inequality (1.5)) can no longer be valid. However, we have the following theorem.

Theorem 2.4

Assume that both $A \in \mathbb{C}^{n\times n}$ *and* $B \in \mathbb{C}^{n\times n}$ are diagonalizable and admit the following decompositions:

$$A = D_1\Lambda_1 D_1^{-1} \quad and \quad B = D_2\Lambda_2 D_2^{-1}, \tag{2.10}$$

where D_1 and D_2 are nonsingular, and $\Lambda_1=\mathrm{diag}(\lambda_1,\lambda_2,\ldots,\lambda_n)$ and $\Lambda_2=\mathrm{diag}(\mu_1,\mu_2,\ldots,\mu_n)$. Then there are permutations π and v of S_n such that

$$\left(\sum_{i=1}^{n} |\lambda_i - \mu_{\pi(i)}|^p\right)^{\frac{1}{p}} \le \left\|(D_1^{-1})^T\right\|_{q,p} \|D_2\|_{q,p} \left\|D_2^{-1}\right\| \|D_1\| \|A-B\|_p, \tag{2.11}$$

$$\left(\sum_{i=1}^{n} |\lambda_i - \mu_{v(i)}|^p\right)^{\frac{1}{p}} \le \left\|(D_2^{-1})^T\right\|_{q,p} \|D_1\|_{q,p} \left\|D_1^{-1}\right\| \|D_2\| \|A-B\|_p, \tag{2.12}$$

where $1<p\le 2$ and $\frac{1}{p}+\frac{1}{q}=1$.

Proof

Using (2.10), we have

$$\|A-B\|_p^p = \left\|D_1\Lambda_1 D_1^{-1} - D_2\Lambda_2 D_2^{-1}\right\|_p^p$$

$$= \left\|D_1\left(\Lambda_1 D_1^{-1}D_2 - D_1^{-1}D_2\Lambda_2\right)D_2^{-1}\right\|_p^p \tag{2.13}$$

and

$$\|A-B\|_p^p = \left\|D_1\Lambda_1 D_1^{-1} - D_2\Lambda_2 D_2^{-1}\right\|_p^p$$

$$= \left\|D_2\left(D_2^{-1}D_1\Lambda_1 - \Lambda_2 D_2^{-1}D_1\right)D_1^{-1}\right\|_p^p. \tag{2.14}$$

We give a proof of (2.11) with the help of (2.13). Similarly one can prove (2.12) using (2.14). Applying (2.8) and (2.9) to (2.13) we obtain

$$\|A-B\|_p^p \ge \frac{\|\Lambda_1 D_1^{-1}D_2 - D_1^{-1}D_2\Lambda_2\|_p^p}{\|(D_1^{-1})^T\|_{q,p}^p \|D_2\|_{q,p}^p}.$$

Using inequality (2.3), there is a permutation π of S_n such that

$$\left\|\Lambda_1 D_1^{-1}D_2 - D_1^{-1}D_2\Lambda_2\right\|_p^p \ge s_n^p(D_1^{-1}D_2)\sum_{i=1}^{n} |\lambda_i - \mu_{\pi(i)}|^p.$$

So we have

$$\sum_{i=1}^{n} |\lambda_i - \mu_{\pi(i)}|^p \le \frac{\|(D_1^{-1})^T\|_{q,p}^p \|D_2\|_{q,p}^p}{s_n^p(D_1^{-1}D_2)} \|A-B\|_p^p.$$

We use the relations

$$s_n^{-1}(D_1^{-1}D_2) = \left\|(D_1^{-1}D_2)^{-1}\right\| \le \|D_1\| \left\|D_2^{-1}\right\|$$

to get the inequality in (2.11).

Theorem 2.5

Under the hypotheses of Theorem2.4,there are permutations$\pi$and$v$of$S_n$such that

$$\left(\sum_{i=1}^{n}|\lambda_i-\mu_{\pi(i)}|^q\right)^{\frac{1}{q}} \geq \frac{\|A-B\|_q}{\|(D_1)^T\|_{p,q}\|D_2^{-1}\|_{p,q}\|D_1^{-1}\|\|D_2\|}, \tag{2.15}$$

$$\left(\sum_{i=1}^{n}|\lambda_i-\mu_{v(i)}|^q\right)^{\frac{1}{q}} \geq \frac{\|A-B\|_q}{\|(D_2)^T\|_{p,q}\|D_1^{-1}\|_{p,q}\|D_1\|\|D_2^{-1}\|}, \tag{2.16}$$

where$q\geq 2$and $\frac{1}{p}+\frac{1}{q}=1$.

Proof

Using (2.10), we have

$$\begin{aligned}\|A-B\|_q^q &= \left\|D_1\Lambda_1D_1^{-1}-D_2\Lambda_2D_2^{-1}\right\|_q^q\\ &= \left\|D_1\left(\Lambda_1D_1^{-1}D_2-D_1^{-1}D_2\Lambda_2\right)D_2^{-1}\right\|_q^q\end{aligned} \tag{2.17}$$

and

$$\begin{aligned}\|A-B\|_q^q &= \left\|D_1\Lambda_1D_1^{-1}-D_2\Lambda_2D_2^{-1}\right\|_q^q\\ &= \left\|D_2\left(D_2^{-1}D_1\Lambda_1-\Lambda_2D_2^{-1}D_1\right)D_1^{-1}\right\|_q^q.\end{aligned} \tag{2.18}$$

Applying (2.6) and (2.7) to (2.17) we obtain

$$\|A-B\|_q^q \leq \left\|(D_1)^T\right\|_{p,q}^q\left\|\Lambda_1D_1^{-1}D_2-D_1^{-1}D_2\Lambda_2\right\|_q^q\left\|D_2^{-1}\right\|_{p,q}^q.$$

Using inequality (2.4), there exists a permutationπofS_nSnsuch that

$$\left\|\Lambda_1D_1^{-1}D_2-D_1^{-1}D_2\Lambda_2\right\|_q^q \leq s_1^q\left(D_1^{-1}D_2\right)\sum_{i=1}^{n}|\lambda_i-\mu_{\pi(i)}|^q,$$

so we have

$$\sum_{i=1}^{n}|\lambda_i-\mu_{\pi(i)}|^q \geq \frac{\|A-B\|_q^q}{\|(D_1)^T\|_{p,q}^q\|D_2^{-1}\|_{p,q}^q s_1^q(D_1^{-1}D_2)}.$$

We use the relations

$$s_1\left(D_1^{-1}D_2\right)=\left\|D_1^{-1}D_2\right\| \leq \left\|D_1^{-1}\right\|\|D_2\|$$

to get the inequality in (2.15).

The proof of inequality (2.16) is similar to the proof of (2.15) and is omitted here.

For $1\leq p\leq 2$, it is well known [2] that the scalar function (1.1) of a matrix A is a submultiplicative matrix norm. However, it is true for $0<p\leq 2$. Actually, according to the Cauchy-Schwarz inequality, we have

$$\|AB\|_p^p = \sum_{i=1}^{n}\sum_{j=1}^{n}\left|\sum_{k=1}^{n} a_{ik}b_{kj}\right|^p$$

$$\leq \sum_{i=1}^{n}\sum_{j=1}^{n}\left(\left(\sum_{k=1}^{n}|a_{ik}|^2\right)^{\frac{1}{2}}\left(\sum_{k=1}^{n}|b_{kj}|^2\right)^{\frac{1}{2}}\right)^p.$$

Since for fixed vector $x=(x_1,x_2,\ldots,x_n)$, the function $p \to (|x_1|^p + |x_2|^p + \cdots + |x_n|^p)^{\frac{1}{p}}$ is decreasing on $(0,\infty)$,

$$\|AB\|_p^p \leq \sum_{i=1}^{n}\sum_{j=1}^{n}\left(\left(\sum_{k=1}^{n}|a_{ik}|^p\right)\left(\sum_{k=1}^{n}|b_{kj}|^p\right)\right)$$

$$= \left(\sum_{i=1}^{n}\sum_{k=1}^{n}|a_{ik}|^p\right)\left(\sum_{j=1}^{n}\sum_{k=1}^{n}|b_{kj}|^p\right)$$

$$= \|A\|_p^p\|B\|_p^p.$$

That is,

$$\|AB\|_p \leq \|A\|_p\|B\|_p. \tag{2.19}$$

If B is nonsingular, then

$$\|A\|_p = \left\|ABB^{-1}\right\|_p \leq \|AB\|_p\left\|B^{-1}\right\|_p.$$

So we have

$$\frac{\|A\|_p}{\|B^{-1}\|_p} \leq \|AB\|_p. \tag{2.20}$$

Similarly, when A is nonsingular, then

$$\|B\|_p = \left\|A^{-1}AB\right\|_p \leq \left\|A^{-1}\right\|_p\|AB\|_p.$$

That is,

$$\frac{\|B\|_p}{\|A^{-1}\|_p} \leq \|AB\|_p. \tag{2.21}$$

Theorem 2.6

Under the assumptions of Theorem 2.4, there are permutations π and v of S_n such that

$$\left(\sum_{i=1}^{n}|\lambda_i-\mu_{\pi(i)}|^p\right)^{\frac{1}{p}} \le \left\|D_1^{-1}\right\|_p\|D_2\|_p\left\|D_2^{-1}\right\|\|D_1\|\|A-B\|_p, \tag{2.22}$$

$$\left(\sum_{i=1}^{n}|\lambda_i-\mu_{\nu(i)}|^p\right)^{\frac{1}{p}} \le \left\|D_2^{-1}\right\|_p\|D_1\|_p\left\|D_1^{-1}\right\|\|D_2\|\|A-B\|_p, \tag{2.23}$$

where$0<p\le 2$.

Proof

The proof is similar to the proof of Theorem2.4and is omitted here.

Remark 2.7

Since

$$\|A\|_{q,p}=\left(\sum_{j=1}^{n}\left(\sum_{k=1}^{n}|a_{kj}|^q\right)^{\frac{p}{q}}\right)^{\frac{1}{p}} \le \left(\sum_{j=1}^{n}\left(\sum_{k=1}^{n}|a_{kj}|^p\right)^{\frac{p}{p}}\right)^{\frac{1}{p}}=\|A\|_p$$

and

$$\left\|A^T\right\|_{q,p}\le\left\|A^T\right\|_p=\|A\|_p$$

for$1<p\le 2$and $\frac{1}{p}+\frac{1}{q}=1$, the bounds in (2.11) and (2.12) are always sharper than those in (2.22) and (2.23), respectively.

When $p=q=2$, then $\|AB\|_2\le\min\{\|A\|_2\|B\|,\|A\|\|B\|_2\}$. We obtain

$$\left(\sum_{i=1}^{n}|\lambda_i-\mu_{\pi(i)}|^2\right)^{\frac{1}{2}} \le \left\|D_1^{-1}\right\|\|D_2\|\left\|D_2^{-1}D_1\right\|\|A-B\|_2$$

$$\le \left\|D_1^{-1}\right\|\|D_2\|\left\|D_2^{-1}\right\|\|D_1\|\|A-B\|_2,$$

$$\left(\sum_{i=1}^{n}|\lambda_i-\mu_{\nu(i)}|^2\right)^{\frac{1}{2}} \le \left\|D_2^{-1}\right\|\|D_1\|\left\|D_1^{-1}D_2\right\|\|A-B\|_2$$

$$\le \left\|D_2^{-1}\right\|\|D_1\|\left\|D_1^{-1}\right\|\|D_2\|\|A-B\|_2.$$

Since

$$\|A\|_{p,q}=\left(\sum_{j=1}^{n}\left(\sum_{k=1}^{n}|a_{kj}|^p\right)^{\frac{q}{p}}\right)^{\frac{1}{q}} \le \left(\sum_{j=1}^{n}\left(\sum_{k=1}^{n}|a_{kj}|^p\right)^{\frac{p}{p}}\right)^{\frac{1}{p}}=\|A\|_p$$

and

$\|A^T\|_{p,q} \leq \|A^T\|_p = \|A\|_p$

for $q \geq 2$ and $\frac{1}{p} + \frac{1}{q} = 1$, we have the following corollary.

Corollary 2.8

Under the same conditions as in Theorem2.4,there are permutations π and v of S_n such that

$$\left(\sum_{i=1}^{n} |\lambda_i - \mu_{\pi(i)}|^q\right)^{\frac{1}{q}} \geq \frac{\|A - B\|_q}{\|D_1\|_p \|D_2^{-1}\|_p \|D_1^{-1}\| \|D_2\|},$$

$$\left(\sum_{i=1}^{n} |\lambda_i - \mu_{v(i)}|^q\right)^{\frac{1}{q}} \geq \frac{\|A - B\|_q}{\|D_1^{-1}\|_p \|D_2\|_p \|D_1\| \|D_2^{-1}\|},$$

where $q \geq 2$ and $\frac{1}{p} + \frac{1}{q} = 1$.

Remark 2.9

When $p=q=2$, we obtain

$$\left(\sum_{i=1}^{n} |\lambda_i - \mu_{\pi(i)}|^2\right)^{\frac{1}{2}} \geq \frac{\|A - B\|_2}{\|D_1\| \|D_2^{-1}\| \|D_1^{-1} D_2\|} \geq \frac{\|A - B\|_2}{\|D_1\| \|D_2^{-1}\| \|D_1^{-1}\| \|D_2\|},$$

$$\left(\sum_{i=1}^{n} |\lambda_i - \mu_{v(i)}|^2\right)^{\frac{1}{2}} \geq \frac{\|A - B\|_2}{\|D_1^{-1}\| \|D_2\| \|D_2^{-1} D_1\|} \geq \frac{\|A - B\|_2}{\|D_1^{-1}\| \|D_2\| \|D_1\| \|D_2^{-1}\|}.$$

PERTURBATION BOUNDS FOR SINGULAR VALUES

For brevity we only consider square matrices. The generalizations from square matrices to rectangular matrices are obvious, and usually problems on singular values of rectangular matrices can be converted to the case of square matrices by adding zero rows or zero columns.

For a Hermitian matrix $G \in \mathbb{C}^{n \times n}$, we always denote $\lambda(G) = (\lambda_1(G), \lambda_2(G), \ldots, \lambda_n(G))$, where $\lambda_1(G) \geq \lambda_2(G) \geq \cdots \geq \lambda_n(G)$ are the eigenvalues of G in decreasing order.

Lemma 3.1

(Lidskii [14], Lemma3.18 [9])

If $G, H \in \mathbb{C}^{n \times n}$ are Hermitian matrices,then

$\lambda(G) - \lambda(H) \prec \lambda(G - H).$

Lemma 3.2

([9], p.18)

Letf(t)be a convex function, $x = (x_1, x_2, \ldots, x_n), y = (y_1, y_2, \ldots, y_n) \in \mathbb{R}^n$ .Then

$$x \prec y \quad \Longrightarrow \quad (f(x_1), f(x_2), \ldots, f(x_n)) \prec_w (f(y_1), f(y_2), \ldots, f(y_n)).$$

Theorem 3.3

Let $\sigma_1(A) \geq \sigma_2(A) \geq \cdots \geq \sigma_n(A)$, $\sigma_1(B) \geq \sigma_2(B) \geq \cdots \geq \sigma_n(B)$ and $\sigma_1(A - B) \geq \sigma_2(A - B) \geq \cdots \geq \sigma_n(A - B)$ be the singular values of the complex matricesA=(a_{ij}),B=(b_{ij})andA−B,respectively.Then

$$\sum_{i=1}^{n} |\sigma_i(A) - \sigma_i(B)|^p \leq \sum_{i,j=1}^{n} |a_{ij} - b_{ij}|^p, \tag{3.1}$$

$$\sum_{i=1}^{n} |\sigma_i(A) - \sigma_i(B)|^q \geq \sum_{i=1}^{n} \sigma_i^q(A - B), \tag{3.2}$$

where1≤p≤2,0<q≤1.

Proof

Let $\varphi(A) \triangleq \begin{pmatrix} 0 & A^* \\ A & 0 \end{pmatrix}$, $\varphi(B) \triangleq \begin{pmatrix} 0 & B^* \\ B & 0 \end{pmatrix}$. Then

$$\varphi(A - B) = \begin{pmatrix} 0 & (A - B)^* \\ A - B & 0 \end{pmatrix}.$$

$\varphi(A), \varphi(B), \varphi(A - B)$ are three Hermitian matrices. Assume that $U_1^* A V_1 = \operatorname{diag}(\sigma_1(A), \sigma_2(A), \ldots, \sigma_n(A))$, $U_2^* B V_2 = \operatorname{diag}(\sigma_1(B), \sigma_2(B), \ldots, \sigma_n(B))$ and $U_3^*(A - B) V_3 = \operatorname{diag}(\sigma_1(A - B), \sigma_2(A - B), \ldots, \sigma_n(A - B))$ are singular value decompositions with U_1, U_2, U_3, V_1, V_2, V_3 unitary. Then

$$Q_1 \triangleq \frac{1}{\sqrt{2}} \begin{pmatrix} V_1 & V_1 \\ U_1 & -U_1 \end{pmatrix}, \quad Q_2 \triangleq \frac{1}{\sqrt{2}} \begin{pmatrix} V_2 & V_2 \\ U_2 & -U_2 \end{pmatrix} \quad \text{and} \quad Q_3 \triangleq \frac{1}{\sqrt{2}} \begin{pmatrix} V_3 & V_3 \\ U_3 & -U_3 \end{pmatrix}$$

are unitary matrices and

$$Q_1^* \varphi(A) Q_1 = \operatorname{diag}(\sigma_1(A), \sigma_2(A), \ldots, \sigma_n(A), -\sigma_1(A), -\sigma_2(A), \ldots, -\sigma_n(A)),$$

$$Q_2^* \varphi(B) Q_2 = \operatorname{diag}(\sigma_1(B), \sigma_2(B), \ldots, \sigma_n(B), -\sigma_1(B), -\sigma_2(B), \ldots, -\sigma_n(B))$$

and

$$Q_3^* \varphi(A - B) Q_3 = \operatorname{diag}(\sigma_1(A - B), \sigma_2(A - B), \ldots, \sigma_n(A - B), \\ -\sigma_1(A - B), -\sigma_2(A - B), \ldots, -\sigma_n(A - B)).$$

By Lemma3.1, we have

$$(\sigma_1(A)-\sigma_1(B),\sigma_2(A)-\sigma_2(B),\ldots,\sigma_n(A)-\sigma_n(B),$$
$$\sigma_n(B)-\sigma_n(A),\ldots,\sigma_2(B)-\sigma_2(A),\sigma_1(B)-\sigma_1(A))$$
$$\prec(\sigma_1(A-B),\sigma_2(A-B),\ldots,\sigma_n(A-B),$$
$$-\sigma_n(A-B),\ldots,-\sigma_2(A-B),-\sigma_1(A-B)). \tag{3.3}$$

First consider the case1≤p≤2. Since the function $f(t)=|t|^p$ is convex on(−∞,+∞), applying Lemma3.2withf(t)to the majorization (3.3) yields

$$(|\sigma_1(A)-\sigma_1(B)|^p,|\sigma_2(A)-\sigma_2(B)|^p,\ldots,|\sigma_n(A)-\sigma_n(B)|^p,$$
$$|\sigma_n(B)-\sigma_n(A)|^p,\ldots,|\sigma_2(B)-\sigma_2(A)|^p,|\sigma_1(B)-\sigma_1(A)|^p)$$
$$\prec_w(\sigma_1^p(A-B),\sigma_2^p(A-B),\ldots,\sigma_n^p(A-B),\sigma_n^p(A-B),\ldots,\sigma_2^p(A-B),\sigma_1^p(A-B)).$$

In particular,

$$(|\sigma_1(A)-\sigma_1(B)|^p,|\sigma_2(A)-\sigma_2(B)|^p,\ldots,|\sigma_n(A)-\sigma_n(B)|^p)$$
$$\prec_w(\sigma_1^p(A-B),\sigma_2^p(A-B),\ldots,\sigma_n^p(A-B)). \tag{3.4}$$

According to Theorem1 of [15] or Theorem3.32 of [9], we have

$$\sum_{i=1}^{n}\sigma_i^p(A-B)\le\sum_{i,j=1}^{n}|a_{ij}-b_{ij}|^p \tag{3.5}$$

for1≤p≤2. Combining (3.4) and (3.5), we obtain (3.1).

When0<q≤1, by considering the convex function $g(t)=-|t|^q$, on $(-\infty,+\infty)$, applying Lemma3.2withg(t)g(t)to the majorization (3.3) yields

$$(-|\sigma_1(A)-\sigma_1(B)|^q,\ldots,-|\sigma_n(A)-\sigma_n(B)|^q,-|\sigma_n(B)-\sigma_n(A)|^q,\ldots,-|\sigma_1(B)-\sigma_1(A)|^q)$$
$$\prec_w(-\sigma_1^q(A-B),\ldots,-\sigma_n^q(A-B),-\sigma_n^q(A-B),\ldots,-\sigma_1^q(A-B)).$$

In particular,

$$(-|\sigma_1(A)-\sigma_1(B)|^q,-|\sigma_2(A)-\sigma_2(B)|^q,\ldots,-|\sigma_n(A)-\sigma_n(B)|^q)$$
$$\prec_w(-\sigma_1^q(A-B),-\sigma_2^q(A-B),\ldots,-\sigma_n^q(A-B)). \tag{3.6}$$

From (3.6), we get (3.2).

Remark 3.4

From inequality (3.1), ifp=2, we get the inequality (1.6). So the inequality (3.1) generalizes the inequality (3.5.33) of [10], p.215, and Theorem3.12 of [11], p.199.

ACKNOWLEDGEMENTS

This work is supported by the National Natural Science Foundation of China (No. 11501126, No. 11471122), the Youth Natural Science Foundation of Jiangxi Province (No. 20151BAB211011), the Science Foundation of Jiangxi Provincial Department of Education (No. GJJ150979), and the Supporting the Development for Local Colleges and Universities Foundation of China - Applied Mathematics Innovative team building.

AUTHORS' CONTRIBUTIONS

All authors conceived of the study, participated in its design and coordination, drafted the manuscript, participated in the sequence alignment, and read and approved the final manuscript.

REFERENCES

1. Li, R: Norms of certain matrices with applications to variations of the spectra of matrices and matrix pencils. Linear Algebra Appl.182, 199-234 (1993)
2. Ostrowski, A: Über Normen von Matrizen. Math. Z.63, 2-18 (1955)
3. Elsner, L, Friedland, S: Singular value, doubly stochastic matrices, and applications. Linear Algebra Appl.220, 161-169 (1995)
4. Chen, X: On perturbation bounds of generalized eigenvalues for diagonalizable pairs. Numer. Math.107, 79-86 (2007)
5. Li, R: Spectral variations and Hadamard products: some problems. Linear Algebra Appl.278, 317-326 (1998)
6. Li, W, Sun, W: Combined perturbation bounds: I. Eigensystems and singular value decompositions. SIAM J. Matrix Anal. Appl.29, 643-655 (2007)
7. Ikramov, KhD: Some estimates for the eigenvalues of matrices. Zh. Vychisl. Mat. Mat. Fiz.10(1), 172-177 (1970)
8. Hoffman, AJ, Wielandt, HW: The variation of the spectrum of a normal matrix. Duke Math. J.20, 37-39 (1953)
9. Zhan, X: Matrix Inequalities. LNM, vol.1790. Springer, Berlin (2002)
10. Horn, RA, Johnson, CR: Topics in Matrix Analysis. Cambridge University Press, Cambridge (1991)
11. Sun, JG: Matrix Perturbation Analysis. Science Press, Beijing (2001)
12. Birkhoff, G: Tres observaciones sobre el Algebra lineal. Rev. Univ. Nac. Tucumán Ser. A, Mat. Fis. Teor.5, 147-151 (1946)
13. Horn, RA, Johnson, CR: Matrix Analysis. Cambridge University Press, Cambridge (1985)
14. Lidskii, VB: On the characteristic numbers of a sum and product of symmetric matrices. Dokl. Akad. Nauk SSSR75, 769-772 (1950)
15. Ikramov, KhD: A simple proof of the generalized Schur inequality. Linear Algebra Appl.199, 143-149 (1994)

Chapter 3

New Iterative Methods for Generalized Singular-value Problems

A. H. Refahi Sheikhani and S. Kordrostami

Department of Applied Mathematics, Faculty of Mathematical Sciences, Lahijan Branch, Islamic Azad University, Lahijan, Iran

ABSTRACT

This paper presents two new iterative methods to compute generalized singular values and vectors of a large sparse matrix. To reach acceleration in the convergence process, we have used a different inner product instead of the common one, Euclidean one. Furthermore, at each restart, a different inner product has been chosen by the researchers. A number of numerical experiments illustrate the performance of the above-mentioned methods.

Citation (APA): Sheikhani, A. R., & Kordrostami, S. (2017). New iterative methods for generalized singular-value problems. Mathematical Sciences, 11(4), 257-265. (9 pages), DOI: https://doi.org/10.1007/s40096-017-0223-3

Keywords: Generalized singular value, Krylov subspace, Iterative, Sparse

INTRODUCTION

There are a number of applications for generalized singular-value decomposition (GSVD) in the literature including the computation of the Kronecker form of the matrix pencilA−λB[5], solving linear matrix equations [1], weighted least squares [2], and linear discriminant analysis [6] to name but a few. In a number of applications like the generalized total least squares problem, the matricesAandBare large and sparse, so in such cases, only a few of the generalized singular vectors corresponding to the smallest or largest generalized singular values are needed. There is a kind of close connection between the GSVD problem and two different generalized eigenvalue problems. In fact, there are many efficient numerical methods to solve generalized eigenvalue problems [8,9,10,11]. In this paper, we will examine the Jacobi–Davidson-type subspace method which is related to the Jacobi–Davidson for the SVD [5], which in turn is inspired by the Jacobi–Davidson method to solve the eigenvalue problem [4]. The main step in Jacobi–Davidson-type method for the (GSVD) is solving the correction equations in an exact manner requiring the solution of linear systems of original size at each iteration. In general, these systems are considered as large, sparse, and nonsymmetrical. For this matter, we use the weighted Krylov subspace process to solve the correction equations in an exact manner, and we show that our proposed method has the feature of asymptotic quadratic convergence. The paper is organized as follows. In "Preparations", we will remind the readers of basic definitions of the generalized singular-value decomposition problems and their elementary properties. "A new iterative method for GSVD" introduces our new numerical methods to solve generalized eigenvalue problems together with an analysis of the convergence of these methods. Several numerical examples are presented in "Numerical experiments". Finally, the conclusions are given in the last section.

PREPARATIONS

Definition 2.1

Supposes that $A \in R^{m\times n}$ and $B \in R^{p\times n}$. The generalized singular values of the pair(A,B)are presented as

$$\sum(A,B)=\{\sigma|\sigma\geq 0,\ \det(A^TA-\sigma^2B^TB)=0\}.$$

Definition 2.2

A generalized singular value is called simple if $\sigma_i\neq\sigma_j$, for all $i\neq j$.

Theorem 2.3

Suppose $A\in R^{m\times n}$, $B\in R^{p\times n}$, *and* m≥n. *Here, taking the previous theorem into consideration, we see that there are orthogonal matrices* $U_{m\times m}$, $V_{p\times p}$ *and a nonsingular matrix* $X_{n\times n}$, *such that*

$$U^TAX=\sum\nolimits_1=\mathrm{diag}(\alpha_1,\ldots,\alpha_n)\quad \alpha_i\geq 0,$$

$$V^TBX=\sum\nolimits_2=\mathrm{diag}(\beta_1,\ldots,\beta_n)\quad \beta_i\geq 0, \tag{1}$$

where $q=\min\{p,n\},\qquad r=\mathrm{rank}(B)$

, *and* $\beta_1\geq\cdots\geq\beta_r>\beta_{r+1}=\cdots=\beta_q=0$. *If* $\alpha_j=0$ *for any* $j,r+1\leq j\leq n,$ *then* $\sum(A,B)=\{\sigma|\sigma\geq 0\}$. *Otherwise,*

$$\sum(A,B)=\left\{\frac{\alpha_i}{\beta_i}|i=1,\ldots,r\right\}.$$

Proof

Refer to [3].

Theorem 2.4

Let $A\in R^{n\times n}$, $B\in R^{n\times n}$ *have the GSVD:*

$$U^TAX=\sum\nolimits_1=\mathrm{diag}(\alpha_i),\quad V^TBX=\sum\nolimits_2=\mathrm{diag}(\beta_i);$$

furthermore, consider it as nonsingular. Here, then, the matrix pencil

$$\begin{pmatrix}0 & A\\ A^T & 0\end{pmatrix}-\lambda\begin{pmatrix}I & 0\\ 0 & B^TB\end{pmatrix} \tag{2}$$

has eigenvalues $\lambda_j=\pm\alpha_j/\beta_j,\ j=1,\ldots,n$ *which corresponds to the eigenvectors:*

$$\begin{pmatrix}u_j\\ \pm x_j/\beta_j\end{pmatrix},\quad j=1,\ldots,n \tag{3}$$

where u_j *is the ith column of* U *and* x_j *is the ith column of* X.

Proof

Refer to [3].

Let D be a diagonal matrix, that is, $D=\mathrm{diag}(d_1,d_2,\ldots,d_n)$. If u and v are two vectors of R^n, we define the D-scalar product of $(u,v)_D=v^TDu$. which is well defined if and only if the matrix D is positively definite or to say $d_i>0, i=1,\ldots,n$. The norm associated with this inner product is the DD-norm $\|\cdot\|_D$ which is defined as $\|u\|_D = \sqrt{(u,u)_D} = \sqrt{u^{\mathrm{T}}Du}\ \forall u \in R^n$.

As assumptionBhas full rank, $(x,y)_{(B^{\mathrm{T}}B)^{-1}} := y^{\mathrm{T}}(B^{\mathrm{T}}B)^{-1}x$ is an inner product, and due to this, the corresponding norm satisfies $\|x\|^2_{(B^{\mathrm{T}}B)^{-1}} := (x,x)_{(B^{\mathrm{T}}B)^{-1}}$. Inspired by the equality $\|Z\|_F^2 = \mathrm{trace}(Z^{\mathrm{T}}Z)$ for a real matrixZ, we define the $(B^{\mathrm{T}}B)^{-1}$ -Frobenius norm ofZby

$$\|Z\|^2_{(B^{\mathrm{T}}B)^{-1},F} = \mathrm{trace}(Z^{\mathrm{T}}(B^{\mathrm{T}}B)^{-1}Z). \tag{4}$$

A new iterative method for GSVD

We will advance different extraction methods here which are often more appropriate for small generalized singular values than the standard one from "A new iterative method for GSVD". Before dealing with these new methods, we should refer to our main idea which is developed considering Krylov subspace methods.

Theorem 3.1

Assume that(σ,u,v)*is a generalized singular triple:* $Aw = \sigma u$ *and* $A^{\mathrm{T}}u = \sigma B^{\mathrm{T}}Bw$, *where*σ*is a simple nontrivial generalized singular value, and* $\|u\| = \|Bw\| = 1$*, and suppose that the correction equations*

$$P = \begin{pmatrix} I - \tilde{u}\tilde{u}^{\mathrm{T}} & 0 \\ 0 & I - B^{\mathrm{T}}B\tilde{w}\tilde{w}^{\mathrm{T}} \end{pmatrix}, \tag{5}$$

are solved exactly in every step. Provided that the initial vectors $(\tilde{u}, \tilde{w})$ *are close enough to*(u,w)*the sequence of approximations* $(\tilde{u}, \tilde{w})$ *converges quadratically to*(u,w).

Proof

Refer to [4].

Lemma 3.2

*Having in mind the Theorem*3.1*, now suppose that*mm*steps of the weighted Arnoldi process*[7]*have been performed on the following matrix:*

$$\begin{pmatrix} I - uu^{\mathrm{T}} & 0 \\ 0 & I - B^{\mathrm{T}}Bww^{\mathrm{T}} \end{pmatrix}\begin{pmatrix} -\theta I & A \\ A^{\mathrm{T}} & -\theta B^{\mathrm{T}}B \end{pmatrix}. \tag{6}$$

Furthermore, consider the matrix $\widetilde{H}_m$ *as the Hessenberg matrix, whose nonzero entries are the scalars* $\tilde{h}_{i,j}$*, constructed by the Weighted Arnoldi process. Here, we notice that the basis* $\widetilde{V}_m = [\tilde{v}_1, \ldots, \tilde{v}_m]$ *constructed by this* *algorithm is*D*D-orthonormal and we have*

$$\widetilde{V}_m^{\mathrm{T}} D \widetilde{V}_m = I_m, \qquad (7)$$

$$\begin{pmatrix} I - uu^{\mathrm{T}} & 0 \\ 0 & I - B^{\mathrm{T}}Bww^{\mathrm{T}} \end{pmatrix} \begin{pmatrix} -\theta I & A \\ A^{\mathrm{T}} & -\theta B^{\mathrm{T}}B \end{pmatrix} \widetilde{V}_m = \widetilde{V}_{m+1} \begin{pmatrix} \widetilde{H}_m \\ h_{m+1,m} e_m^{\mathrm{T}} \end{pmatrix}. \qquad (8)$$

Proof

See [4].

We know that similar to Krylov methods, the*m*th(m≥1)iterate $x_m = [s_m, t_m]^t$ of the weighted-FOM and weighted-GMRES methods belong to the affine Krylov subspace:

$$\begin{pmatrix} s_0 \\ t_0 \end{pmatrix} + \kappa_m \left(\begin{pmatrix} I - uu^{\mathrm{T}} & 0 \\ 0 & I - B^{\mathrm{T}}Bww^{\mathrm{T}} \end{pmatrix} \times \begin{pmatrix} -\theta I & A \\ A^{\mathrm{T}} & -\theta B^{\mathrm{T}}B \end{pmatrix}, \begin{pmatrix} r_0^{(s)} \\ r_0^{(t)} \end{pmatrix} \right). \qquad (9)$$

Now, it is the time to prove our main theorem.

Theorem 3.3

*Considering Theorem*3.1,mm*steps of the weighted Arnoldi process have been run on*(7)*. Here, the iterate* $x_m = [s_m, t_m]^t$ *is the exact solution of the correction equation:*

$$P \begin{pmatrix} -\theta I & A \\ A^{\mathrm{T}} & -\theta B^{\mathrm{T}}B \end{pmatrix} \begin{pmatrix} s \\ t \end{pmatrix} = -r, \quad s \perp \tilde{u}, \quad t \perp \tilde{w}. \qquad (10)$$

Proof

The iterate x_m^{WF} of the weighted-FOM method is selected, because its residual isD-orthonormal or

$$\begin{pmatrix} r_m^{(s)} \\ r_m^{(t)} \end{pmatrix}^{\mathrm{WF}} \perp_D \kappa_m \left(\begin{pmatrix} I - uu^{\mathrm{T}} & 0 \\ 0 & I - B^{\mathrm{T}}Bww^{\mathrm{T}} \end{pmatrix} \times \begin{pmatrix} -\theta I & A \\ A^{\mathrm{T}} & -\theta B^{\mathrm{T}}B \end{pmatrix}, \begin{pmatrix} r_0^{(s)} \\ r_0^{(t)} \end{pmatrix} \right). \qquad (11)$$

The iterate x_m^{WG} of the weighted-GMRES method is selected to lessen the residualD-norm in (9). Here, we notice that it is the solution of the least squares problem:

$$\text{minimize}_{[s,t]^t\in(4.4)}\left\|\begin{pmatrix} A\tilde{w}-\theta\tilde{u} \\ A^{\mathrm{T}}\tilde{u}-\theta B^{\mathrm{T}}B\tilde{w}\end{pmatrix} -P\begin{pmatrix}-\theta I & A \\ A^{\mathrm{T}} & -\theta B^{\mathrm{T}}B\end{pmatrix}\begin{pmatrix}s\\t\end{pmatrix}\right\|_D. \tag{12}$$

In these methods, we use theD-inner product and theD-norm to calculate the solution in the affine subspace (9) and we create aD-orthonormal basis of the Krylov subspace:

$$K_m\left(\begin{pmatrix} I-uu^{\mathrm{T}} & 0 \\ 0 & I-B^{\mathrm{T}}Bww^{\mathrm{T}}\end{pmatrix}\begin{pmatrix}-\theta I & A \\ A^{\mathrm{T}} & -\theta B^{\mathrm{T}}B\end{pmatrix},\begin{pmatrix} r_0^{(s)} \\ r_0^{(t)}\end{pmatrix}\right). \tag{13}$$

by the weighted Arnoldi process. An iterate$x_m$of these two methods can be transcribed as

$$\begin{pmatrix}s_m\\t_m\end{pmatrix}=\begin{pmatrix}s_0\\t_0\end{pmatrix}+\widetilde{V}_m\begin{pmatrix}y_m^{(s)}\\y_m^{(t)}\end{pmatrix},$$

where $y_m \in R^m$.

Therefore, the matching residual $r_m=\left[r_m^{(s)},r_m^{(t)}\right]^t$ satisfies

$$\begin{aligned}
\begin{pmatrix}r_m^{(s)}\\r_m^{(t)}\end{pmatrix} &= \begin{pmatrix} A\tilde{w}-\theta\tilde{u} \\ A^{\mathrm{T}}\tilde{u}-\theta B^{\mathrm{T}}B\tilde{w}\end{pmatrix}-\begin{pmatrix} I-uu^{\mathrm{T}} & 0 \\ 0 & I-B^{\mathrm{T}}Bww^{T}\end{pmatrix}\begin{pmatrix}-\theta I & A \\ A^{\mathrm{T}} & -\theta B^{\mathrm{T}}B\end{pmatrix}\begin{pmatrix}s_m\\t_m\end{pmatrix}\\
&= \begin{pmatrix} A\tilde{w}-\theta\tilde{u} \\ A^{\mathrm{T}}\tilde{u}-\theta B^{T}B\tilde{w}\end{pmatrix}-P\begin{pmatrix}-\theta I & A \\ A^{\mathrm{T}} & -\theta B^{\mathrm{T}}B\end{pmatrix}\left(\begin{pmatrix}s_0\\t_0\end{pmatrix}+\widetilde{V}_m\begin{pmatrix}y_m^{(s)}\\y_m^{(t)}\end{pmatrix}\right),\\
&= \begin{pmatrix}r_0^{(s)}\\r_0^{(t)}\end{pmatrix}-P\begin{pmatrix}-\theta I & A \\ A^{\mathrm{T}} & -\theta B^{\mathrm{T}}B\end{pmatrix}\widetilde{V}_m\begin{pmatrix}y_m^{(s)}\\y_m^{(t)}\end{pmatrix},\\
&= \widetilde{V}_{m+1}\left(\beta e_1-\begin{pmatrix}\widetilde{H}_m\\h_{m+1,m}e_m^{\mathrm{T}}\end{pmatrix}\begin{pmatrix}y_m^{(s)}\\y_m^{(t)}\end{pmatrix}\right),
\end{aligned}$$

where $\beta=\|r_0\|_D$, $r_0=\left[r_0^{(s)},r_0^{(t)}\right]^t$,and$e_1$is the first vector of the canonical basis. At this point, the weighted-FOM method entails finding the vector $y_m^{\mathrm{WF}}=\left[y_m^{(s)},y_m^{(t)}\right]^t$ solution of the problem:

$$\widetilde{V}_m^{\mathrm{T}}D\widetilde{V}_{m+1}(\beta e_1-\widetilde{H}_m y_m^{\mathrm{WF}})=0,$$

which is equal to solve

$$\widetilde{H}_m y_m^{\mathrm{WF}}=\beta e_1. \tag{14}$$

To the extent that the weighted-GMRES method is considered, the matrix$\widetilde{V}_{m+1}$isD-orthonormal, so we have

$$\|r_m\|_D^2=\left\|\widetilde{V}_{m+1}(\beta e_1-\widetilde{H}_m y_m)\right\|_D^2=\left\|\beta e_1-\widetilde{H}_m y_m\right\|_2^2,$$

and problem (12) is condensed to find the vector y_m^{WG} solution of the minimization problem:

$$\text{minimize}_{y \in R^m} \left\| \beta e_1 - \widetilde{H}_m y \right\|_2. \tag{15}$$

We can reach the solution of (14) and (15) with the use of the**QR**decomposition of the matrix$\widetilde{H}_m$, as for the FOM and GMRES algorithms. Whenmmis equal to the degree of the minimal polynomial of

$$\begin{pmatrix} I - uu^{\mathrm{T}} & 0 \\ 0 & I - B^{\mathrm{T}}Bww^{\mathrm{T}} \end{pmatrix} \begin{pmatrix} -\theta I & A \\ A^{\mathrm{T}} & -\theta B^{\mathrm{T}}B \end{pmatrix}$$

for $r_0 = [r_0^{(s)}, r_0^{(t)}]^t$, the Krylov subspace (13) will be invariant. Therefore, the iterate $x_m = [s_m, t_m]^t$ gained by both methods is the exact solution of the correction Eq.(10).

It is time to write the main algorithm in this paper now. The following algorithm applies FOM, GMRES, weighted-FOM, and weighted-GMRES processes to solve the correction Eq.(10) and as a final point to solve the generalized singular-value decomposition problem. They are represented as F-JDGSVD, G-JDGSVD, WF-JDGSVD, and WG-JDGSVD.

Algorithm 3.1 F-JDGSVD, G-JDGSVD, WF-JDGSVD and WG-JDGSVD

Input: *Starting vectors* u_1 *and* w_1, *and a tolerance* ε, *parameter* m , *and initial vector* x_0

Output: *An approximate triple* (θ, u, w) *for the largest generalized singular triple satisfying*

$$\left\| \begin{pmatrix} Aw - \theta u \\ A^T u - \theta B^T Bw \end{pmatrix} \right\| \leq \varepsilon$$

1. $s = u_1, t = w_1, U_0 = [\ ], W_0 = [\ ]$

   for $k = 1, 2, \ldots$ **do**

2. $U_k = RGS\left(U_{k-1}, s\right)$

   $W_k = RGS_{B^T B}\left(W_{k-1}, t\right)$

3. *Compute the kth column of* AW_k, $A^T U_k$, *and* $B^T BW_k$

   Compute the kth row and column of $H_k = U_k^T A W_k$

4. *Compute the approximately largest generalized singular triple* (θ, c, d),

   of the projected system using the standard extraction technique.

5. $u = U_k c,\ w = W_k d$

6. $r = \begin{pmatrix} Aw - \theta u \\ A^T u - \theta B^T Bw \end{pmatrix}$

7. *Stop if* $\|r\| \le \varepsilon$

8. *Compute*

$$\mathsf{A} = \begin{pmatrix} I - uu^T & 0 \\ 0 & I - B^T Bww^T \end{pmatrix} \begin{pmatrix} -\theta I & A \\ A^T & -\theta B^T B \end{pmatrix}$$

9. *for solving the system* $\mathsf{A}\begin{pmatrix} s \\ t \end{pmatrix} = -r$

compute $r_0 = b - Ax_0$

for $l = 1, 2, \ldots$ **do**

for solving the system by FOM or GMRES methods:

Compute $\beta = \|r_0\|_2$ *and* $v_1 = r_0 / \beta$.

Construct the orthonormal basis V_m *by the Arnoldi process, starting with* v_1

Form the approximate solution:

FOM: *Solve the system* $H_m y_m = \beta e_1$ *by the QR factorization of* H_m *and set*

$x_m = x_0 + V_m y_m$, $r_m = b - Ax_m$.

GMRES: *Compute* $y_m = \operatorname{argmin}_{y \in R^m} \|\beta e_1 - H_m y\|_2$, *by the QR factorization of* H_m *and set*

$x_m = x_0 + V_m y_m$, $r_m = b - Ax_m$

for solving the system by WFOM or WGMRES methods:

Choose the vector d *such as* $\|d\|_2 = \sqrt{n}$, *and set* $D = diag(d)$

Compute $\tilde{\beta} = \|r_0\|_D$ *and* $\tilde{v}_1 = r_0 / \tilde{\beta}$.

Construct the D *-orthonormal basis* V_m *by the weighted Arnoldi process, starting with* $\tilde{v}_1$

Form the approximate solution:

WFOM: *Solve the system* $H_m y_m = \tilde{\beta} e_1$ *by the QR factorization of* H_m *and set*

$x_m = x_0 + V_m y_m$, $r_m = b - Ax_m$.

WGMRES: *compute* $y_m = \operatorname{argmin}_{y \in R^m} \|\tilde{\beta} e_1 - \overline{H}_m y\|_2$, *by the QR factorization of* $\overline{H}_m$,

set $x_m = x_0 + V_m y_m$, $r_m = b - Ax_m$.

If $\|r_m\|_2 \le \varepsilon_1$ *stop else set* $x_0 = x_m$, $r_0 = r_m$.

As Algorithm 3.1 displays, there are two loops in this algorithm. One of them computes the largest generalized singular value called the outer iteration, and the other called the inner iteration solves the system of linear equation at each iteration. Numerical tests indicate that there is a significant relation between parametermand the norm of residual vector and the computational time.

Convergence

We will now demonstrate that the method we have proposed has asymptotically quadratic convergence to generalized singular values when the correction equations are solved in an exact manner and tend toward linear convergence when they are solved with a sufficiently small residual reduction.

Theorem 3.4

*Having in mind Theorem3.3, suppose that*mm*steps of the weighted Arnoldi process have been performed on*(6)*and*$x_m = [s_m, t_m]^{\mathrm{T}}$ *is the exact solution of the correction Eq.*(10).*Provided that he initial vectors*$(\tilde{u}, \tilde{w})$ *are close enough to*(u,w)*, the sequence of approximations*$(\tilde{u}, \tilde{w})$*converges quadratically to*(u,w).

Proof

Suppose

$$\mathrm{A} = \begin{pmatrix} 0 & A \\ A^T & 0 \end{pmatrix}, \quad \mathrm{B} = \begin{pmatrix} I & 0 \\ 0 & B^T B \end{pmatrix}$$

andPare like what you have seen in (5). Let $[s_m, t_m]^T$ with $s_m \perp \tilde{u}$ and $t_m \perp \tilde{w}$ be the exact solution to the correction equation:

$$P(\mathrm{A} - \theta \mathrm{B}) \begin{pmatrix} s_m \\ t_m \end{pmatrix} = -r. \tag{16}$$

Besides, let $\alpha u = \tilde{u} + s, \ \ s \perp \tilde{u},$ and $\beta w = \tilde{w} + t, \ \ t \perp \tilde{w},$ for certain scalarsαandβ, satisfy (15); note that these decompositions are possible meanwhile $u^{\mathrm{T}}\tilde{u} \neq 0$ and $w^{\mathrm{T}}\tilde{w} \ / = 0$ because of the assumption that the vectors$(\tilde{u}, \tilde{w})$are close to(u,w). Projecting (16) yields

$$P(\mathrm{A} - \theta \mathrm{B}) \begin{pmatrix} s \\ t \end{pmatrix} = -r + P \begin{pmatrix} (\mu_1 - \theta)s \\ (\mu_2 - \theta)B^{\mathrm{T}}Bt \end{pmatrix}. \tag{17}$$

Subtracting (16) from (17) gives

$$P(\mathrm{A} - \theta \mathrm{B}) \begin{pmatrix} s - s_m \\ t - t_m \end{pmatrix} = P \begin{pmatrix} (\mu_1 - \theta)s \\ (\mu_2 - \theta)B^{\mathrm{T}}Bt \end{pmatrix}.$$

Thus for$(\tilde{u}, \tilde{w})$close enough to(u,w),P(A−θB)is a bijection from $\tilde{u}^{\perp} \times \tilde{w}^{\perp}$onto itself. Together with

$\mu_1 = \tilde{u}^{\mathrm{T}} A(\tilde{w} + t) = \theta + O(\|t\|),$

$\mu_2 = (\tilde{w} + t)^{\mathrm{T}} A^{\mathrm{T}} (\tilde{u} + s) / \|B(\tilde{w} + t)\|^2 = \theta + O(\|s\| + \|t\|),$

this implies asymptotic quadratic convergence:

$$\left\| \begin{pmatrix} \alpha u - (\tilde{u} + s_m) \\ \beta w - (\tilde{w} + t_m) \end{pmatrix} \right\| = \left\| \begin{pmatrix} s - s_m \\ t - t_m \end{pmatrix} \right\| = O\left(\left\| \begin{pmatrix} s \\ t \end{pmatrix} \right\|^2 \right).$$

NUMERICAL EXPERIMENTS

In this section, we look for the largest generalized singular value, using the following default options of the proposed method:

Maximum dimension of search spaces	3030
Maximum iterations to solve correction equation	1010
Fix target until‖r‖≤ε	0.010.01
Initial search spaces	Random

Example 4.1

The matrix pair(A,B)(A,B)is constructed, such that that they are similar to experiments as [7]. We choose two diagonal matrices of dimensionn=1000. Forj=1,2,…,1000

$$C = \mathrm{diag}(c_j), \quad c_j = (n - j + 1)/2n, \quad S = \sqrt{1 - C^2},$$

$$D = \mathrm{diag}(d_j), \quad d_j = \lceil j/250 \rceil + r_j$$

where the$r_j$uniformly distributed on the interval(0,1)and⌈·⌉denotes the ceil function. We take

$$A = Q_1 C D Q_2, \quad B = Q_1 S D Q_2$$

where$Q_1$and$Q_2$are two random orthogonal matrices. The estimated condition numbers ofA andBare4.4e2and5.7e0, respectively (Table1).

Table 1: Implementation of Algorithm 3.1 for(A,B)with different values ofmm

mm	F-JDGSVD			G-JDGSVD			WF-JDGSVD			WG-GSVD		
	σ_{max}	$\|r\|2$	Time	σ_{max}	$\|r\|2$	Time	σ_{max}	$\|r\|2$	Time	σ_{max}	$\|r\|2$	Time
4	0.5766	0.0084	23.95	0.5767	0.0062	28.35	0.5773	9.22e–5	31.13	0.5770	8.88e–6	22.08
6	0.5773	0.0052	19.82	0.5770	0.0043	23.32	0.5772	4.82e–5	28.76	0.5768	4.01e–6	17.51
8	0.5773	0.0023	16.10	0.5771	0.0028	19.30	0.5773	7.92e–6	23.66	0.5772	1.00e–6	14.70
10	0.5772	0.0058	14.85	0.5772	0.0014	16.31	0.5773	2.81e–6	17.99	0.5772	9.94e–7	12.04

We can see that by increasing the value ofm, the number of outer and inner iterations decreases. Therefore, the consuming time also decreases. But not that ifmis very large, the number of iterations increases because of loosing the orthogonality property. This example is given to show the improvement brought by the weighted methodsWF-JDGSVDWF-JDGSVDandWG-JDGSVDWG-JDGSVDis simultaneously on the relative error and on the computational time (Fig.1).

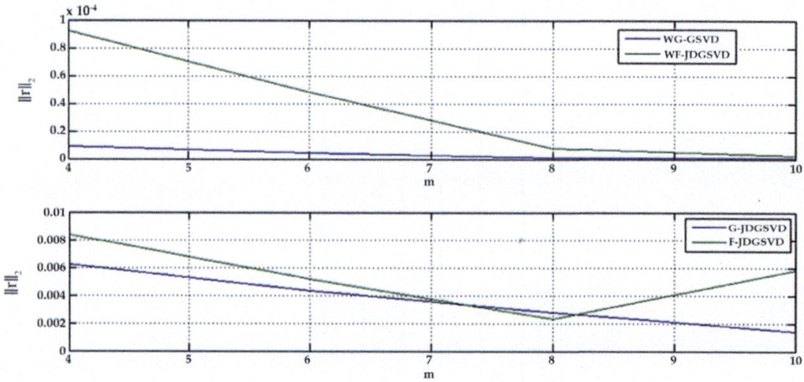

Figure1: *Errors plot*created by F-JDGSVD, G-JDGSVD, WF-JDGSVD, and WG-GSVD.

From figure one, we can see that the suggested method WG-JDGSVD is more accurate form the other methods.

Example 4.2

In this experiment, we takeA=CDandB=SDof various dimensionn=400,800,1000,1200.

This example is given to show the performance of two new methods on the large sparse problems. In this test, we have difficulties in computing the largest singular value for ill-conditioned matricesAandB. We note that in this experiments, due to the ill-conditioning ofAandB, it turned out to be advantageous to turn of the Krylov option.

Example 4.3

Consider the matrix pair(A,B), whereAis selected from the university of Florida sparse matrix collection [8] as*lp*-ganges. This matrix arises from a linear programming problem. Its size is$1309\times1706$and it has a total ofNz=6937nonzero elements. The estimated condition number is2.1332e4, andBis the1309×1706identity matrix (Tables2,3).

Table2: Implementation of Algorithm 3.1 for(A,B)with various dimensions andm=6

n	F-JDGSVD		G-JDGSVD		WF-JDGSVD		WG-GSVD		κ(A)	κ(B)
	‖r‖2	Time	‖r‖2	Time	‖r‖2	Time	‖r‖2	Time		
400	8.82e–4	7.03	0.0098	6.08	2.47e–8	11.85	2.14e–9	11.78	3.5e2	3.2e0
800	0.0085	19.59	0.0063	21.89	9.19e–8	26.09	4.44e–8	22.25	3.6e2	5.6e0
1200	0.0034	27.83	0.0073	29.35	6.74e–6	41.18	5.19e–7	42.35	4.8e2	6.6e0
1600	0.0075	38.65	0.0084	35.89	1.19e–5	49.09	4.99e–5	58.17	6.0e2	8.9e0

Table3: Implementation of Algorithm 3.1 for(A,B)with different values ofmm

m	F-JDGSVD			G-JDGSVD			WF-JDGSVD			WG-GSVD		
	σ_{max}	‖r‖2	Time	σ_{max}	‖r‖2	Time	σ_{max}	‖r‖2	Time	σ_{max}	‖r‖2	Time
10	3.9889	0.0075	52.57	3.9865	0.0079	48.86	2.7297	0.00034	63.59	3.9890	0.00015	55.36
20	3.9907	0.0054	46.63	3.9889	0.0035	42.84	2.7298	0.00098	56.99	3.9890	0.00041	47.39
30	2.7298	0.0016	39.78	3.9889	0.0097	36.08	3.9907	0.00043	48.74	3.9888	0.00040	39.65
40	3.9897	0.0091	33.17	3.9888	0.0052	30.89	2.7298	0.00027	38.37	3.9887	0.00014	32.68

We should mention that, for all considered Krylov subspaces sizes, each weighted method converges in less iterations and less time than its corresponding standard method. The convergence of F-JDGSVD and G-JDGSVD is slow, and we have linear asymptotic convergence. However, the two WF-JDGSVD and WG-JDGSVD methods have quadratic asymptotic convergence, because the correction Eq.(10) is solved exactly.

Remark 4.4

From the above examples and tables, we can see that the two suggested methods are more accurate than G-JDGSVD and F-JDGSVD for the same value*m*, but its computational times are often a little longer than G-JDGSVD and F-JDGSVD. Therefore, we can use WF-JDGSVD and WG-GSVD if the computational time is less important.

Remark 4.5

The algorithm we have described finds the largest generalized singular triple. We can compute multiple generalized singular triples of the pair(A,B)using a deflation technique. Suppose that$U_f=[u_1,\ldots,u_f]$and$W_f=[w_1,\ldots,w_f]$contain the already found generalized singular vectors, whereBW_fhas orthonormal columns. We can check that the pair of deflated matrices

$$\hat{A} := (I - U_f U_f^{\mathrm{T}})A(I - W_f W_f^{\mathrm{T}} B^{\mathrm{T}} B) \quad \text{and}$$
$$\hat{B} := B(I - W_f W_f^{\mathrm{T}} B^{\mathrm{T}} B) \qquad (18)$$

has the same generalized singular values and vectors as the pair(A,B) (see [3]).

Example 4.6

In generalized singular-value decomposition, ifB=I_n, then×nidentity matrix, we get the singular value ofA.SVDhas important applications in image and data compression. For example, consider the following image. This image is represented by a1185×1917matrixA. Which we can then decompose via the singular-value decomposition asA=U∑ V^TwhereUis1185×1185,∑is1185×1917, andVis1917×1917. The matrixA, however, can also be written as a sum of rank 1 matrices $A = \sum_{j=1}^{r} \sigma_j u_j v_j^{\mathrm{T}}$, where $\sigma_1 \geq \sigma_2 \geq \cdots \geq \sigma_r > 0$ are thernonzero singular value ofA. In digital image processing, any matrixAof orderm×n(m≥n) generally has a large number of small singular values. Suppose there are(n−k)small singular values ofAAthat can be neglected (Fig.2).

Figure2: Original image.

Then, the matrix $A_k = \sigma_1 u_1 v_1^{\mathrm{T}} + \sigma_2 u_2 v_2^{\mathrm{T}} + \cdots + \sigma_k u_k v_k^{\mathrm{T}}$ is a very good approximation ofA, and such an approximation can be adequate. Even whenkis chosen much less thenn, the digital image corresponding to$A_k$can be very close to the original image. Below are the subsequent approximations using various numbers of singular values.

5 singular values

10 singular values

The observation on those examples, we found whenk≤20, the images are blurry but with the increase of singular values, when their numbers are about50, we have a good approach to the original image.

20 singular values

30 singular values

50 singular values

100 singular values

CONCLUSIONS

In this paper, we have suggested two new iterative methods, namely, WF-JDGSVD and WG-JDGSVD, for the computation of some of the generalized singular values and corresponding vectors. Various examples studied illustrate these methods. To accelerate the convergence, we applied the Krylov subspace method for solving the correction equations in large sparse problems. In our methods, we see the existence of asymptotically quadratic convergence, because the correction equations are solved exactly. In the meantime, the correction equations in F-JDGSVD and G-JDGSVD methods are solved inexactly for large sparse problems, so we have linear convergence.

As the amount of the WF-JDGSVD and WG-JDGSVD methods is not much larger than that of the F-JDGSVD and G-JDGSVD methods, and as the weighted methods need less iterations to convergence, the parallel version of the weighted methods seems very interesting. From the tables and the figures, we see that when$m$increases, the suggested methods are more accurate than the previous methods; moreover, by increasing the dimension of the matrix, two suggested methods are applicable; this results are supported by convergence theorem which shows the asymptotically quadratic convergence to generalized singular values.

REFERENCES

1. Betcke, T.: The generalized singular value decomposition and the method of particular solutions. SIAM. Sci. Comput.30, 1278–1295 (2008)
2. Hochstenbach, M.E.: Harmonic and refined extraction methods for the singular value problem, with applications in least square problems. BIT44, 721–754 (2004)
3. Hochstenbach, M.E.: A Jacobi–Davidson type method for the generalized singular value problem. Linear Algebra Appl.431, 471–487 (2009)
4. Hochstenbach, M.E., Sleijpen, G.L.C.: Two-sided and alternating Jacobi–Davidson. Linear Algebra Appl.358(1–3), 145–172 (2003)
5. Kagstrom, B.: The generalized singular value decomposition and the general$A-\lambda B$problem. BIT24, 568–583 (1984)
6. Park, C.H., Park, H.: A relationship between linear discriminant analysis and the generalized minimum squared error solution. SIAM J. Matrix Anal. Appl.27, 474–492 (2005)
7. Saad, Y.: Krylov subspace methods for solving large unsymmetrical linear systems. Math. Comput.37, 105–126 (1981)
8. Saberi Najafi, H., Refahi Sheikhani, A.H: A new restarting method in the Lanczos algorithm for generalized eigenvalue problem. Appl. Math. Comput.184, 421–428 (2007)
9. Saberi Najafi, H., Refahi Sheikhani, A.H.: FOM-inverse vector iteration method for computing a few smallest, (largest) eigenvalues of pair (A, B). Appl. Math. Comput.188, 641–647 (2007)
10. Saberi Najafi, H., Refahi Sheikhani, A.H., Akbari, M.: Weighted FOM-inverse vector iteration method for computing a few smallest (largest) eigenvalues of pair (A, B). Appl. Math. Comput.192, 239–246 (2007)
11. Saberi Najafi, H., Edalatpanah, S.A., Refahi Sheikhani, A.H.: Convergence analysis of modified iterative methods to solve linear systems. Mediterr. J. Math.11(3), 1019–1032 (2014)

Chapter 4

Blind Distributed Estimation Algorithms for Adaptive Networks

Muhammad O Bin Saeed, Azzedine Zerguine and Salam A Zummo

Electrical Engineering Department, King Fahd University of Petroleum & Minerals, Dhahran 31261, Saudi Arabia

ABSTRACT

Until recently, a lot of work has been done to develop algorithms that utilize the distributed structure of an ad hoc wireless sensor network to estimate a certain parameter of interest. However, all these algorithms assume that the input regressor data is available to the sensors, but this data is not always available to the sensors. In such cases, blind estimation of the required parameter is needed. This work formulates two newly developed blind block-recursive algorithms based on singular value decomposition (SVD) and Cholesky factorization-based techniques. These adaptive algorithms are

Citation (APA): Saeed, M. O. B., Zerguine, A., & Zummo, S. A. (2014). Blind distributed estimation algorithms for adaptive networks. EURASIP Journal on Advances in Signal Processing, 2014(1), 136. (20 pages), DOI: https://doi.org/10.1186/1687-6180-2014-136

then used for blind estimation in a wireless sensor network using diffusion of data among cooperative sensors. Simulation results show that the performance greatly improves over the case where no cooperation among sensors is involved.

Keywords: Blind estimation, Diffusion, Adaptive networks

INTRODUCTION

This work studies the problem of blind distributed estimation over an ad-hoc wireless sensor network (WSN). WSNs have recently generated a great deal of renewed interest in distributed computing. New research avenues have opened up in the fields of estimation and tracking of parameters of interest, in applications requiring a robust, scalable and low-cost solution. To appreciate such applications, consider a set ofNsensor nodes spread over a geographic area as shown in Figure1. Sensor measurements are taken at each node at every time instant. The objective of the sensor is to estimate a certain unknown parameter of interest using these sensed measurements.

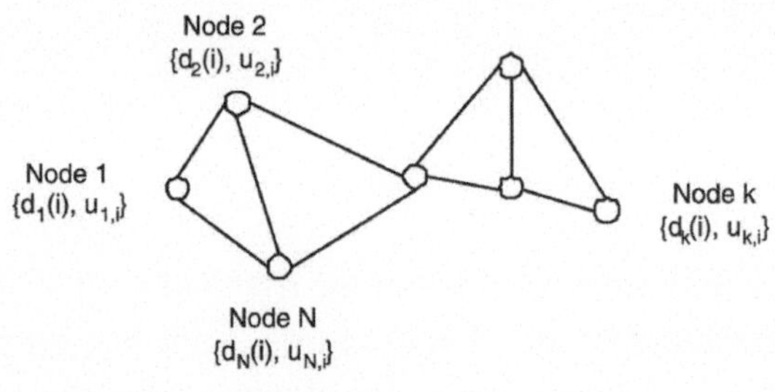

Figure 1: Adaptive network ofNnodes.

Several algorithms have been devised in the literature for distributed estimation [1,2,3,4,5]. The work in [1] introduces a distributed estimation approach using the recursive least squares algorithm. Other algorithms involving the least-mean-square (LMS) approach have also been suggested [2,3,4,5].

However, all these algorithms assume that the input regressor data,$\mathbf{u}_{k,i}$, is available at the sensors. If this information is not available, then the said problem becomes a blind estimation problem. Blind algorithms have been a topic of interest ever since Sato devised a blind algorithm [6] in the context of equalization [7]. Since then, several algorithms have been derived for blind estimation [8,9,10,11,12,13,14,15]. The work in [8] summarizes the second-order statistics-based approaches for blind identification. These include multichannel as well as single-channel blind estimation methods such as the works in [9] and [10]. The work in [11] is one of the most cited blind estimation techniques for a single-input-single-output (SISO) model. However, unlike in [11], it is shown in [12] that the technique of [11] can be improved upon using only two blocks of data. A key idea in [12] is used in [13] to devise an algorithm that does indeed show improvement over the algorithm of [11]. However, the computational complexity of this new algorithm (in [13]) is very demanding. A generalized algorithm is devised in [14], improving upon both algorithms developed in [12,13]. In [15], a Cholesky factorization-based least squares solution is suggested that simplifies the work of [11,13,14]. Although the performance of the algorithm developed in [15] is not as good as that of the previous algorithms, it nevertheless provides an excellent trade-off between performance level and computational complexity. However, in systems where less complexity is required and performance can be compromised to some extent, this algorithm would provide a good substitute to the algorithms developed in [12,13].

As mentioned above, for the case where the input regressor data is not available to the WSN environment used, then blind estimation techniques become mandatory. In this case, since blind estimation techniques have not yet been developed for this field, blind block-recursive least squares algorithms would have to be devised, inspired from the works in [11] and [15], and then implemented in a distributed WSN environment using the diffusion approach suggested in [1].

The following notation has been used here. Boldface letters are used for vectors/matrices and normal font for scalar quantities. Matrices are defined by capital letters and small letters are used for vectors. The notation $(.)^T$stands for transposition for vectors and matrices and expectation operation is denoted byE[.]. Any other mathematical operators used in this paper will be defined as and when introduced in the paper.

The paper is divided as follows: Section 2 defines the problem statement. Section 3 gives a brief overview of the blind estimation algorithms taken into consideration in this work. Section 4 proposes the newly developed recursive forms of the two algorithms, as well as their diffusion counterparts, to be used in wireless sensor networks. Section 5 studies the computational complexity of all the algorithms. The simulation results are discussed in detail in section 6. Finally, the paper is concluded in section 7.

PROBLEM STATEMENT

Consider a network of K sensor nodes deployed over a geographical area, to estimate an (M×1) unknown parameter vector $\mathbf{w}^o$ as shown in Figure 1. Each node k has access to a time realization of a scalar measurement $d_k(i)$ that is given by

$$d_k(i) = \mathbf{u}_{k,i}\mathbf{w}^o + v_k(i), \qquad 1 \le k \le K \tag{1}$$

where $\mathbf{u}_{k,i}$ is a (1×M) input regressor vector, v_k is a spatially uncorrelated zero-mean additive white Gaussian noise with variance $\sigma^2_{v_k}$ and i denotes the time index. The input data is assumed to be Gaussian. The aim of this work is to estimate the unknown vector $\mathbf{w}^o$ using the sensed data $d_k(i)$ without knowledge of the input regressor vector. The estimate of the unknown vector can be denoted by an (M×1) vector $\mathbf{w}_{k,i}$. Assuming that each node k cooperates only with its neighbors and k has access to updates $\mathbf{w}_{l,i}$, from its K_k neighboring nodes at every time instant i, where $\{l \in K_k, l \neq k\}$, in addition to its own estimate, $\mathbf{w}_{k,i}$. The adapt-then-combine (ATC) diffusion scheme [16] first updates the local estimate at each node using the adaptive algorithm and then fuses together the estimates from the K_k neighboring nodes. This scheme will be used in this work for the development of our distributed algorithm. Note that, even though this work is designed for a fixed topology, it can be extended to a dynamic one.

BLIND ESTIMATION ALGORITHM

In this work, the input regressor data, $\mathbf{u}_k(i)$ is assumed to be not available to the sensors and the unknown vector $\mathbf{w}^o$ is estimated using only the sensed values, $d_k(i)$. Since the data considered here is Gaussian, a method using second-order statistics only is sufficient for such an estimation problem as it will capture all the required data statistics. Even for the case of non-Gaussian data, such an approach would provide a suboptimal yet accurate enough estimate. The work in [11] uses the second-order statistics in an intelligent manner to create a null space with respect to the unknown vector $\mathbf{w}^o$. At the receiver end, this null space is then exploited to estimate the unknown vector. The authors in [15] further simplify the algorithm of [11] by proposing a new algorithm that reduces complexity but at a cost of performance degradation. These two algorithms are taken into consideration in this work as one provides excellent results whereas the other provides a computationally tractable solution.

Singular Value Decomposition-based Blind Algorithm

The work in [11] uses redundant filterbank precoding to construct data blocks that have trailing zeros. These data blocks are then collected at the receiver and used for blind channel identification. In this work, however, there is no precoding required. The trailing zeros will still be used though, for estimation purposes. Let the unknown parameter vector be of size (M×1). Suppose the input vector is a (P×1) vector with (P−M) trailing zeros

$$\mathbf{s}_k(i) = \left\{s_{k,0}(i), s_{k,1}(i), \ldots, s_{k,M-1}(i), 0, \ldots, 0\right\}^T, \tag{2}$$

wherePandMare related throughP= 2M−1. The unknown parameter vector can be written in the form of a convolution matrix given by

$$\mathbf{W} = \begin{bmatrix} w(0) & \cdots & w(M-1) & 0 & \cdots & 0 \\ 0 & \ddots & \ddots & \ddots & \ddots & \vdots \\ \vdots & \ddots & \ddots & \ddots & \ddots & 0 \\ 0 & \cdots & 0 & w(0) & \cdots & w(M-1) \end{bmatrix}^T, \tag{3}$$

where$\mathbf{w}^o$=[**w**(**0**),**w**(**1**),...,**w**(**M**−**1**)] is the unknown parameter vector. The output data block can now be written as

$$\mathbf{d}_k(i) = \mathbf{W}\mathbf{s}_k(i) + \mathbf{v}_k(i), \tag{4}$$

where$\mathbf{d}_k$(i) is the ((2M−1)×1) output data block and$\mathbf{v}_k$is the added noise. The output blocks {$\mathbf{d}_k$(i)} are collected together to form the following matrix

$$\mathbf{D}_{k,N} = (\mathbf{d}_k(0), \mathbf{d}_k(1), \ldots, \mathbf{d}_k(N-1)), \tag{5}$$

whereNis greater than the minimum number of data blocks required for the input blocks to have a full rank. The singular value decomposition (SVD) of the autocorrelation of$\mathbf{D}_{k,N}$gives a set of null eigenvalues. Thus, the eigendecomposition

$$\mathbf{D}_{k,N}\mathbf{D}_{k,N}^T = \left(\bar{\mathbf{U}}_k \; \tilde{\mathbf{U}}_k\right) \begin{pmatrix} \Sigma_{M\times M} & \mathbf{0}_{M\times(M-1)} \\ \mathbf{0}_{(M-1)\times M} & \mathbf{0}_{(M-1)\times(M-1)} \end{pmatrix} \times \begin{pmatrix} \bar{\mathbf{U}}_k^T \\ \tilde{\mathbf{U}}_k^T \end{pmatrix}, \tag{6}$$

where$\bar{\mathrm{U}}_k$is the (P×M) matrix of eigenvectors and$\tilde{\mathrm{U}}_k$is a (P× (M−1)) matrix whose columns form the null space for$\mathbf{D}_{k,N}$. This implies

$$\tilde{\mathbf{U}}_k^T \mathbf{W} = \mathbf{0}, \tag{7}$$

which can also be written as

$$\tilde{\mathbf{u}}_{k,m}^T \mathbf{W} = \mathbf{0}^T, \tag{8}$$

where $m = 1,\ldots,M-1$ and $\tilde{\mathbf{u}}_{k,m}$ is simply the m_{th} vector of $\tilde{\mathbf{U}}_k$. This equation denotes convolution since, as pointed earlier, $\mathbf{W}$ is essentially a convolution matrix. Since convolution is a commutative operation, Equation 8 can also be written as

$$\mathbf{w}^T \mathcal{U}_k := \mathbf{w}^T \left(\mathcal{U}_{k,1} \ldots \mathcal{U}_{k,M-1}\right) = \mathbf{0}^T, \tag{9}$$

where $\mathbf{w}$ is a component vector of the convolution matrix $\mathbf{W}$ and $U_{k,m}$ is an (M×M) Hankel matrix given by

$$\mathcal{U}_{k,m} = \begin{bmatrix} \tilde{\mathbf{u}}_{k,m}(0) & \tilde{\mathbf{u}}_{k,m}(1) & \cdots & \tilde{\mathbf{u}}_{k,m}(P-M) \\ \tilde{\mathbf{u}}_{k,m}(1) & \tilde{\mathbf{u}}_{k,m}(2) & \cdots & \tilde{\mathbf{u}}_{k,m}(P-M+1) \\ \vdots & \vdots & \vdots & \vdots \\ \tilde{\mathbf{u}}_{k,m}(M-1) & \tilde{\mathbf{u}}_{k,m}(M) & \cdots & \tilde{\mathbf{u}}_{k,m}(P-1) \end{bmatrix}. \tag{10}$$

The final parameter estimate is given by the unique solution (up to a constant factor) of Equation 9. It is important to note here that due to the presence of noise, the final estimate is not accurate.

Cholesky Factorization-based Blind Algorithm

The work in [15] describes a method that replaces the SVD operation with the Cholesky factorization operation to blindly estimate the channel. Again, the received block data matrix can be written as (5). Taking the autocorrelation of $\mathbf{D}_{k,N}$ and assuming the input data regressors to be white Gaussian with variance $\sigma^2_{\mathbf{s},k}$, we get

$$\mathbf{R}_{\mathbf{d},k} = \mathrm{E}\left[\mathbf{D}_N \mathbf{D}_N^T\right] = \sigma^2_{\mathbf{s},k} \mathbf{W}\mathbf{W}^T + \sigma^2_{\mathbf{v},k}\mathbf{I}. \tag{11}$$

Now if the second-order statistics of both the input regressor data as well as the additive noise are known, then the correlation matrix for the unknown vector can be written as

$$\mathbf{R}_{\mathbf{w}} = \mathbf{W}\mathbf{W}^T = \left(\mathbf{R}_{\mathbf{d},k} - \sigma^2_{\mathbf{v},k}\mathbf{I}\right) / \sigma^2_{\mathbf{s},k}. \tag{12}$$

However, this information, particularly the information about the input regressor data, is not always known and cannot be easily estimated either. Therefore, the correlation matrix of the unknown parameter vector has to be approximated by the correlation matrix of the received/sensed data. Now the algorithm in [15] uses the Cholesky factor of this correlation matrix to provide a least squares estimate of the unknown parameter vector.

The method given in [15] is summarized here. Since the correlation matrix is not available at the receiver, an approximate matrix is calculated using K-blocks of data. So the correlation matrix is given by

$$\hat{\mathbf{R}}_{\mathbf{d},k} = \frac{1}{K}\sum_{i=1}^{K} \mathbf{d}_k(i)\mathbf{d}_k^T(i). \tag{13}$$

As the second-order statistics of the noise are not known, the noise variance is estimated and then subtracted from the data correlation matrix. Thus, we have

$$\hat{\mathbf{R}}_{\mathbf{w},k} = \hat{\mathbf{R}}_{\mathbf{d}_k} - \hat{\sigma}_{\mathbf{v},k}^2 \mathbf{I}_K = \frac{1}{K}\sum_{i=1}^{K} \mathbf{d}_k(i)\mathbf{d}_k^T(i) - \hat{\sigma}_{\mathbf{v},k}^2 \mathbf{I}_K. \tag{14}$$

Taking the Cholesky factor of this matrix gives us the upper triangular matrix $\hat{\mathbf{G}}_k$

$$\hat{\mathbf{G}}_k = \text{chol}\left\{\hat{\mathbf{R}}_{\mathbf{w},k}\right\} \tag{15}$$

Next, we use thevecoperator to get an ($M^2\times$ 1) vector$\hat{\mathbf{g}}_k$

$$\hat{\mathbf{g}}_k = \text{vec}\left\{\hat{\mathbf{G}}_k\right\}. \tag{16}$$

It is given in [15] that the vectors**g**and**w**°are related through

$$\mathbf{g} = \mathbf{Q}\mathbf{w}^o, \tag{17}$$

where**Q**is an ($M^2\times M$) selection matrix given by $[\mathbf{J}_1^T \mathbf{J}_2^T \cdots \mathbf{J}_M^T]^T$, and the ($M\times M$) matrices**J**$_c$are defined as

$$\mathbf{J}_c = \begin{cases} 1, & \text{if } a+b=c-1 \\ 0, & \text{otherwise}, \end{cases} \tag{18}$$

where {a,b,c} = 0,...,M−1. So the least squares estimate of the unknown parameter vector is given by [15]

$$\hat{\mathbf{w}}_k = \left(\mathbf{Q}^T\mathbf{Q}\right)^{-1}\mathbf{Q}^T\hat{\mathbf{g}}_k. \tag{19}$$

The work in [15] also gives a method for estimating the noise variance that is on the whole adequate except it may not provide correct estimates of the noise variance at low SNRs. As a result, subtracting the estimated variance from the autocorrelation matrix may not yield a positive-definite matrix. In such cases, the use of Cholesky factorization may not be justified. However, neglecting the noise variance estimate altogether may lead to a poor estimate of the parameter vector. Despite this shortcoming, the main advantage of this method remains its very low computational complexity. Whereas the method of [11] requires the singular value decomposition of the autocorrelation matrix followed by the building of Hankel matrices using the null eigenvectors and then finding a unique solution to an over-determined set of linear equation, this method [15] simply evaluates the Cholesky factor (upper triangular matrix) of the autocorrelation matrix and then uses it to directly find the required estimate. Computational complexity is, thus, greatly reduced but at the cost of a performance degradation.

Both of the above-mentioned methods require several blocks of data to be stored before estimation can be performed. Although the least squares approximation gives a good estimate, a sensor network requires an algorithm that can be deployed in a distributed manner, which is possible only with recursive algorithms. Therefore, the first step would be to make both algorithms in [11] and [15] recursive in order to utilize them in a WSN setup.

PROPOSED RECURSIVE BLIND ESTIMATION ALGORITHMS

In the ensuing, the previously mentioned blind estimation algorithms are made recursive and applied over a wireless sensor network.

Blind Block Recursive SVD Algorithm

Here, we show how the algorithm from [11] can be made into a blind block-recursive algorithm. Since the algorithm requires a complete block of data at each processing instant, we therefore base our iterative process on data blocks as well. So instead of the matrix$\mathbf{D}_k$, we have the block data vector$\mathbf{d}_k$. The autocorrelation matrix for the first data block is defined in as

$$\hat{\mathbf{R}}_{\mathbf{d},k}(1) = \mathbf{d}_k(1)\mathbf{d}_k^T(1). \tag{20}$$

The matrix is expanded for two blocks in the original algorithm as

$$\begin{aligned}\hat{\mathbf{R}}_{\mathbf{d},k}(2) &= \mathbf{D}_{k,2}\mathbf{D}_{k,2}^T \\ &= \begin{bmatrix} \mathbf{d}_k(1) & \mathbf{d}_k(2) \end{bmatrix}\begin{bmatrix} \mathbf{d}_k^T(1) \\ \mathbf{d}_k^T(2) \end{bmatrix} \\ &= \mathbf{d}_k(1)\mathbf{d}_k^T(1) + \mathbf{d}_k(2)\mathbf{d}_k^T(2) \\ &= \hat{\mathbf{R}}_{\mathbf{d},k}(1) + \mathbf{d}_k(2)\mathbf{d}_k^T(2).\end{aligned} \tag{21}$$

Thus, a generalization of (21) can be written as

$$\hat{\mathbf{R}}_{\mathbf{d},k}(i) = \hat{\mathbf{R}}_{\mathbf{d},k}(i-1) + \mathbf{d}_k(i)\mathbf{d}_k^T(i). \tag{22}$$

The first few iterations may not give a good estimate and the error may even seem to be increasing as the matrix will be rank deficient at this early stage. However, as more data blocks are processed, the rank becomes gradually full and the estimate then begins to gradually improve. The next step is to get the eigendecomposition of the autocorrelation matrix. Applying the

SVD on $\mathbf{R}_{d,k}$ yields the eigenvector matrix $\mathbf{U}_k$, from which we get the (M−1 ×M) matrix $\tilde{\mathbf{U}}_k$ that forms the null space of the autocorrelation matrix. From $\tilde{\mathbf{U}}_k$, we then form the M Hankel matrices of size (M×M+ 1) each, which are then concatenated to give the matrix $\mathcal{U}_k(i)$ from which the estimate $\tilde{\mathbf{w}}_k(i)$ is finally derived. This sequential derivation process is depicted below in (23):

$$\text{SVD}\left\{\mathbf{R}_{\mathbf{d},k}(i)\right\} \Rightarrow \mathbf{U}_k(i) \Rightarrow \tilde{\mathbf{U}}_k(i) \Rightarrow \mathcal{U}_k(i) \Rightarrow \tilde{\mathbf{w}}_k(i). \tag{23}$$

The update for the estimate of the unknown parameter vector is then given by

$$\hat{\mathbf{w}}_k(i) = \lambda_k \hat{\mathbf{w}}_k(i-1) + (1-\lambda_k)\,\tilde{\mathbf{w}}_k(i), \tag{24}$$

It can be seen from (23) that the recursive algorithm does not become computationally less complex. However, it does require lesser memory compared to the original algorithm of [11] and the result improves with an increase in the number of data blocks processed. The performance almost matches that of the algorithm of [11].

Algorithm ?? describes the steps of the blind block recursive SVD (RS) algorithm. The forgetting factor is fixed in this case. If the forgetting factor value were to be changed to $\lambda_k(i) = 1 - \frac{1}{i}$, the algorithm would then become the variable forgetting factor RS (VFFRS) algorithm. However, simulations show that the VFFRS algorithm converges more slowly than its fixed forgetting factor counterpart. The simulation results show that if the forgetting factor is small for the fixed forgetting factor case, the algorithm converges faster even though it gives a higher error floor at steady-state. While for the VFFRS algorithm, the forgetting factor increases with time resulting in slow convergence even though the steady-state error is lower compared to the fixed forgetting factor case.

Algorithm 1: Summary of blind block recursive SVD algorithm.

Step 1. Form auto-correlation matrix for iteration i from equation (22).
Step 2. Get $\mathbf{U}_k(i)$ from SVD of $\hat{\mathbf{R}}_{\mathbf{d},k}(i)$.
Step 3. Form $\tilde{\mathbf{U}}_k(i)$ from the null eigenvectors of $\mathbf{U}_k(i)$.
Step 4. Form Hankel matrices of size $(L \times M - 1)$ from individual vectors of $\tilde{\mathbf{U}}_k(i)$.
Step 5. Form $\mathcal{U}_k(i)$ by concatenating the Hankel matrices.
Step 6. The null eigenvector from the SVD of $\mathcal{U}_k(i)$ is the estimate $\tilde{\mathbf{w}}_k(i)$.
Step 7. Use $\tilde{\mathbf{w}}_k(i)$ in equation (24) to get the update $\hat{\mathbf{w}}_k(i)$.

Blind Block Recursive Cholesky Algorithm

In this section, we show how the algorithm of [15] can be converted into a blind block recursive solution.

Equation 14 can be rewritten as

$$\hat{\mathbf{R}}_{\mathbf{w},k}(i) = \frac{1}{i}\sum_{n=1}^{i}\mathbf{d}_k(n)\mathbf{d}_k^T(n) - \hat{\sigma}_{\mathbf{v},k}^2\mathbf{I}_K$$

$$= \frac{1}{i}\mathbf{d}_k(i)\mathbf{d}_k^T(i) + \frac{1}{i}\sum_{n=1}^{i-1}\mathbf{d}_k(n)\mathbf{d}_k^T(n) - \hat{\sigma}_{\mathbf{v},k}^2\mathbf{I}_K$$

$$= \frac{1}{i}\left(\mathbf{d}_k(i)\mathbf{d}_k^T(i) - \hat{\sigma}_{\mathbf{v},k}^2\mathbf{I}_K\right) + \frac{i-1}{i}\hat{\mathbf{R}}_{\mathbf{w},k}(i-1). \quad (25)$$

Similarly, we have

$$\hat{\mathbf{G}}_k(i) = chol\left\{\hat{\mathbf{R}}_{\mathbf{w},k}(i)\right\}, \quad (26)$$

$$\hat{\mathbf{g}}_k(i) = vec\left\{\hat{\mathbf{G}}_k(i)\right\}. \quad (27)$$

Letting $\mathbf{Q}_A = (\mathbf{Q}^T\mathbf{Q})^{-1}\mathbf{Q}^T$, we have

$$\bar{\mathbf{w}}_k(i) = \mathbf{Q}_A\hat{\mathbf{g}}_k(i). \quad (28)$$

We further apply a smoothing step to get the final estimate:

$$\hat{\mathbf{w}}_k(i) = \lambda_k(i)\hat{\mathbf{w}}_k(i-1) + (1-\lambda_k(i))\,\bar{\mathbf{w}}_k(i), \quad (29)$$

where $\lambda_k(i) = 1 - \frac{1}{i}$ is a variable forgetting factor.

Letting $\lambda_k(i) = 1 - \frac{1}{i}$, the blind block recursive Cholesky algorithm is summarised in Algorithm ??. This table defines the Blind Block Recursive Cholesky algorithm with variable forgetting factor (VFFRC). If the forgetting factor is fixed then the algorithm can simply be called Blind Block Recursive Cholesky (RC) algorithm. Simulation results show that the VFFRC algorithm converges to the least squares solution obtained through the algorithm given in [15]. The RC algorithm can also achieve the same result if the value of the forgetting factor is extremely close to 1. However, the convergence speed of the RC algorithm is slower than that of the VFFRC algorithm even though it requires lesser memory and is slightly less computationally complex. There are two issues with the recursive algorithm. Firstly, the Cholesky factorization cannot be applied until at leastMblocks of data have been received as the data correlation matrix needs to be first positive definite before the Cholesky method can be correctly applied. The second issue involves the variance of the additive noise. In [15], it is shown that if the noise variance can be estimated, the estimate of the unknown vector will improve. However, using the noise variance in the recursive algorithm can make the resulting matrix have zero or negative eigenvalues before a sufficient number of data blocks were processed, thus making the use of Cholesky factorization unjustifiable. However, neglecting the noise variance altogether will lead to a performance degradation of this algorithm even though it will be computationally less complex than the SVD approach. One approach is to estimate the noise variance after a certain number of blocks have been received and then use that value for the remainder of the iterations.

Algorithm 2: Summary of blind block recursive Cholesky (RC) algorithm.

Step 1. Let forgetting factor be defined as $\lambda_k(i) = 1 - \frac{1}{i}$.
Step 2. Form auto-correlation matrix for iteration i using $\lambda_k(i)$ in equation (25) to get

$$\hat{\mathbf{R}}_{\mathbf{w},k}(i) = (1 - \lambda_k(i)) \left(\mathbf{d}_k(i)\mathbf{d}_k^T(i) - \hat{\sigma}_{\mathbf{v},k}^2 \mathrm{I}_K \right) + \lambda_k(i)\, \hat{\mathbf{R}}_{\mathbf{w},k}(i-1)$$

Step 3. Get $\hat{\mathbf{G}}(i)$ as the Cholesky factor of $\hat{\mathbf{R}}_{\mathbf{w},k}(i)$.
Step 4. Apply the *vec* operator to get $\hat{\mathbf{g}}_k(i)$.
Step 5. Use $\lambda_k(i)$ in equation (29) to get the final update

$$\hat{\mathbf{w}}_k(i) = \mathbf{Q}_A \left(\hat{\mathbf{g}}_k(i) - \lambda_i \hat{\mathbf{g}}_k(i-1) \right) + \lambda_k(i)\hat{\mathbf{w}}_k(i-1).$$

Diffusion Blind Block Recursive Algorithms

In a wireless sensor network, a distributed algorithm is required, through which nodes can interact with each other and improve their individual estimates as well as the overall performance of the network. In such a scenario, a recursive algorithm is required. This is one major reason for requiring a recursive blind algorithm. Each node can individually update its estimate and then collaborate with the neighboring nodes to improve that estimate. A comparison of different distributed schemes has shown that theAdapt-Then-Combine(ATC) diffusion strategy provides the best performance [16]. Therefore, we also implement our distributed algorithms using the ATC scheme.

For the diffusion-based RS algorithm, all nodes evaluate their own autocorrelation matrix updates and then perform the SVD operation. This is followed by a preliminary estimate of the unknown vector. The preliminary results are then combined with those of the neighboring nodes. As a result of this cooperation, the result of the network improves, as will be shown in the simulations. Similarly, for the diffusion RC algorithm, each node performs the recursion and reaches a preliminary estimate of the unknown vector which is then combined with those of the neighboring nodes. These are summarized in Algorithms ?? and ??, where the subscriptkdenotes the node number,$N_k$is the set of neighbors of nodek, $\hat{\mathbf{h}}_k$is the intermediate estimate for nodekand$c_{lk}$is the weight connecting nodekto its neighboring nodel$\in N_k$, where$N_k$includes nodek, and$\sum_{clk}=1$.

Algorithm 3: Summary of diffusion blind block recursive SVD algorithm.

Step 1. Form auto-correlation matrix for iteration i from equation (22) for each node k.

$$\hat{\mathbf{R}}_{\mathbf{d},k}(i) = \mathbf{d}_{k,i}\mathbf{d}_{k,i}^T + \hat{\mathbf{R}}_{\mathbf{d},k}(i-1)$$

Step 2. Get $\mathbf{U}_k(i)$ from SVD of $\hat{\mathbf{R}}_{\mathbf{d},k}(i)$.

Step 3. Form $\tilde{\mathbf{U}}_k(i)$ from the null eigenvectors of $\mathbf{U}_k(i)$.

Step 4. Form Hankel matrices of size $(L \times M - 1)$ from individual vectors of $\tilde{\mathbf{U}}_k(i)$.

Step 5. Form $\mathcal{U}_k(i)$ by concatenating the Hankel matrices.

Step 6. The null eigenvector from the SVD of $\mathcal{U}_k(i)$ is the estimate $\tilde{\mathbf{w}}_{k,i}$.

Step 7. Use $\tilde{\mathbf{w}}_{k,i}$ in equation (24) to get the intermediate update $\hat{\mathbf{h}}_{k,i}$.

$$\hat{\mathbf{h}}_{k,i} = \lambda\hat{\mathbf{w}}_{k,i-1} + (1-\lambda)\tilde{\mathbf{w}}_{k,i}$$

Step 8. Combine estimates from neighbors of node k to get $\hat{\mathbf{w}}_{k,i}$.

$$\hat{\mathbf{w}}_{k,i} = \sum_{l \varepsilon N_k} c_{lk}\hat{\mathbf{h}}_{l,i}$$

Algorithm 4: Summary of diffusion blind block recursive Cholesky algorithm.

Step 1. Let forgetting factor be defined as $\lambda_{k,i} = 1 - \frac{1}{i}$.
Step 2. Form auto-correlation matrix for iteration k from

$$\hat{\mathbf{R}}_{\mathbf{w},k}(i) = (1 - \lambda_{k,i})\left(\mathbf{d}_{k,i}\mathbf{d}_{k,i}^T - \hat{\sigma}_{v,k}^2 \mathrm{I}_K\right) + \lambda_{k,i}\hat{\mathbf{R}}_{\mathbf{w},k}(i-1)$$

Step 3. Get $\hat{\mathbf{G}}_k(i)$ as the Cholesky factor of $\hat{\mathbf{R}}_{\mathbf{w},k}(i)$.
Step 4. Apply the *vec* operator to get $\hat{\mathbf{g}}_{k,i}$.
Step 5. The intermediate update is then given as

$$\hat{\mathbf{h}}_{k,i} = \mathbf{Q}_A\left(\hat{\mathbf{g}}_{k,i} - \lambda_{k,i}\hat{\mathbf{g}}_{k,i-1}\right) + \lambda_{k,i}\hat{\mathbf{w}}_{k,i-1}.$$

Step 6. The final update is the weighted sum of the estimates of all neighbors of node k

$$\hat{\mathbf{w}}_{k,i} = \sum_{l \varepsilon N_k} c_{lk}\hat{\mathbf{h}}_{l,i}$$

COMPLEXITY OF THE RECURSIVE ALGORITHMS

In order to fully understand the variation in performance of these two algorithms, it is necessary to look at their computational complexity as it will allow us to estimate the loss in performance that would result from a reduction in computational load. We first analyze the complexity of the original algorithms and then deal with that of their recursive versions.

Blind SVD Algorithm

The length of the unknown parameter vector is$M$and the data block size is$K$. Since a total number of$N$data blocks are required for the estimation of the unknown parameter vector, where$N \geq K$, the resulting data matrix will therefore be of size$K \times N$. The data correlation matrix will thus be of size$K \times K$and this function will require$K^2(2N-1)$ calculations (including both multiplications and additions) for its computation. The next step is singular value decomposition (SVD), done using the QR decomposition algorithm. This algorithm requires a total of$\left[\frac{4}{3}K^3 + \frac{3}{2}K^2 + \frac{19}{6}K - 6\right]$calculations. Then the null eigenvectors are separated and each eigenvector is used to form a Hankel matrix with all the Hankel matrices then stacked together to form a matrix of size$M \times (K-M)(M-1)$. The unique null vector of this new matrix gives the estimate of the unknown vector. To find this eigenvector requires another $\left[\left(2K + \frac{7}{3}\right)M^3 - 2M^4 + (1 - 4K)\frac{M^2}{2} + \frac{19}{6}M - 6\right]$calculations. So the overall computational load required for the algorithm can be given as

$$T_{C,\text{SVD}} = \frac{4}{3}K^3 + \left(2N + \frac{1}{2}\right)K^2 + \frac{19}{6}(K+M)$$
$$+ \left(2K + \frac{7}{3}\right)M^3 - 2M^4 + (1-4K)\frac{M^2}{2} - 12. \tag{30}$$

Blind Cholesky Algorithm

Like the SVD algorithm, here also the unknown vector length isMand the data block size isK. For computational purposes and for the blind SVD algorithm, the total number of data blocks is taken asN. The correlation process is the same except for the final averaging step which results in an extra division so the total number of calculations becomes$K^2(2N-1) + 1$. The next step is to estimate the noise variance, which requires an SVD decomposition and therefore an extra number of calculations given by $\left[\frac{4}{3}K^3 + \frac{3}{2}K^2 + \frac{19}{6}K - 6\right]$. After the SVD decomposition, only one further division is required to estimate the noise variance. The noise variance is then subtracted from the diagonal of the correlation matrix, resulting in anotherK calculations. After that, the Cholesky factorization is performed, which requires$\left[\frac{1}{3}\left(M^3 + 3M^2 + M\right)\right]$calculations. Finally, the last step is to get the estimate of the unknown vector through the pseudo-inverse of Cholesky-factorized data correlation matrix and this step requires further $[M(2M^2-1)]$ calculations. Thus, the total number of calculations required is given as

$$T_{C,\text{Chol}} = \frac{4}{3}K^3 + \left(2N + \frac{1}{2}\right)K^2 + \frac{19}{6}K - 4$$
$$+ \frac{1}{3}\left(7M^3 + 3M^2 - M\right). \tag{31}$$

Blind Block Recursive SVD Algorithm

When the blind SVD algorithm is made recursive, we notice that this will involve only a slight change in the overall algorithm but will really halve the total computational load. Since the correlation matrix is only being updated at each iteration, the number of calculations required for the first step are now only $2K^2$instead of$K^2(2N-1)$. However, an extra (M+ 2) calculations are required for the final step. The overall number of calculations is thus given as

$$T_{C,\mathrm{SRLS}} = \frac{4}{3}K^3 + \frac{7}{2}K^2 + \frac{19}{6}K + \left(2K + \frac{7}{3}\right)M^3 - 2M^4$$
$$+ (1 - 4K)\frac{M^2}{2} + \frac{25}{6}M - 10 = O\left(K^3\right). \quad (32)$$

Blind Block Recursive Cholesky Algorithm

Similarly, the number of calculations for the first step for this algorithm is reduced to ($2K^2$+ 2) from the ($K^2(2N-1)$ + 1) calculations required by its non-recursive counterpart. The final step includes an extra (K^2+M+ 2) calculations. Thus, the total number of calculations is now given as

$$T_{C,\mathrm{CRLS}} = \frac{4}{3}K^3 + \frac{7}{2}K^2 + \frac{19}{6}K + \frac{1}{3}\left(7M^3 + 3M^2 + 2M\right)$$
$$= O\left(K^3\right). \quad (33)$$

However, it should be noted here that the estimation of the noise variance need not be repeated at each iteration. After a few iterations, the number of which can be fixeda priori, the noise variance can be estimated once only and then this same estimated value can be used in the remaining iterations. The number of calculations, thus, reduces to

$$T_{C,\mathrm{CRLS}} = 2K^2 + \frac{1}{3}\left(7M^3 + 3M^2 + 2M\right) + 4 = O\left(K^2\right). \quad (34)$$

Comparison of All Algorithms

Here, we compare all of the algorithms discussed in the previous sections, using specific scenarios where the value forMis fixed to 4, that ofKis varied for all algorithms and the value ofNis varied between 10 and 20 for the least squares algorithms. The number of calculations for the two recursive algorithms discussed before are shown for one iteration only. Recall that in the second algorithm, i.e. the blind block recursive Cholesky algorithm, the noise variance is calculated only once, after a pre-selected number of iterations have occurred, and then kept constant for the remaining iterations. Tables1and2below summarize the results.

Table 1: Computations for original least squares algorithms under different settings

M= 4	***N*= 10**	***N*= 10**	***N*= 20**	***N*= 20**	***N*= 20**
	***K*= 8**	***K*= 10**	***K*= 8**	***K*= 10**	***K*= 20**
SVD	2,434	4,021	3,714	6,021	28,496

Chol	2,180	3,575	3,460	5,575	27,090

Table 2: Computations for recursive algorithms under different settings

M= 4	*K*= 8	*K*= 10	*K*= 20
RS	1,352	2,327	13,702
RC	1,100	1,883	12,298
RCNV	300	372	972

Table1lists the number of computations for the original algorithms, showing that the Cholesky-based method requires fewer computations than the SVD-based method and so the trade-off between performance and complexity is justified. If the number of blocks is small, then the Cholesky-based method may even perform better than the SVD-based method as shown in [15]. Here it is assumed that the exact length of the unknown vector is known. Generally, an upper bound of this value is known and that value is used instead of the exact value, resulting in an increase in computations. This assumption is made for both algorithms here to make their comparative study fair.

Table2lists the computations-per-iteration for the recursive versions of these two algorithms. RS and RC give the number of computations for the recursive SVD algorithm and the recursive Cholesky algorithm respectively. RCNV lists the number of computations when the noise variance is estimated only once in the recursive Cholesky algorithm. This shows how the complexity of the algorithm can be reduced by an order of magnitude by adopting an extra implicit assumption regarding the wide-sense stationarity of the noise and hence the constancy of its variance from one iteration to the next. Although the performance does suffer slightly, the gain in the reduction of computational complexity more than compensates for this loss.

It is important to note here that even though the SVD and Cholesky factorization operations are being run at every iteration, there is a significant gain achieved in the calculation of the autocorrelation function. While each batch processing algorithm would require a total ofP^2N^2multiplications, where ($P \times N$) is the size of the data block matrix, the recursive algorithms only requireP^2Nmultiplications. Thus, there is a reduction in the number of multiplications by a factor of$N$, which becomes significant when the number of blocks$N$is large.

SIMULATIONS AND RESULTS

Here we compare results for the newly developed recursive algorithms. Results are shown for a network of 20 nodes, shown in Figure2. The forgetting factor is both varied as well as kept fixed in order to study its impact on the performance of each algorithm. The two algorithms are compared with each other and in different scenarios. The forgetting factor, data block size and network size are changed one at a time while all other variables are kept constant in order to closely monitor the impact of each of these three varying parameters on the performance of each algorithm.

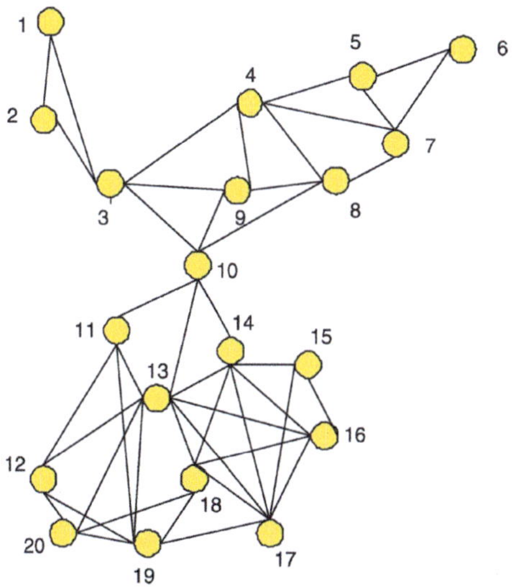

Figure 2: Network of 20 nodes.

Performance of the SVD and Cholesky Algorithms

Initially, the two algorithms are used to identify an unknown parameter vector of length$M = 4$ in an environment with the two signal-to-noise ratios (SNR) of 10 and 20 dB. The two forgetting factors used are fixed at$\lambda = \{0.9, 0.99\}$. The block size is taken as$K = 8$. The results for both algorithms are shown in Figures3,4,5and6, for both diffusion (DRC, DRS) and no cooperation (NRC, NRS) cases. As can be seen from these figures, the Cholesky algorithm does not perform well with the smaller forgetting factor. However, the performance improves appreciably with an increase in the forgetting factor but its speed of convergence decreases significantly as well. However, the one main positive attribute of the Cholesky algorithm remains to be its low computational complexity. For the SVD algorithm, the performance improves slightly with an increase in forgetting factor but at a significant loss of convergence speed. These remarks have prompted us to further analyze the impact of the forgetting factor on the performance on these two algorithms, as discussed next.

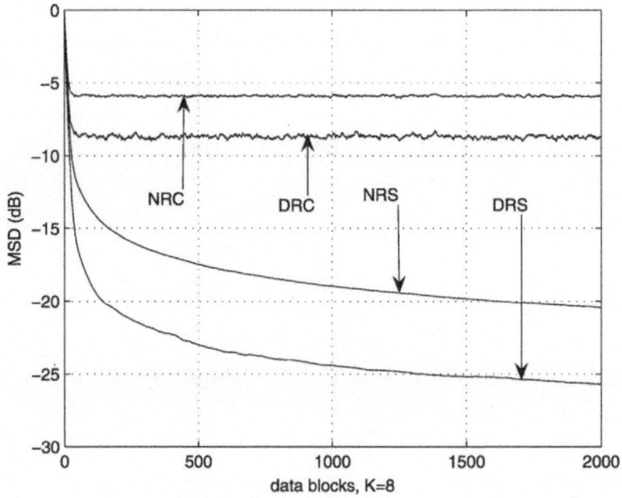

Figure 3: MSD at SNR = 10 dB andλ= 0.9.

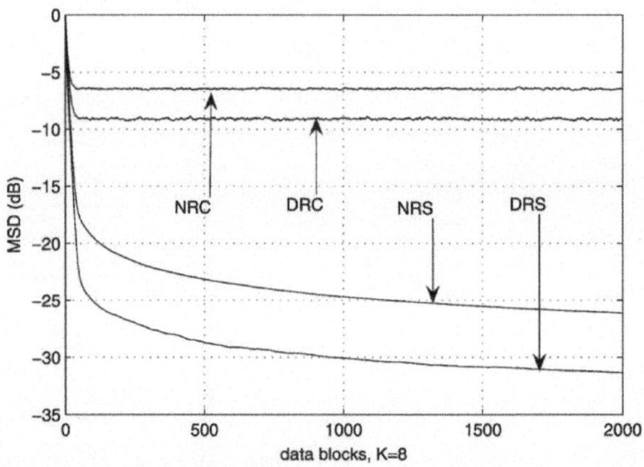

Figure 4: MSD at SNR = 20 dB andλ= 0.9.

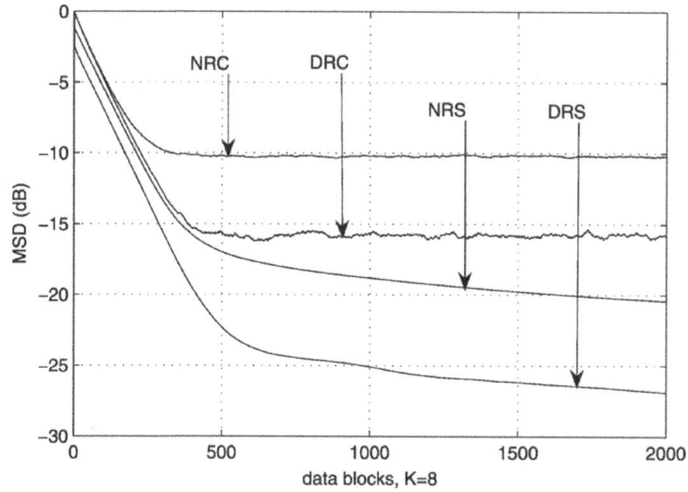

Figure 5: MSD at SNR = 10 dB andλ= 0.99.

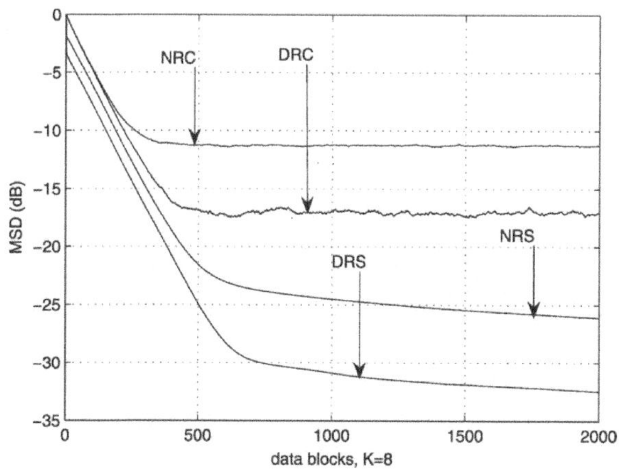

Figure 6: MSD at SNR = 20 dB andλ= 0.99.

Further Simulation-based Analysis of the Effect of Forgetting Factor

Next, the performance of each algorithm is separately studied for different values of the forgetting factor. For the fixed forgetting factor case, the values taken areλ= {0.9,0.95,0.99} and the results are compared with those of the variable forgetting factor case. The SNR is chosen as 20 dB and the network size is taken to be 20 nodes. Figure7shows the results for the Cholesky factorization-based RC algorithm. It is seen that the performance improves as the forgetting factor is increased but the convergence slows down. The algorithm performs best when the forgetting factor is variable. The results for the SVD-based RS algorithm are shown in Figures8,9and10. Figure8shows the results for all three fixed forgetting factors as well as those for its variable one. However, as there is not much difference in performance between the four cases, Figure8is then zoomed in to see more clearly the algorithm's transient and near-steady-state behavior. Figure9shows the result of this zooming effect. The speed of convergence is fastest forλ= 0.9 and slowest forλ= 0.99. For the variable forgetting factor (VFF) case, the speed is fast initially but then slows down with time. Figure10shows the behavior of the algorithm near steady-state. It is evident that the fixed case ofλ= 0.99 would yield the lowest steady-state error whereas the VFF case would take the longest to reach the steady-state. Although the steady-state performance of the variable forgetting factor may be as good as for the case ofλ=0.99 or even better, its speed of convergence is too slow.

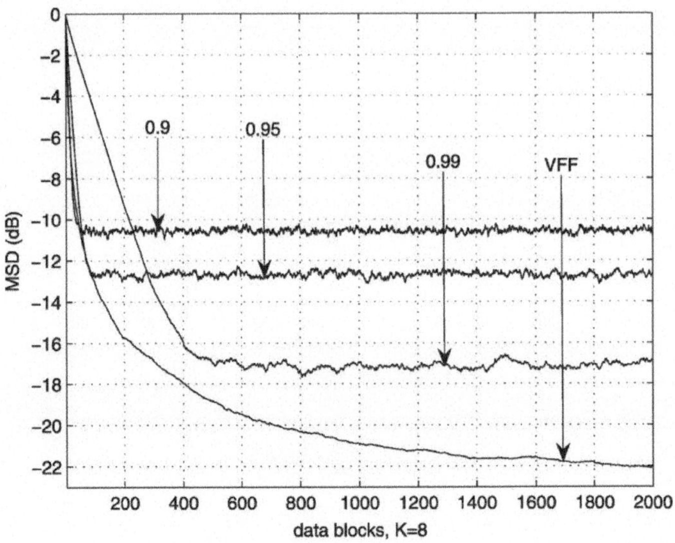

Figure 7: MSD at SNR = 20 dB for RC with different forgetting factors.

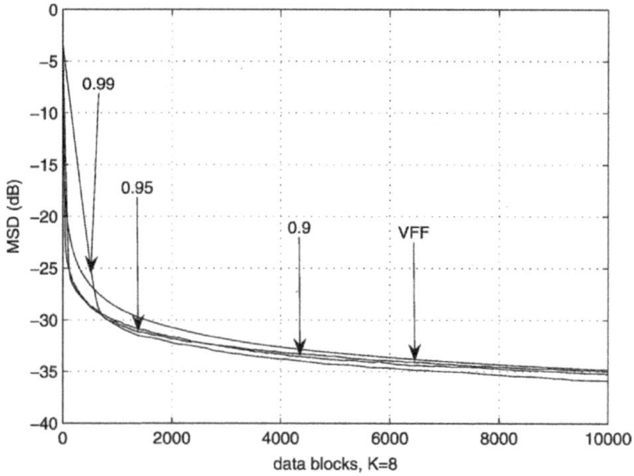

Figure 8: MSD at SNR = 20 dB for RS with different forgetting factors.

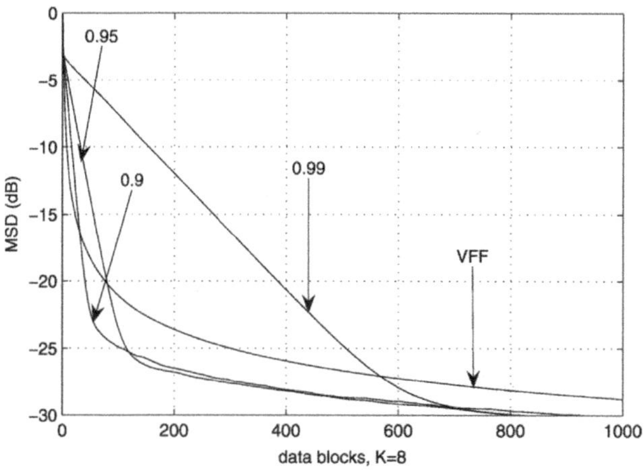

Figure 9: MSD at SNR = 20 dB for RS with different forgetting factors (Transient behavior).

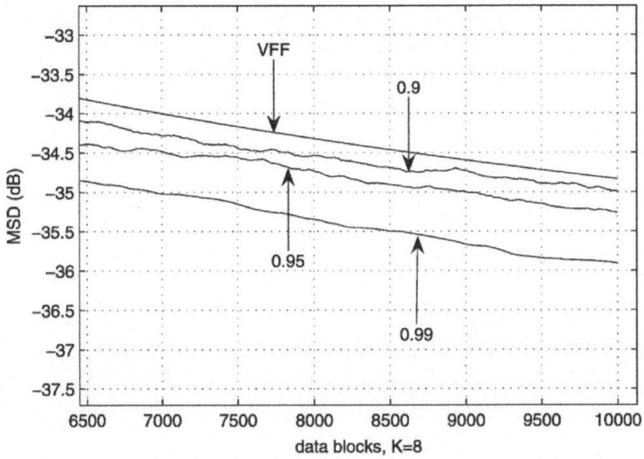

Figure 10: MSD at SNR = 20 dB for RS with different forgetting factors (Near steady-state behavior).

Performance of the Two Algorithms using an Optimal Forgetting Factor

From the results of Figures7,8,9and10, it can easily be inferred that the Cholesky factorization-based approach yields the best results when the forgetting factor is varied whereas the SVD-based algorithm performs best if the forgetting factor is fixed. In order to have a fair performance comparison, the two algorithms need to be compared under conditions in which they both perform best. Figures11and12give the best performance results of the two algorithms, respectively. As can be seen from these two figures, at an SNR of 10 dB, the Cholesky-based DRC algorithm performs slightly better than the SVD-based RS algorithm without diffusion, whereas both SVD-based algorithms outperform the Cholesky-based algorithms at an SNR of 20 dB. However, the RC algorithm remains computationally less complex than the RS one. A final choice of either of these two algorithms will have to be based on a trade-off between their complexity and performance.

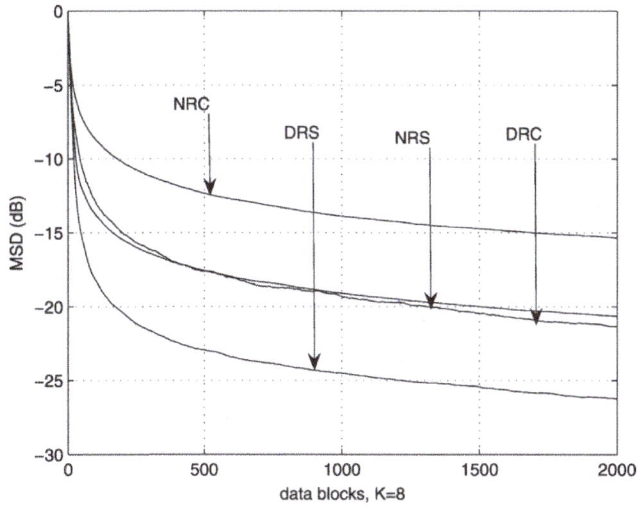

Figure 11: MSD at SNR = 10 dB under best performance conditions.

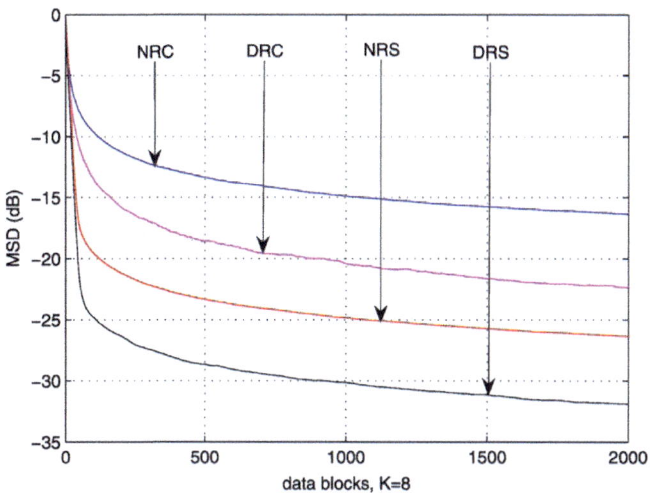

Figure 12: MSD at SNR = 20 dB under best performance conditions.

It would be only fair to compare the performance of these algorithms with the original batch processing algorithms from [11] and [15]. At an SNR of 10 dB, the MSE value for the SVD-based algorithm of [11] is −25.69 dB while that of the Cholesky-based algorithm from [15] is −17.79 dB. The corresponding numbers at an SNR of 20 are −31.38 and −18.36 dB, respectively. Comparing these results with the figures, we see that both recursive algorithms perform similar to their batch-processing counterparts.

Furthermore, the diffusion algorithms perform better than the batch processing algorithms. The comparison results are tabulated in Table3.

Table 3: Performance comparison with the batch processing algorithms (all results are in dB)

SNR	CHOL [[15]]	NRC	DRC	SVD [[11]]	NRS	DRS
10 dB	−17.79	−15.36	−21.34	−25.69	−20.63	−26.23
20 dB	−18.36	−16.34	−22.35	−31.38	−26.35	−31.90

Effect of Block Size

Since it has been stated in [11] and [15] that the block size can affect the performance of the algorithm, the performance of our algorithms is also tested here for various block sizes. The block size is varied asK= {5,8,10,15,20} and the SNR is set to 20 dB. These settings are applied to both algorithms separately. Here it is important to note that as the size of the data block increases, the total amount of data required for the same number of blocks also increases. Figures13and14show the results for the RC algorithm and clearly demonstrate that the algorithm fails badly forK= 5. However, for the remaining block sizes, the algorithm's performance remains almost unaffected by the block size changes. The convergence speeds are nearly the same (see Figure13) and the performance at steady-state is similar as well for the remaining block sizes, with only a slight difference (see Figure14). From Figure14it can be inferred that the best result, in every respect, is achieved when the block size is just large enough to achieve a full rank input data matrix (K=8 in this case), as expected. Thus, it is essential to estimate a tight upper bound for the size of the unknown vector in order to achieve good performance. Figures15and16show the results for the RS algorithm. Here, the performance improves gradually with an increase in block size. However, the speed of convergence is slow for a large block size (see Figure15) even though a larger block size gives better performance at steady-state (see Figure16). Again it can be inferred that it is best to keep the block size reasonably small in order to achieve a good trade off between performance and speed of convergence, especially when taking into account the fact that a larger block size would mean sensing more data for the same number of blocks.

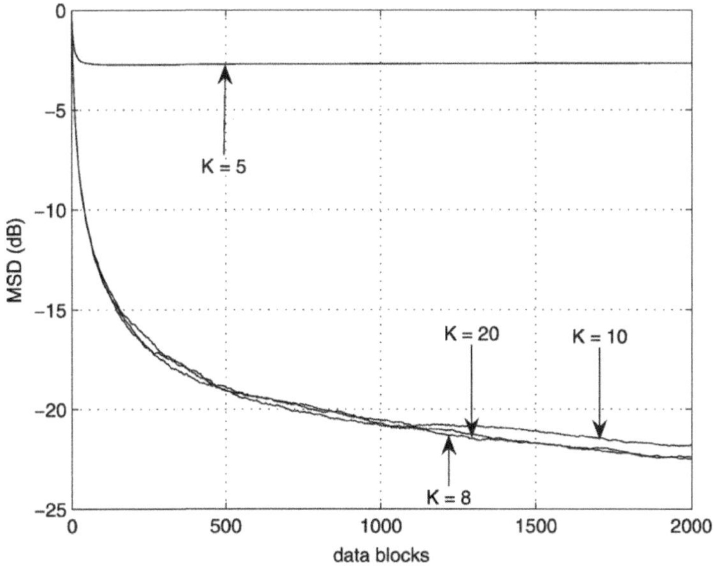

Figure 13: MSD at SNR = 20 dB for varying K for RC.

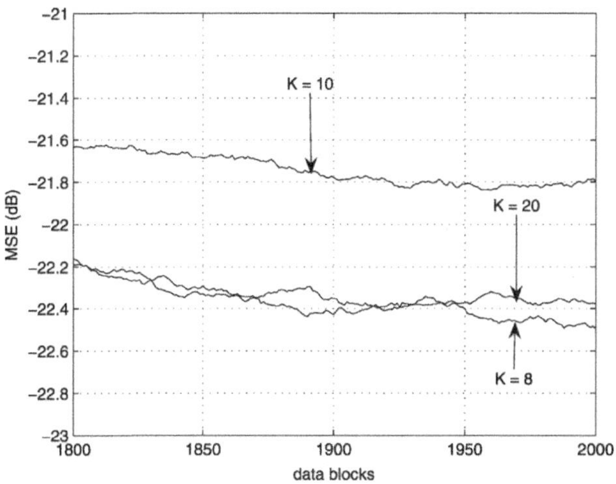

Figure 14: MSD at SNR = 20 dB for varying K for RC (last 200 runs).

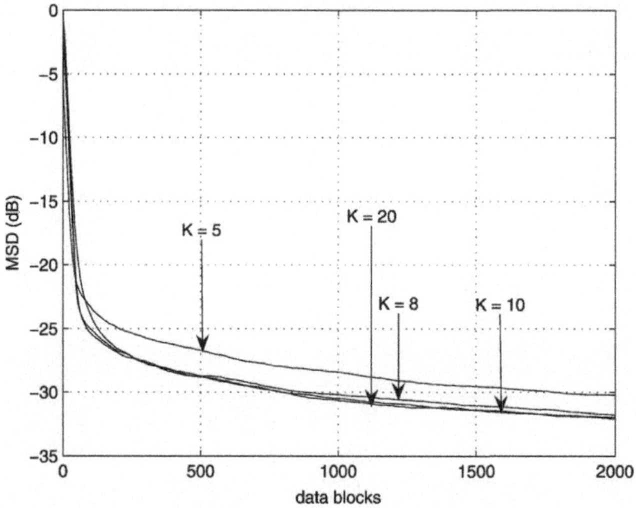

Figure 15: MSD at SNR = 20 dB for varying K for RS.

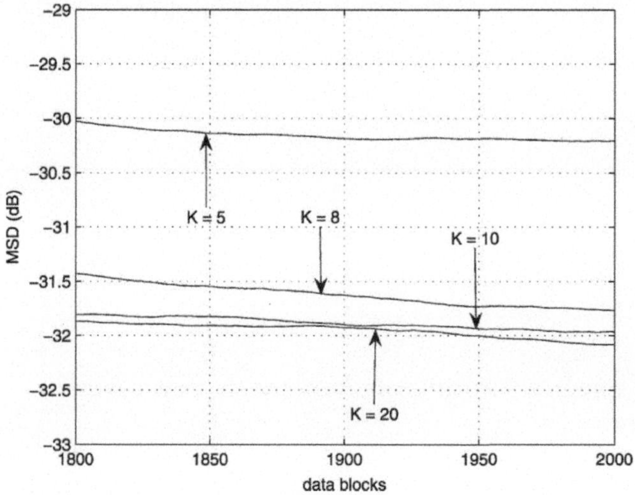

Figure 16: MSD at SNR = 20 dB for varying K for RS (last 200 runs).

Effect of Network Size

Here the effect of the size of the network on the performance of the algorithms is discussed. For this purpose, the size of the network is varied over the rangeN= {10−50} while the forgetting factor is kept fixed atλ= 0.9 for the RS algorithm and made variable for the RC algorithm. The block size is taken asK= 8. This performance comparison is also carried out for both algorithms separately. The number of neighbors for each node is increased gradually as the size of the network is increased. Figures17and18show results for the RC algorithm. The performance is poor forN= 10 but improves asNincreases. The initial speed of convergence is similar for various network sizes as can be seen in Figure17but, near steady-state, the networks with large sizes show a slight improvement in performance, as shown in Figure18. Figures19and20show the results for the RS algorithm. Here the trend is slightly different. It can be seen that the initial speed of convergence improves with an increase inN(see Figure19) but the improvement in performance is slightly smaller near steady-state (see Figure20). Also, the difference in performance is smaller for larger networks, which is as expected.

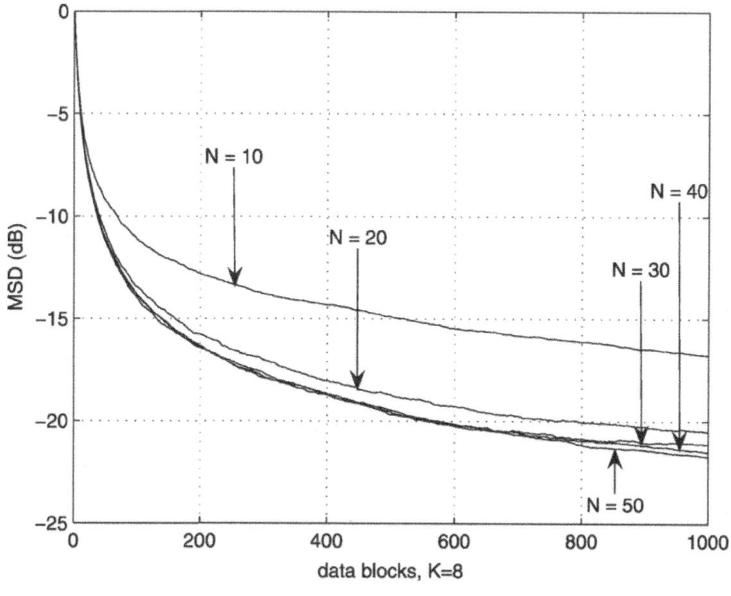

Figure 17: MSD at SNR = 20 dB for varying network sizes for RC.

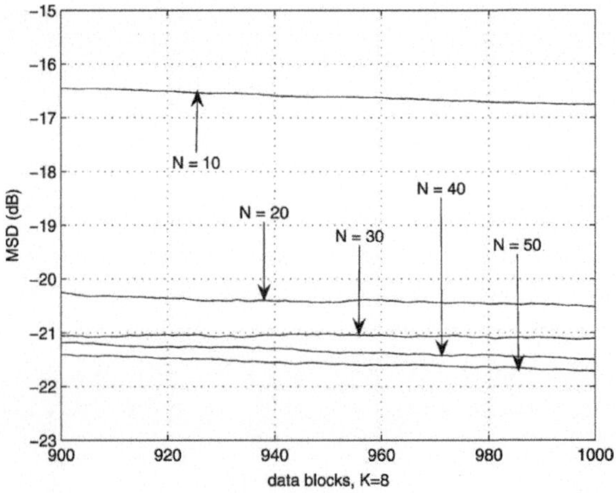

Figure 18: MSD at SNR = 20 dB for varying network sizes for RC (last 100 runs).

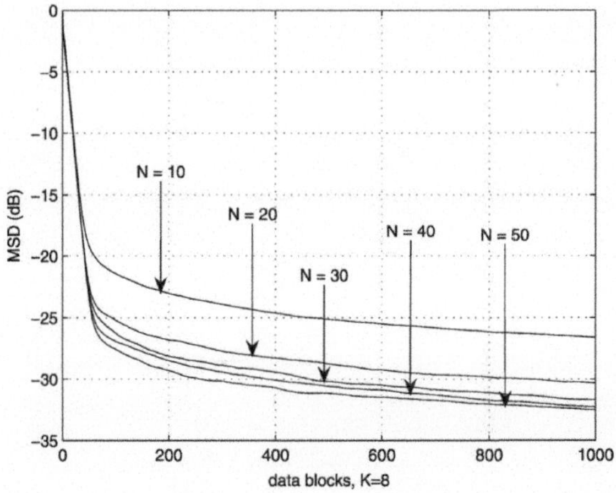

Figure 19: MSD at SNR = 20 dB for varying network sizes for RS.

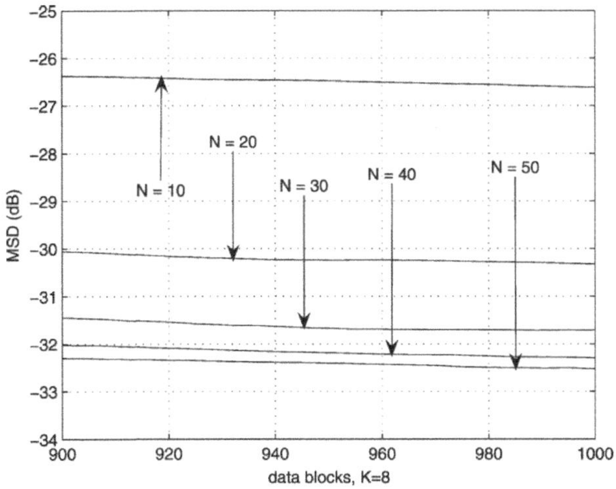

Figure 20: MSD at SNR = 20 dB for varying network sizes for RS (last 100 runs).

Effect of Node Malfunction

Finally, it is shown how the performance can be affected if one or more nodes malfunction. Two different network sizes are chosen in two different SNR scenarios to show how performance gets affected by node malfunction or failure. First, a network of 20 nodes is used and five nodes are switched off. Here, switching off a node means that it stops to participate in any further estimation process. The functioning nodes then re-calibrate the weights for the remaining neighbors while the weights for the failed nodes are set to zero. The nodes with the maximum number of neighbors are switched off to see how seriously the network performance might be affected. Results are shown for both SNRs, 10 and 20 dB, in Figures21and22respectively. The network size is then increased to 50 nodes with about a quarter of the nodes (13) switched off. The corresponding results are shown in Figures23and24. The RC algorithm's performance is worst affected at SNR = 10 dB but remains almost unaffected at SNR = 20 dB, with the small difference in performance getting even smaller when the network size is increased. Under similar test conditions as for the RC algorithm, the degradation of the SVD-based algorithm's performance was found to be similar to that of the RC's in all test cases. This clearly shows that the SVD-based algorithm is also strongly dependent on the connectivity of the nodes. As expected, the overall performance improves with an increase in network size. The effect of switched-off nodes on the performance of both algorithms, however, is similar when the ratio of switched-off nodes with maximum numbers of neighbor nodes to the total number nodes is the same.

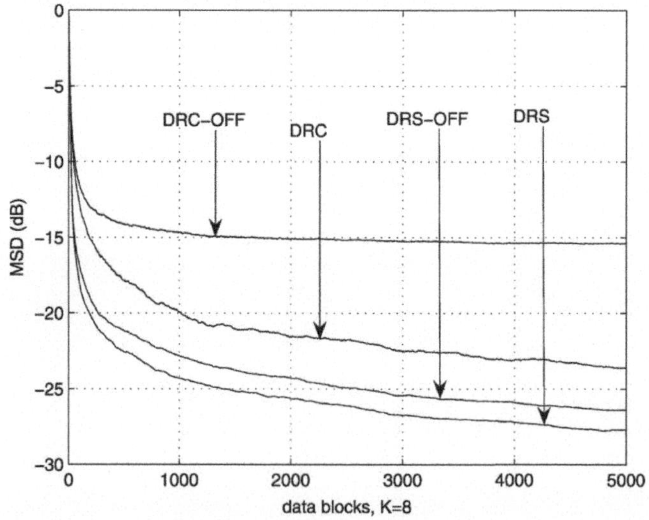

Figure 21: MSD at SNR = 10 dB andN= 20 nodes when five most connected nodes are switched off.

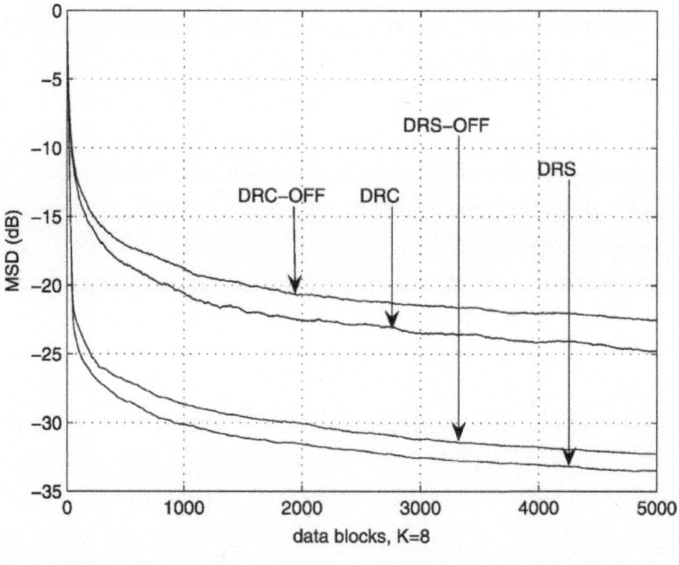

Figure 22: MSD at SNR = 20 dB andN= 20 nodes when five most connected nodes are switched off.

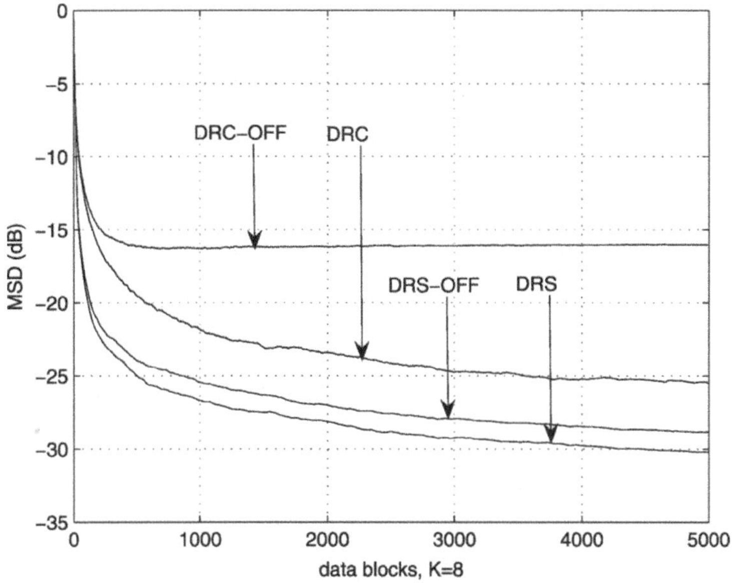

Figure 23: MSD at SNR = 10 dB andN= 50 nodes when 13 most connected nodes are switched off.

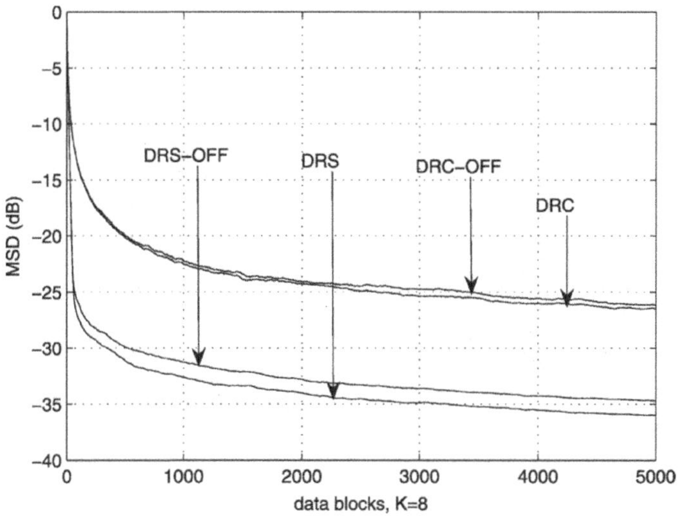

Figure 24: MSD at SNR = 20 dB andN=50 nodes when 13 most connected nodes are switched off.

CONCLUSION

This work develops blind block recursive least squares algorithms based on Cholesky factorization and singular value decomposition (SVD). The algorithms are then used to estimate an unknown vector of interest in a wireless sensor network using cooperation between neighboring sensor nodes. Incorporating the algorithms in the sensor networks creates new diffusion-based algorithms, which are shown to perform much better than their non-diffusion-based counterparts. The new algorithms have been tested using both a variable as well as a fixed forgetting factor. The two developed algorithms are named diffusion blind block recursive Cholesky (DRC) and diffusion blind block recursive SVD (DRS) algorithms. Extensive simulation work comparing the two algorithms under different scenarios revealed that the DRS algorithm performs much better than the DRC algorithm but at the cost of a higher computational complexity. Also, of the two algorithms, the DRC algorithm performs better when the forgetting factor is variable whereas the DRS algorithm gives better results with a fixed forgetting factor. In the case of DRS, the value of the forgetting factor does not effect the overall performance a great deal except for a slight variation in convergence speed and steady-state performance. It was also seen that the size of the data block has an effect on the performance of the two algorithms. The speed of convergence slows down with an increasing block size which means an increasing amount of data to be processed. A block size increase, however, does not necessarily improve performance. It was found that, in general, a small block size gives a better performance. Therefore, it is essential to estimate a very low upper bound to the size of the unknown vector so that the data block size to be used is not unnecessarily large. Next, it was noticed that an increase in the network size improves performance but the improvement gradually decreases with an increasing network size. Moreover, it was shown that switching off some nodes with the largest neighborhoods can slightly degrade the performance of the algorithm. Finally at low SNRs, the Cholesky-based algorithm suffers from a severe degradation, whereas the SVD-based one only experiences a slight degradation.

ACKNOWLEDGMENTS

The authors acknowledge the support provided by the Deanship of Scientific Research (DSR) at KFUPM under Research Grants RG1216, RG1112, SB101024, and FT100012.

REFERENCES

1. Sayed AH, Lopes CG: Distributed recursive least-squares strategies over adaptive networks. InProceedings of the 40th Asilomar Conference on Signals, Systems, Computers. Monterey, CA; 2006:233-237.
2. Lopes CG, Sayed AH: Incremental adaptive strategies over distributed networks.IEEE Trans. Signal Process2007, 55: 4064-4077.
3. Lopes CG, Sayed AH: Diffusion least-mean squares over adaptive networks: formulation and performance analysis.IEEE Trans. Signal Process2008, 56(7):3122-3136.
4. Schizas ID, Mateos G, Giannakis GB: Distributed LMS for consensus-based in-network adaptive processing.IEEE Trans. Signal Process2009, 57(6):2365-2382.
5. Bin Saeed MO, Zerguine A, Zummo SA: A variable step-size strategy for distributed estimation over distributed networks.Eur. J. Adv. Signal Process2013, 2013: 135. 10.1186/1687-6180-2013-135
6. Sato Y: A method of self-recovering equalization for multilevel amplitude-modulation.IEEE Trans. Commun1975, COM-23(6):679-682.
7. Proakis J:Digital Communications. McGraw-Hill, New York; 2000.
8. Tong L, Perreau S: Multichannel blind identification: from subspace to maximum likelihood methods.Proc. IEEE1998, 86(10):1951-1968. 10.1109/5.720247
9. Xu G, Liu H, Tong L, Kailath T: A least-squares approach to blind channel identification.IEEE Trans. Signal Process1995, 43(12):2982-2993. 10.1109/78.476442
10. Abed-Meraim K, Qiu W, Hua Y: Blind system identification.IEEE Trans. Signal Process1997, 45(3):770-773. 10.1109/78.558501
11. Scaglione A, Giannakis GB, Barbarossa S: Redundant filterbank precoders and equalizers part II: blind channel estimation, synchronization, and direct equalization.IEEE Tran. Signal Proc1999, 47(7):2007-2022. 10.1109/78.771048
12. Manton JH, Neumann WD: Totally blind channel identification by exploiting guard intervals.Syst. Control Lett2003, 48(2):113-119. 10.1016/S0167-6911(02)00278-5
13. Pham DH, Manton JH: A subspace algorithm for guard interval based channel identification and source recovery requiring just two received

blocks. InProceedings of the IEEE ICASSP '03. Hong Kong; 2003:317-320.

14. Su B, Vaidyanathan PP: A generalized algorithm for blind channel identification with linear redundant precoders.Eur. J. Adv. Signal Proc2007, 2007: 1-13.
15. Choi J, Lim C-C: A cholesky factorization based approach for blind FIR channel identification.IEEE Tran. Signal Proc2008, 56(4):1730-1735.
16. Cattivelli F, Sayed AH: Diffusion LMS strategies for distributed estimation.IEEE Trans. Signal Process2010, 58(3):1035-1048.

Chapter 5

A DFT-based Approximate Eigenvalue and Singular Value Decomposition of Polynomial Matrices

Mahdi Tohidian[1], **Hamidreza Amindavar**[1] **and Ali M Reza**[2]

[1]Department of Electrical Engineering, Amirkabir University of Technology, Tehran, Iran

[2]Electrical Engineering and Computer Science, University of Wisconsin-Milwaukee, Milwaukee, WI 53201–0784, USA.

ABSTRACT

In this article, we address the problem of singular value decomposition of polynomial matrices and eigenvalue decomposition of para-Hermitian matrices. Discrete Fourier transform enables us to propose a new algorithm based on uniform sampling of polynomial matrices in frequency domain. This formulation of polynomial matrix decomposition allows for controlling spectral properties of the decomposition. We set up a nonlinear quadratic

Citation (APA): Tohidian, M., Amindavar, H., & Reza, A. M. (2013). A DFT-based approximate eigenvalue and singular value decomposition of polynomial matrices. EURASIP Journal on Advances in Signal Processing, 2013(1), 93. (16 pages), DOI: https://doi.org/10.1186/1687-6180-2013-93

minimization for phase alignment of decomposition at each frequency sample, which leads to a compact order approximation of decomposed matrices. Compact order approximation of decomposed matrices makes it suitable in filterbank and multiple-input multiple-output (MIMO) precoding applications or any application dealing with realization of polynomial matrices as transfer function of MIMO systems. Numerical examples demonstrate the versatility of the proposed algorithm provided by relaxation of paraunitary constraint, and its configurability to select different properties.

Keywords: Singular Value Decomposition, Discrete Fourier Transform, Singular Vector, Polynomial Matrix, Polynomial Matrice

INTRODUCTION

Polynomial matrices have been used for a long time for modeling and realization of multiple-input multiple-output (MIMO) systems in the context of control theory [1]. Nowadays, polynomial matrices have a wide spectrum of applications in MIMO communications [2,3,4,5,6], source separation [7], and broadband array processing [8]. They also have a dominant role in development of multirate filterbanks [9].

More recently, there have been much interest in polynomial matrix decomposition such as QR decomposition [10,11,12], eigenvalue decomposition (EVD) [13,14], and singular value decomposition (SVD) [5,11]. Lambert [15] has utilized Discrete Fourier transform (DFT) domain to change the problem of polynomial EVD to pointwise EVD. Since EVD is obtained at each frequency separately, eigenvectors are known at each frequency up to a scaling factor. Therefore, this method requires many frequency samples to avoid abrupt changes in adjacent eigenvectors.

Although, many methods of designing principle component filterbanks have been developed that are equivalent to EVD of pseudo circulant polynomial matrices [16,17], the next pioneering work on polynomial matrix EVD is presented by McWhirter et al. [13]. They use an extension of Jacobi algorithm known as SBR2 for EVD of para-Hermitian polynomial matrices which guarantees exact paraunitarity of eigenvector matrix. Since final goal of SBR2 algorithm is to have strong decorrelation, the decomposition does not necessarily satisfy spectral majorization property. SBR2 algorithm has also been modified for QR decomposition and SVD [10,11]. Jacobi-type algorithms are not the only proposed methods for polynomial matrix decomposition. Another iterative method for spectrally majorized EVD is presented in [14] which is based on the maximization of zeroth-order diagonal energy. Spectral majorization property of this algorithm is verified via simulation. Followed by the work of [6], a DFT-based approximation of polynomial SVD is also proposed in [18] which uses model order truncation by phase optimization.

In this article, we present polynomial EVD and SVD based on DFT formulation. It transforms the problem of polynomial matrix decomposition to the problem of, pointwise in frequency, constant matrix decomposition. At first it seems that applying inverse DFT on the decomposed matrices leads to polynomial EVD and SVD of the corresponding polynomial matrix. However, we will show later in this article that in order to have compact order decomposition, phase alignment of decomposed constant matrices in DFT domain results in polynomial matrices with considerably lower order. For this reason, a quadratic nonlinear minimization problem is set up to minimize the decomposition error for a given finite order constraint. Consequently, the required number of frequency samples and computational complexity of decomposition reduce dramatically. The algorithm provides compact order matrices as an approximation of polynomial matrix decomposition for an arbitrary polynomial order. This is suitable in MIMO communications and filterbank applications, where we deal with realization of MIMO linear time invariant systems. Moreover, formulation of polynomial EVD and SVD in DFT domain enables us to select the property of decomposition. We show that if eigenvalues (singular values) intersect at some frequencies in frequency domain, smooth decomposition, and spectrally majorized decomposition are distinct. The proposed algorithm is able to reach to either of these properties.

The remainder of this article is organized as follows. The relation between polynomial matrix decomposition and DFT matrix decomposition is formulated in Section 2. In Section 3, two important spectral properties of decomposition, namely spectral majorization and smooth decomposition, are provided using appropriate arrangement of singular values (eigenvalues) and corresponding singular vectors (eigenvectors). The equality of polynomial matrix and dft matrix decomposed matrices decompositions are guaranteed via the finite duration constraint, which is investigated in Section 4. The finite duration constraint imposes the phase angles of singular vector (eigenvector) to minimize a nonlinear quadratic function. A solution for this problem is proposed in Section 5. Section 6 presents the results of some computer simulations which are considered to demonstrate performance of the proposed decomposition algorithm.

Notation

Some notational conventions are as follows: constant values, vectors, and matrices are in regular character lower case, lower case over-arrow, and upper case, respectively. Coefficients of polynomial (scalar, vector, and matrix) are with indeterminate variable n in the square brackets. Any polynomial (scalar, vector, and matrix) is distinguished by bold character and indeterminate variable z in the parenthesis and its DFT by bold character and indeterminate variable k in the brackets.

PROBLEM FORMULATION

Denote a $p \times q$ polynomial matrix $\mathbf{A}(z)$ such that each element of $\mathbf{A}(z)$ is a polynomial. Equivalently, we can indicate this type of matrix by coefficient matrix $A[n]$,

$$\mathbf{A}(z) = \sum_{n=N_{\min}}^{N_{\max}} A[n] z^{-n} \tag{1}$$

where $A[n]$ is only non-zero in the interval $[N_{\min}, N_{\max}]$. Define the effective degree of $\mathbf{A}(z)$ as $N_{\max} - N_{\min}$ (or the length of $A[n]$ as $N_{\max} - N_{\min} + 1$).

The polynomial matrix multiplication of a $p \times q$ matrix $\mathbf{A}(z)$ and a $q \times t$ matrix $\mathbf{B}(z)$ is defined as

$$\begin{aligned} \mathbf{C}(z) &= \mathbf{A}(z)\mathbf{B}(z) \\ \mathbf{c}_{ij}(z) &= \textstyle\sum_{k=1}^{q} \mathbf{a}_{ik}(z)\mathbf{b}_{kj}(z). \end{aligned}$$

We can obtain the coefficient matrix of product by matrix convolution of $A[n]$ and $B[n]$, that is defined as

$$\begin{aligned} C[n] &= A[n] * B[n] \\ c_{ij}[n] &= \textstyle\sum_{k=1}^{q} a_{ik}[n] * b_{kj}[n] \end{aligned}$$

where $*$ denotes the linear convolution operator.

Denote para-conjugate of a polynomial matrix as

$$\tilde{\mathbf{A}}(z) = \mathbf{A}_*^T(z^{-1}) = \sum_{N_{\min}}^{N_{\max}} A^H[n] z^n.$$

in which, $*$ as a subscript denotes the complex conjugate of coefficients in the polynomial matrix $\mathbf{A}(z)$.

A matrix is said to be para-Hermitian if $\tilde{\mathbf{A}}(z) = \mathbf{A}(z)$ or equivalently $A[n] = A^H[-n]$.

We call a polynomial matrix paraunitary if $\tilde{\mathbf{U}}(z)\mathbf{U}(z) = \mathbf{I}$, where $\mathbf{I}$ is a $q \times q$ identity matrix.

Thin EVD of a $p \times p$ para-Hermitian polynomial matrix $\mathbf{A}(z)$ is of the form

$$\mathbf{A}(z) = \mathbf{U}(z)\mathbf{\Lambda}(z)\tilde{\mathbf{U}}(z), \tag{2}$$

and thin SVD of a $p \times q$ arbitrary polynomial matrix is of the form,

$$\mathbf{A}(z) = \mathbf{U}(z)\mathbf{\Sigma}(z)\tilde{\mathbf{V}}(z) \tag{3}$$

where $\mathbf{U}(z)$ and $\mathbf{V}(z)$ are $p \times r$ and $q \times r$ paraunitary matrices, respectively. $\mathbf{\Lambda}(z)$ and $\mathbf{\Sigma}(z)$ represent $r \times r$ diagonal matrices where r is the rank of $\mathbf{A}(z)$.

We can equivalently write EVD of a para-Hermitian matrix and SVD of a polynomial matrix in coefficient matrix form

$$A[n] = U[n] * \Lambda[n] * U^H[-n] \tag{4}$$

$$A[n] = U[n] * \Sigma[n] * V^H[-n] \tag{5}$$

in which, $U[n]$, $V[n]$, $\Lambda[n]$, and $\Sigma[n]$ are the coefficient matrices corresponding to $\mathbf{U}(z)$, $\mathbf{V}(z)$, $\mathbf{\Lambda}(z)$, and $\mathbf{\Sigma}(z)$.

In general, EVD and SVD of a finite-order polynomial matrix are not finite order. As an example, suppose EVD of para-Hermitian polynomial matrix

$$A(z) = \begin{bmatrix} 2 & z^{-1}+1 \\ z+1 & 2 \end{bmatrix}. \tag{6}$$

Eigenvalues and eigenvectors of the polynomial matrix in (6) are neither of finite order nor rational

$$\Lambda(z) = \begin{bmatrix} 2+(z^{-1}+2+z)^{1/2} & 0 \\ 0 & 2-(z^{-1}+2+z)^{1/2} \end{bmatrix}$$

$$\mathbf{U}(z) = \frac{1}{\sqrt{2}}\begin{bmatrix} (z^{-1}+1)(z^{-1}+2+z)^{-1/2} & z^{-1} \\ 1 & (z^{-1}+1)(z^{-1}+2+z)^{-1/2} \end{bmatrix}.$$

The same results can be found for polynomial QR decomposition in [12].

We mainly explain the proposed algorithm for polynomial SVD, yet wherever it seems necessary we explain the result for both decomposition.

The decomposition in (3) can also be approximated by samples of discrete-time Fourier transform, yields a decomposition off the form

$$\mathbf{A}[k] = \mathbf{U}'[k]\,\mathbf{\Sigma}'[k]\,\mathbf{V}'^H[k], \quad k = 0, 1, \ldots, K-1. \tag{7}$$

Such a decomposition can be obtained by taking the K-point DFT of coefficient matrix $A[n]$,

$$\mathbf{A}[k] = \mathbf{A}(z)|_{z=w_K^k} = \sum_{N_{\min}}^{N_{\max}} A[n]\, w_K^{kn}, \quad k = 0, 1, \ldots, K-1, \tag{8}$$

where $w_K = \exp(-j2\varpi/K)$.

DFT formulation plays an important role in decomposition of polynomial matrices because it replaces the problem of polynomial SVD that involves many protracted steps with K conventional SVD that are pointwise in frequency. It also enables us to control spectral properties of the decomposition. However, it causes two inherent drawbacks:

- Regardless of what is the trajectory of polynomial singular values in frequency domain, conventional SVD order singular values irrespectively of the ordering in neighboring frequency samples.
- In frequency domain, samples of polynomial singular vectors are known up to a scalar complex exponential by using the SVD at each frequency sample, which yields to discontinuous variation between neighboring frequency samples.

The first issue is directly dealt with the spectral properties of the decomposition. In Section 3, we would explain why arranging singular values in decreasing order yields to approximate spectral majorization, while smooth decomposition requires rearrangement of singular values and their corresponding singular vectors.

For the second issue, suppose conventional SVD of an arbitrary constant matrix A. If the pair $\vec{u}$ and $\vec{v}$ are the left and right singular vectors corresponding to a non-zero singular value, for an arbitrary scalar phase angle θ, the pair $e^{j\theta}\vec{u}$ and $e^{j\theta}\vec{v}$ are also left and right singular vectors corresponding to the same singular value. Although this non-uniqueness is trivial in conventional SVD, it plays a crucial role in polynomial SVD. When we perform SVD at each frequency of DFT matrix as in (7), these non-uniquenesses in phase exist at each frequency regardless of other frequency samples.

Denote $\vec{\mathbf{u}}_i[k]$ and $\vec{\mathbf{v}}_i[k]$ the ith column vector of the desired matrices $\mathbf{U}(z)$ and $\mathbf{V}(z)$. Then all the vectors of the form

$$\begin{cases} \vec{\mathbf{u}}_i'[k] = e^{j\theta_i[k]}\vec{\mathbf{u}}_i[k] \\ \vec{\mathbf{v}}_i'[k] = e^{j\theta_i[k]}\vec{\mathbf{v}}_i[k], \quad i = 1, 2, \ldots, r \\ \sigma_i'[k] = \sigma_i[k] \end{cases} \tag{9}$$

have the chance to appear as the ith column of $\mathbf{U}'[k]$ and $\mathbf{V}'[k]$, and ith diagonal element of$\Sigma'[k]$, respectively. Moreover, in many applications, specially those which are related to MIMO precoding, we can relax constraints of the problem by letting singular values to be complex (see applications of polynomial SVD in [4,18])

$$\begin{cases} \vec{\mathbf{u}}_i'[k] = e^{j\theta_i^u[k]}\vec{\mathbf{u}}_i[k] \\ \vec{\mathbf{v}}_i'[k] = e^{j\theta_i^v[k]}\vec{\mathbf{v}}_i[k] \\ \sigma_i'[k] = e^{j(\theta_i^v[k]-\theta_i^u[k])}\sigma_i[k] \end{cases}, \quad i = 1,2,\ldots,r. \tag{10}$$

Given this situation, singular values have not all their conventional meaning. For instance, the greatest singular value is conventionally 2-norm of the corresponding matrix, which is not true for complex singular values. The process of compensating singular vectors for these phases is what we call*phase alignment*and is developed in Section 4.

Based on what was mentioned above, Algorithm 1 gives the descriptive pseudo code for DFT-based SVD. Modifications of the algorithm for EVD of para-Hermitian matrices are straightforward. If at each frequency sample all singular values are in decreasing order, REARRANGE function (which is described in Algorithm 2) is only required for smooth decomposition, otherwise for spectral majorization, no further arrangement is required. For the phase alignment, first we need to compute phase angles which is indicated in the algorithm by DOGLEG function and is described in Algorithm 3.

SPECTRAL MAJORIZED DECOMPOSITION VERSUS SMOOTH DECOMPOSITION

Two of the most appealing decomposition properties are smooth decomposition [19] and spectral majorization [13]. These two objectives do not always occur at the same time, hence we should choose which one we are willing to use as our main objective.

In many filterbank applications which are dealt with principle components filterbank, spectral majorization and strong decorrelation are both required [16]. Since smooth decomposition leads to more compact decomposition, in cases that the only objective is strong decorrelation, exploiting smooth decomposition is reasonable. The DFT-based approach of polynomial matrix decomposition is capable of decomposing a matrix with either of these properties with small modification.

Algorithm 1: Approximate SVD

```
[U[n], Σ[n], V[n]] ← ASVD (A[n])
for k = 0,1,···,K−1
    Compute A[k] from (8):
        A[k] = Σ_{N_min}^{N_max} A[n] w_K^{kn}
    Decompose A[k] from (7):
        [U'[k], Σ[k], V'[k]] ← SVD (A[k])
end(for)
If smooth decomposition is required use Algorithm 2:
    [U'[k], Σ'[k], V'[k]] ← REARRANGE (U'[k], Σ'[k], V'[k])
for i = 1,2,···,r
    Compute phase angles using Algorithm 3:
        [θ_i^u[k], θ_i^v[k]] ← DOGLEG (ũ'_i[k], ṽ'_i[k])
     for k = 0,1,...,K̄−1
        Phase alignment using (9) or (10)
     end(for)
 end(for)
 for n = 0,1,...,M−1
     Compute decomposed polynomial matrices:
         U[n] ← Σ_{k=0}^{K} U[k] W^{−kn}
         V[n] ← Σ_{k=0}^{K} V[k] W^{−kn}
         Σ[n] ← Diagonal elements of U^H[−n]*A[n]*V[n]
 end(for)
```

Polynomial EVD of a para-Hermitian matrix is said to have spectral majorization property if [13,16]

$$\boldsymbol{\lambda}_1(e^{j\omega}) \geq \boldsymbol{\lambda}_2(e^{j\omega}) \geq \cdots \geq \boldsymbol{\lambda}_r(e^{j\omega}), \quad \forall \omega.$$

Note that, eigenvalues corresponding to para-Hermitian matrices are real in all frequencies.

We can extend the definition to the polynomial SVD, replacing singular values with eigenvalues in the definition, we have

$$\boldsymbol{\sigma}_1(e^{j\omega}) \geq \boldsymbol{\sigma}_2(e^{j\omega}) \geq \cdots \geq \boldsymbol{\sigma}_r(e^{j\omega}), \quad \forall \omega.$$

If we let singular values to be complex, we can replace absolute value of singular values in the definition.

A polynomial matrix have no discontinuity in frequency domain, hence we modify definition of smooth decomposition presented in [19] to fit with our problem and avoid unnecessary discussions.

Polynomial EVD (SVD) of a matrix is said to possess smooth decomposition if eigenvectors (singular vectors) have no discontinuity in frequency domain, that is

$$\left|\frac{d}{d\omega}\mathbf{u}_{il}(e^{j\omega})\right| < \infty, \quad \forall\omega \text{ and } \begin{array}{l} i = 1,2,\ldots,r \\ l = 1,2,\ldots,p, \end{array} \tag{11}$$

where $\mathbf{u}_{il}$ is the lth element of $\vec{\mathbf{u}}_i$.

If eigenvalues (singular values) of a polynomial matrix intersect at some frequencies, the spectral majorization and smooth decomposition are not simultaneously realizable. As an example, suppose $\mathbf{A}(z)$ is a polynomial matrix with $\vec{\mathbf{u}}_1(z)$ and $\vec{\mathbf{u}}_2(z)$ are eigenvectors corresponding to distinct eigenvalues $\lambda_1(z)$ and $\lambda_2(z)$, respectively. Lets assume $\vec{\mathbf{u}}_1(e^{j\omega})$ and $\vec{\mathbf{u}}_2(e^{j\omega})$ have no discontinuity in frequency domain, and $\lambda_1(e^{j})$ and $\lambda_2(e^{j})$ intersect at some frequencies. Denote

$$\lambda_1'(e^{j\omega}) = \begin{cases} \lambda_1(e^{j\omega}) & \lambda_1(e^{j\omega}) \geq \lambda_2(e^{j\omega}) \\ \lambda_2(e^{j\omega}) & \lambda_1(e^{j\omega}) < \lambda_2(e^{j\omega}) \end{cases},$$
$$\lambda_2'(e^{j\omega}) = \begin{cases} \lambda_2(e^{j\omega}) & \lambda_1(e^{j\omega}) \geq \lambda_2(e^{j\omega}) \\ \lambda_1(e^{j\omega}) & \lambda_1(e^{j\omega}) < \lambda_2(e^{j\omega}) \end{cases}, \tag{12}$$

Algorithm 2 Rearrangement for smooth decomposition

$[\mathbf{U}'[k], \Sigma[k], \mathbf{V}'[k]] \leftarrow$ **REARRANGE**$(\mathbf{U}'[k], \Sigma[k], \mathbf{V}'[k])$

```
for k = 1,2,...,K
    Define S = {1,2,...,r}
    for i = 1,2,...,r
        ι ← arg max_{j∈S} (|c_ij^{u'}[k]| + |c_ij^{v'}[k]|) / 2
        u''_i[k] ← u'_ι[k]
        v''_i[k] ← v'_ι[k]
        σ''_i[k] ← σ_ι[k]
        S ← S − {ι}
    end(for)
    U'[k] ← U''[k]
    V'[k] ← V''[k]
    Σ[k] ← Σ''[k]
end(for)
```

and

$$\vec{\mathbf{u}}_1'(e^{j\omega}) = \begin{cases} \vec{\mathbf{u}}_1(e^{j\omega}) & \lambda_1(e^{j\omega}) \geq \lambda_2(e^{j\omega}) \\ \vec{\mathbf{u}}_2(e^{j\omega}) & \lambda_1(e^{j\omega}) < \lambda_2(e^{j\omega}) \end{cases},$$
$$\vec{\mathbf{u}}_2'(e^{j\omega}) = \begin{cases} \vec{\mathbf{u}}_2(e^{j\omega}) & \lambda_1(e^{j\omega}) \geq \lambda_2(e^{j\omega}) \\ \vec{\mathbf{u}}_1(e^{j\omega}) & \lambda_1(e^{j\omega}) < \lambda_2(e^{j\omega}) \end{cases}. \tag{13}$$

Obviously, $\vec{\mathbf{u}}_1'(e^{j\omega})$ and $\vec{\mathbf{u}}_2'(e^{j\omega})$ are eigenvectors corresponding to distinct eigenvalues $\lambda'_1(e^{j\omega})$ and $\lambda'_2(e^{j\omega})$, respectively. Note that, $\lambda'_1(e^{j\omega}) \geq \lambda'_2(e^{j\omega})$ for all frequencies, which means $\lambda_1(e^j)$ and $\lambda_2(e^j)$ are spectrally majorized. However, $\lambda_1(e^{j\omega})$ and $\lambda_2(e^{j\omega})$ are discontinuous at intersection frequencies of $\lambda_1(e^j)$ and $\lambda_2(e^j)$, which implies that they are not smooth anymore. In this situation, although $\lambda'_1(e^{j\omega})$, $\lambda'_2(e^{j\omega})$, $\vec{\mathbf{u}}'_1(e^{j\omega})$, and $\vec{\mathbf{u}}'_2(e^{j\omega})$ are not even analytic, we can approximate them with finite order polynomials.

If a decomposition has spectral majorization, its eigenvalues (singular values) are of decreasing order in all frequencies. Therefore, they are in decreasing order in any arbitrary frequency sample set, including DFT frequencies. Obviously the converse is only approximately true. Hence, for polynomial EVD to possess spectral majorization approximately, it suffices to arrange sampled eigenvalues (singular values) of (7) in decreasing order. Since we only justify spectral majorization at DFT frequency samples, the resulting EVD (SVD) may possess the property only approximately. Similar results can be seen in [14,20].

To have smooth singular vectors, we propose an algorithm based on inner product of consecutive frequency samples of singular vectors. We can accumulate smoothing requirement in (11) for all r elements as

$$\left\| \frac{d}{d\omega} \vec{\mathbf{u}}_i(e^{j\omega}) \right\| < \infty, \quad \forall \omega \text{ and } \quad i = 1, 2, \ldots, r. \tag{14}$$

Let B be the upper bound of norm of derivative and $\Re\{\cdot\}$ be the real value of a complex value.

For an arbitrary $\Delta\omega$ we have

$$\begin{aligned} \left\| \vec{\mathbf{u}}_i(e^{j(\omega+\Delta\omega)}) - \vec{\mathbf{u}}_i(e^{j\omega}) \right\|^2 &= 2 - 2\Re\left\{ \vec{\mathbf{u}}_i^H(e^{j(\omega+\Delta\omega)}) \vec{\mathbf{u}}_i(e^{j\omega}) \right\} \\ &< (\Delta\omega B)^2 \quad \forall \omega, \end{aligned} \tag{15}$$

that is, for a smooth singular vector $\Re\{\vec{\mathbf{u}}_i(e^{j(\omega+\Delta\omega)})\ \vec{\mathbf{u}}_i(e^{j\omega})\}$ can be made to be as close to unity as desired by making $\Delta\omega$ sufficiently small. In our problem $\vec{\mathbf{u}}_i(e^{j\omega})$ is sampled uniformly with $\Delta\omega = \frac{2\pi}{K}$. Since EVD is performed at each frequency sample independently, $\vec{\mathbf{u}}_i[k]$ and $\vec{\mathbf{u}}_i[k+1]$ are not necessarily two consecutive frequency samples of a smooth eigenvector. Therefore, we should rearrange eigenvalues and eigenvectors to yield smooth decomposition. This can be done for each sample of eigenvector $\vec{\mathbf{u}}_i[k]$ by seeking for the eigenvector of successor sample $\vec{\mathbf{u}}_j[k+1]$ with the most value of $\Re\{\vec{\mathbf{u}}_i^H[k]\,\vec{\mathbf{u}}_j[k+1]\}$.

Define inner product $c_{ij}^u[k]$ as

$$c_{ij}^{u}[k] = \vec{\mathbf{u}}_i^{H}[k-1]\,\vec{\mathbf{u}}_j[k].$$

Since, $\vec{\mathbf{u}}'_i[k]$ is a scalar phase multiplication of $\vec{\mathbf{u}}_i[k]$, computation of $\Re\{c_{ij}^{u}[k]\}$ is not possible before phase alignment. Due to (15), for sufficiently small $\Delta\omega$, two consecutive samples of a smooth singular vector can be as close as desired and we can approximate

$$\Re\left\{c_{ij}^{u}[k]\right\} \approx |c_{ij}^{u}[k]| = |c_{ij}^{u'}[k]|,$$

which allows us to use inner product of $\vec{\mathbf{u}}'[k]$ instead of $\vec{\mathbf{u}}[k]$. From (12) and (13), it can be seen that before the intersection of eigenvalues, consecutive eigenvectors which are sorted by conventional EVD in decreasing order, are from the same smooth eigenvector and so $|c_{11}^{u'}[k]|$ and $|c_{22}^{u'}[k]|$ are near unity. However, if $k-1$ and k are two frequency sample before and after the intersection, respectively, due to decreasing order of eigenvalues, smoothed eigenvectors are swapped after intersection. Therefore, $|c_{11}^{u'}[k]|$ and $|c_{22}^{u'}[k]|$ are some values near zero, instead $|c_{12}^{u'}[k]|$ and $|c_{21}^{u'}[k]|$ are near unity.

Algorithm 2 describes a simple rearrangement procedure to track eigenvectors (singular vectors) for smooth decomposition.

FINITE DURATION CONSTRAINT

Phase alignment is critical to have compact order decomposition. Another aspect of this fact is revealed in the coefficient's domain perspective of (7). In this domain, the multiplication is replaced by circular convolution

$$\begin{aligned} A[((n))_K] &= U'[((n))_K] \circledast \Sigma'[((n))_K] \circledast V'^{H}[((-n))_K] \\ &= U[((n))_K] \circledast \Sigma[((n))_K] \circledast V^{H}[((-n))_K] \end{aligned} \tag{16}$$

in which $\circledast$ is the circular convolution operator and $((n))_K$ denotes n module K.

Polynomial SVD corresponds to linear convolution in the coefficients domain, however the decomposition obtained from DFT corresponds to circular convolution. Recalling from discrete-time signal processing, it is well known that we can equivalently utilize circular convolution instead of linear convolution if convoluted signals are zero-padded adequately. That is, for $x_1[n]$ and $x_2[2]$ are two signals with the length of N_1 and N_2, respectively, apply zero padding such that zero padded signals have the length N_1+N_2-1 [21]. Hence, if the last $M-1$ coefficients of $U[n]$, $\Sigma[n]$, and $V[n]$, are zero, the following results are hold:

$$A[((n))_K] = U[((n))_K] \circledast \Sigma[((n))_K] \circledast V^H[((-n))_K] \Rightarrow A[n] = U[n] * \Sigma[n] * V^H[-n],$$
$$U[((n))_K] \circledast U^H[((-n))_K] = \delta[((n))_K] I \Rightarrow U[n] * U^H[-n] = \delta[n] I,$$
$$V[((n))_K] \circledast V^H[((-n))_K] = \delta[((n))_K] I \Rightarrow V[n] * V^H[-n] = \delta[n] I. \quad (17)$$

Therefore, the problem is to obtain the phase set$\{\vec{\theta}_i[k]\}$and correcting the singular vectors using (9). The phase set$\{\vec{\theta}_i[k]\}$should be such that the resulting coefficients satisfy (17).

Without loss of generality, let$U[n]$ and$V[n]$ be causal, i.e.,$U[n]=V[n]=0$ for$n<0$.$U[n]$ and$V[n]$ (which are supposed to be of length$M$) should be multi-sequence zero-padded at least with $(M-1)$ zeros.

$$U[n] = \frac{1}{K}\sum_{k=0}^{K-1} \mathbf{U}[k]\, w_K^{-kn} = \mathbf{0} \quad (18)$$

for$n=M, M+1, \ldots, K-1$, in which$K \geq 2M-1$. If these conditions are satisfied, circular convolution can be used instead of linear convolution.

Since the available matrix of singular vectors at each frequency is$\mathbf{U}'[k]$, inserting (9) in (18) for each singular vector separately leads to

$$\sum_{k=0}^{K-1} \mathbf{u}_i'[k]\, e^{-j\theta_i^u[k]} w_K^{-kn} = \mathbf{0} \quad (19)$$

for$n=M, M+1, \ldots, K-1$.

Without loss of generality, let$\theta_i[0]=0$. In a more compact form we can express these $(K-M)$-folded equations in matrix form

$$\mathcal{F}_M(\vec{\mathbf{u}}_i')\vec{\mathbf{x}}(\vec{\theta}_i^u) = -\vec{f}_M(\vec{\mathbf{u}}_i') \qquad i = 1, 2, \ldots, r \quad (20)$$

in which $\vec{\mathbf{x}}(\vec{\theta}_i^u) = [\exp(j\theta_i^u[1]), \ \exp(j\theta_i^u[2]), \ \ldots, \exp(j\theta_i^u[K-1])]^T$, $\vec{f}_M(\vec{\mathbf{u}}_i') = [\vec{\mathbf{u}}_i'^T[0], \vec{\mathbf{u}}_i'^T[0], \ldots, \vec{\mathbf{u}}_i'^T[0]]^T$ is a $p(K-M) \times 1$ vector, and

$$\mathcal{F}_M(\vec{\mathbf{u}}_i') = \begin{bmatrix} \vec{\mathbf{u}}_i'[1] w_K^{-M} & \vec{\mathbf{u}}_i'[2] w_K^{-2M} & \cdots & \vec{\mathbf{u}}_i'[K-1] w_K^{-(K-1)M} \\ \vec{\mathbf{u}}_i'[1] w_K^{-(M+1)} & \vec{\mathbf{u}}_i'[2] w_K^{-2(M+1)} & \cdots & \vec{\mathbf{u}}_i'[K-1] w_K^{-(K-1)(M+1)} \\ \vdots & \vdots & & \vdots \\ \vec{\mathbf{u}}_i'[1] w_K^{-(K-1)} & \vec{\mathbf{u}}_i'[2] w_K^{-2(K-1)} & \cdots & \vec{\mathbf{u}}_i'[K-1] w_K^{-(K-1)^2} \end{bmatrix}.$$

For polynomial EVD, Equation (20) is enough, however, for polynomial SVD we have two options. To approximate SVD with approximately positive singular values, we must augment $\mathcal{F}_M(\vec{\mathbf{u}}_i')$ and $\vec{f}_M(\vec{\mathbf{u}}_i')$ with similar defined matrix and vector for $\mathbf{v}_i'[k]$

$$\mathcal{F}_M(\vec{\mathbf{u}}_i', \vec{\mathbf{v}}_i') = \begin{bmatrix} \mathcal{F}_M(\vec{\mathbf{u}}_i') \\ \mathcal{F}_M(\vec{\mathbf{v}}_i') \end{bmatrix} \text{ and } \vec{f}_M(\vec{\mathbf{u}}_i', \vec{\mathbf{v}}_i') = \begin{bmatrix} \vec{f}_M(\vec{\mathbf{u}}_i') \\ \vec{f}_M(\vec{\mathbf{v}}_i') \end{bmatrix},$$

then solve

$$\mathcal{F}_M(\vec{\mathbf{u}}_i', \vec{\mathbf{v}}_i')\vec{\mathbf{x}}(\theta_i) = -\vec{f}_M(\vec{\mathbf{u}}_i', \vec{\mathbf{v}}_i') \quad i = 1, 2, \ldots, r. \tag{21}$$

An additional degree of freedom is obtained by letting singular values to be complex. However, an straightforward solution which yield to singular values and singular vectors of order$M$is complicated. Instead, we impose the finite duration constraint only two singular vectors

$$\begin{cases} \mathcal{F}_M(\vec{\mathbf{u}}_i')\vec{\mathbf{x}}(\vec{\theta}_i^{\nu}) = -\vec{f}_M(\vec{\mathbf{u}}_i') \\ \mathcal{F}_M(\vec{\mathbf{v}}_i')\vec{\mathbf{x}}(\vec{\theta}_i^{\nu}) = -\vec{f}_M(\vec{\mathbf{v}}_i') \end{cases} \quad i = 1, 2, \ldots, r. \tag{22}$$

If$K \geq 2M-1$, then the last$M-1$ coefficients of resulting polynomial vectors are zero. Therefore, according to (17),$\mathbf{U}(z)$ and$\mathbf{V}(z)$ are paraunitary. On the other hand, if$K \geq 2M+N_{max}-N_{min}-1$, circular convolution relation of coefficient

$$\Sigma[\,((n))_K] = U^H[\,((-n))_K] \circledast A[\,((n))_K] \circledast V[\,((n))_K]$$

results in the linear convolution $\Sigma[n] = U^H[-n] * A[n] * V[n]$. This guarantee that$\mathbf{\Sigma}(z)$ is a diagonal polynomial matrix of order $2M+N_{max}-N_{min}-2$. Obviously, if$\mathbf{U}(z)$ and$\mathbf{V}(z)$ are paraunitary and$\mathbf{\Sigma}(z) = \tilde{\mathbf{U}}(z)\mathbf{A}(z)\mathbf{V}(z)$is a diagonal matrix,$\mathbf{A}(z) = \mathbf{U}(z)\mathbf{\Sigma}(z)\tilde{\mathbf{V}}(z)$is the polynomial SVD of$\mathbf{A}(z)$.

Once the set of phase$\{\theta_i^u[k], \theta_i^v[k]\}$are obtained from (20), (21), or (22), phase alignment of$\vec{\mathbf{u}}_i'[k]$ and $\vec{\mathbf{v}}_i'[k]$can be done using (10) and inverse DFT of$\mathbf{U}[k]$ and$\mathbf{V}[k]$ yield to coefficient matrices$U[n]$ and$V[n]$. For obtaining singular values, we have two options, we can either set$K \geq 2M-1$ and phase align $\sigma_i'[k]$ using (10). After inverse DFT of$\mathbf{\Sigma}[k]$, we should truncate $\Sigma[n]$ to have duration$M$. Another option which yields to more accurate results is by calculating$\tilde{\mathbf{U}}(z)\mathbf{A}(z)\mathbf{V}(z)$and replacing off-diagonal elements with zero.

Next, we provide a minimization approach to determine the unknown set $\{\theta_i[k]\}$.

GRADIENT DESCENT SOLUTION

In general, there may exist no phase vector$\vec{\theta}$ which satisfies (20). Even when there exists a phase vector that satisfies the finite duration constraint, the solution is not straightforward. For these reasons, we can view (20) as a minimization problem [6]. We use energy of the highest order coefficients (the coefficients that we equate to zero in (18)) as the objective to the minimization problem

$$J(\vec{\theta}_i) = \left\| \mathcal{F}_M(\vec{\mathbf{u}}_i')\vec{\mathbf{x}}(\vec{\theta}_i^u) + \vec{f}_M(\vec{\mathbf{u}}_i') \right\|^2, \quad i = 1, 2, \ldots, r. \tag{23}$$

An alternative minimization technique as a solution for this phase optimization problem is proposed in [6], which we describe it in this section.

Throughout this section, we focus on solving $\vec{\theta}_i = \arg\min J(\vec{\theta}_i)$ as a least square solution for a single singular vector $\vec{\mathbf{u}}_i[k]$, so we drop the subscript "i" from the quantity $\vec{\theta}_i$ and use F and $\vec{f}$, instead of $\mathcal{F}_M(\vec{\mathbf{u}}_i')$ and $\vec{f}_M(\vec{\mathbf{u}}_i')$ to simplify the notation. The objective $J(\vec{\theta})$ is intentionally presented as a function of $\vec{\theta}$ to emphasize the fact that our problem is classified as an unconstrained optimization.

We exploit the trusted region strategy for the problem (23). By utilizing the information about the gradient vector and Hessian matrix in each step, trusted region strategy constructs a model function m_k which have a similar behavior close to the current point $\vec{\theta}^{(k)}$. The model m_k is usually defined as the second-order Taylor series expansion (or its approximation) of $J(\vec{\theta} + \vec{\varphi})$ around $\vec{\theta}$, that is

$$m_k(\vec{\varphi}) = J(\vec{\theta}) + \vec{\varphi}^T \nabla J(\vec{\theta}) + \vec{\varphi}^T \nabla^2 J(\vec{\theta})\vec{\varphi},$$

where $\nabla J(\vec{\theta})$ and $\nabla^2 J(\vec{\theta})$ are the gradient vector and the Hessian matrix corresponding to $J(\vec{\theta})$, respectively. The model m_k is designed to be a good approximation of $J(\vec{\theta})$ near the current point and is not trustworthy on regions far from the current point. Consequently, the restriction in minimization of m_k on a region around $\vec{\theta}^{(k)}$ is crucial, that is

$$\vec{\varphi} = \arg\min_{\vec{\varphi}} m_k(\vec{\varphi}) \quad \|\vec{\varphi}\| < R. \tag{24}$$

where R is the trusted region radius.

The decision about shrinking of the trusted region is determined by comparing the actual reduction in objective function and predicted reduction. Given a step $\vec{\varphi}$, the ratio

$$\rho = \frac{J(\vec{\theta}) - J(\vec{\theta} + \vec{\varphi})}{m_k(0) - m_k(\vec{\varphi})} \tag{25}$$

is used as a criterion to indicate if the trusted region is small enough.

Among methods which approximate the solution of the constrained minimization (24) dogleg procedure is the only one which leads to analytical approximation. It also promises to achieve at least as much reduction in m_k as is possible by Cauchy point (the minimizer of m_k along the steepest descent direction $-\nabla J(\vec{\theta})$, subject to the trusted region) [22]. However, this procedure requires Hessian matrix (or an approximation of it) to be positive definite.

Hessian Matrix Modification

The gradient vector and Hessian matrix corresponding to $J(\vec{\theta})$ are as follows

$$\begin{aligned} \vec{g}(\vec{\theta}) &= 2\Im\left\{\mathbf{X}(\vec{\theta})\mathcal{F}^H(\mathcal{F}\vec{\mathbf{x}}(\vec{\theta})+\vec{f})\right\}, \\ H(\vec{\theta}) &= 2\Re\left\{\mathbf{X}(\vec{\theta})\mathcal{F}^H\mathcal{F}\mathbf{X}(\vec{\theta})^H\right\} \\ &\quad -2\Re\left\{\operatorname{diag}\left(\mathbf{X}(\vec{\theta})\mathcal{F}^H(\mathcal{F}\vec{\mathbf{x}}(\vec{\theta})+\vec{f})\right)\right\}, \end{aligned} \tag{26}$$

where $\mathbf{X}(\vec{\theta})$ is a diagonal matrix with the kth diagonal element $\exp(-j\theta[k])$ and $k=0, 1, \ldots, K-1$.

In general, Hessian matrix in (26) does not promise to be always positive definite. Therefore, we should modify Hessian matrix to yield a positive definite approximation.

We provide a simple modification which brings some desirable features by omitting the second term from the Hessian matrix and diagonal loading to guarantee positive definiteness

$$H(\vec{\theta}) \approx 2\Re\left\{\mathbf{X}(\vec{\theta})\mathcal{F}^H\mathcal{F}\mathbf{X}(\vec{\theta})^H\right\} + \alpha I. \tag{27}$$

The term $2\Re\left\{\mathbf{X}(\vec{\theta})\mathcal{F}^H\mathcal{F}\mathbf{X}(\vec{\theta})^H\right\}$ is positive semi-definite and in many situations, it is much more significant than the second term of Hessian matrix in (26). Hence, with diagonal loading αI (I is with conformable size and α is very small), the modified Hessian matrix guarantees (27) to be positive definite and provides the desired properties in contrast to use the exact Hessian matrix.

Dogleg Method

Dogleg method starts with the unconstrained minimization of (24)

$$\phi^h = -H^{-1}\vec{g} \tag{28}$$

When the trusted region radius is so large that $\|\phi^H\| \le R$, it is the exact solution of (24) and we select it as the dogleg method answer. On the other hand, for small R the solution of (24) is $-R\vec{g}/\|\vec{g}\|$. For intermediate values of R, the optimal solution lies on a curved trajectory between these two points [22].

Algorithm 3: Trusted region dogleg algorithm.

Obtain $\vec{\theta}_0$ from (33)
Given $R_0 < \bar{R}$
for $i = 0, 1, 2, \cdots$
    Obtain $\vec{\phi}_h$ from (28)
    if $\left\|\vec{\phi}^h\right\| < R_0$
        $\vec{\phi}_i \leftarrow \vec{\phi}^u$
    else
        Obtain $\vec{\phi}^g$ from (29)
        if $\left\|\vec{\phi}^g\right\| > R_i$
            $\vec{\phi}_i \leftarrow R_i \frac{\vec{\phi}^g}{\|\vec{\phi}^g\|}$
        else
            $\vec{\phi}^d \leftarrow \vec{\phi}^h - \vec{\phi}^g$
            Solve $\left\|\vec{\phi}^d\right\|\tau^2 + 2(\vec{\phi}^g)^H\vec{\phi}^d\tau + \left\|\vec{\phi}^g\right\| - R_i^2 = 0$
            Select the root $\tau > 0$
            $\vec{\phi}_i \leftarrow \vec{\phi}^g + \tau\vec{\phi}^h$
      Evaluate ρ_i from (25)
      if $\rho_i < \frac{1}{4}$
          $R_{i+1} \leftarrow 0.25R_i$
    elseif $\rho > \frac{3}{4}$ and $\|\vec{\phi}_i\| = R_i$
        $R_{i+1} \leftarrow min(2R_i, \bar{R})$
    else
        $R_{i+1} \leftarrow R_i$
    if $\rho_i > 0$
        $\theta_{i+1} \leftarrow \theta_i + \phi_i$
    else
        $\theta_{i+1} \leftarrow \theta_i$
end(for)

The dogleg method approximates this trajectory by a path consisting of two line segments. The first line segment starts from the origin to the unconstrained minimized point along the steepest descent direction

$$\phi^g = -\frac{\vec{g}^T\vec{g}}{\vec{g}^T H\vec{g}}\vec{g}. \tag{29}$$

The second line segment starts fromϕ^gtoϕ^h. These two line segments form an approximate trajectory which its intersection with the sphere$\|\phi\|=R$is the approximate solution of (24) when$\|\phi^h\|>R$.

Alternating Minimization

Another solution of (23) is provided by converting the problem of multivariate minimization to a sequence of single-variate minimization problem via alternating minimization [6]. In each iteration, a series of single-variate minimization is performed, while other parameters are held unchanged.

Each Iteration consists of $k-1$ steps, which at each step one parameter $\theta[k]$ is updated. Suppose we are at step k of ith iteration. At this step $k-1$ first parameters were updated in the current iteration, and $K-k-2$ last parameters were updated from the previous iteration. These parameters are held fixed, while $\theta[k]$ is minimized at the current step,

$$\theta_i[k] = \arg\min_{\theta[k]} J\left(\theta_i[1],\ldots,\theta_i[k-1],\theta[k],\theta_{i-1}[k+1],\ldots,\theta_{i-1}[K-1]\right). \tag{30}$$

The cost function is guaranteed to be non-incremental at each step; however, this method is also converges to a local minima which highly depend on the initial guess of the algorithm. For solving (30) it is suffices to make the kth element of gradient vector in (26) equal to zero. Suppose the calculation are performed for phase alignment of $\vec{\mathbf{u}}'[k]$ k=0,1,…,K−1,

$$\begin{aligned}\frac{\partial J}{\partial\theta[k]} &= \Im\left\{e^{-j\theta[k]}t_i[k]\right\}\\ &= 2\,|t_i[k]|^2\sin(\angle t_i[k]-\theta[k]) = 0,\end{aligned} \tag{31}$$

where $\angle t_u^{(i)}[k]$ is the phase angle of $t_u^{(i)}[k]$ and

$$\begin{aligned}t_i[k] &= \sum_{l=0}^{k-1} e^{j\theta_i[l]}\vec{\mathbf{u}}'^H[k]\,\vec{\mathbf{u}}'[l]\,\frac{w_K^{(k-l)M}-1}{1-w_K^{(k-l)}}\\ &+\sum_{l=k+1}^{K-1} e^{j\theta_{i-1}[l]}\vec{\mathbf{u}}'^H[k]\,\vec{\mathbf{u}}'[l]\,\frac{w_K^{(k-l)M}-1}{1-w_K^{(k-l)}}.\end{aligned}$$

Fortunately, Equation (31) has a closed form solution

$$\theta_i[k] = \begin{cases}\angle t_i[k]\\ \angle t_i[k]+\pi\end{cases}. \tag{32}$$

However, only the second case of (32) has positive second partial deviation. Therefore, the global minima of (30) is

$\theta_i[k] = \angle t_i[k]+\pi.$

Initial Guess

All algorithms of unconstrained minimization require to be supplied by a starting point, which we denoted by $\vec{\theta}_0$. To avoid getting stuck in local minima, we should select a good initial guess. This can be accomplished by minimizing a different but similar cost function denoted by $J'(\vec{\theta})$

$$J'(\vec{\theta}) = \left\| \vec{\mathbf{x}}(\vec{\theta}_i) - \mathcal{F}^{\dagger}\vec{f} \right\|^2$$

in which $\dagger$ represents pseudo inverse.
Solving $J'(\vec{\theta})$ yields to a simple initial guess

$$\vec{\theta}_0 = \angle \mathcal{F}^{\dagger}\vec{f}. \tag{33}$$

Based on what have been mentioned in this section, a pseudo-code description of the trusted region dogleg algorithm is given by Algorithm 3. In this algorithm, we start with the initial guess of (33) and a trusted region radius upper bound $\bar{R}$. Then we continue the trusted region minimization procedure as described in this section.

SIMULATION RESULTS

In this section, we present some examples to demonstrate the performance of the proposed algorithm. For the first example, our algorithm is applied to a polynomial matrix example from [11]

$$\mathbf{A}(z) = \begin{bmatrix} 2 & 0 & 2z^{+1} \\ z^{+1} & 1 & 0 \\ 0 & z^{-1} & 1 \end{bmatrix}. \tag{34}$$

Frequency behavior of singular values can be seen in Figure1. There is no intersection of singular values, so the setup of the algorithm either for spectral majorization or frequency smoothness leads to identical decomposition.

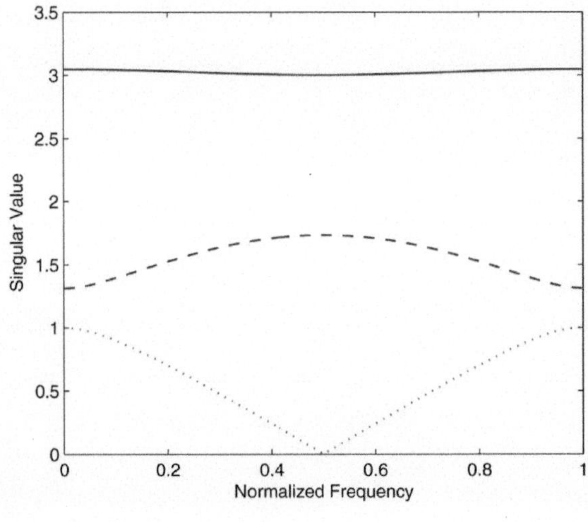

Figure 1: Singular values versus frequency.

For having approximately positive singular values, we use (21). Define the average energy of highest order coefficients for the pair of polynomial singular vectors $\vec{\mathbf{u}}_i$ and $\vec{\mathbf{v}}_i$ as $E_i^{u,v} = J(\vec{\theta}_i)/(K - M)$ (we expect energy of highest order coefficients to be zero or at least minimized). A plot of E_i versus iteration for each pair of singular vectors is depicted in Figure2. The decomposition length is $M=9$ (order is 8) and we use $K=2M+(N_{max}-N_{min})=20$ number of DFT points.

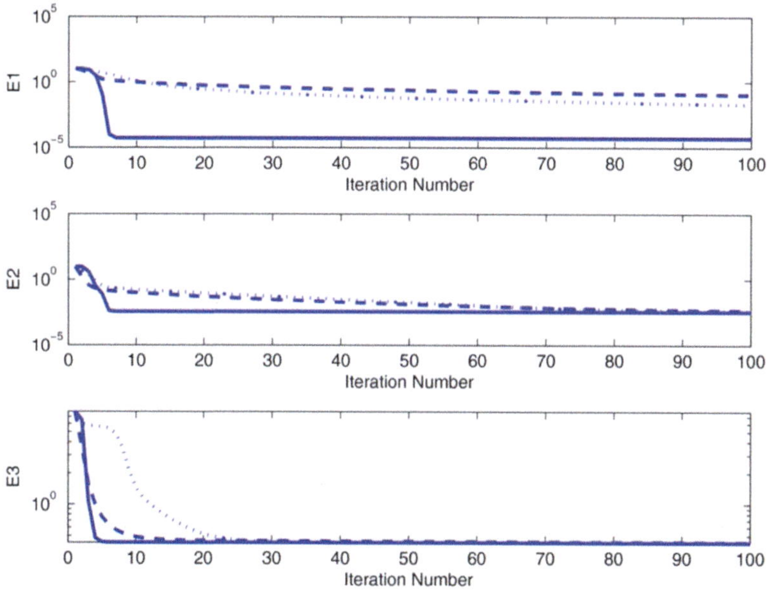

Figure 2: Average highest order coefficients energy E_i versus iteration number for a decomposition with approximately positive singular values.Dotted line: Cauchy points. Dashed line: Alternative minimization. Solid Line: proposed algorithm.

As it is seen, the use of dogleg method with approximate Hessian matrix leads to a fast convergence in contrast with using alternative minimization and Cauchy-point (which is always selected along the gradient direction). Of course we should consider that due to matrix inversion, computational complexity of Dogleg method is $O(K^3)$ while computational complexity of alternative minimization and Cauchy point is $O(K^2)$.

The final value of average highest order coefficient for three pair of singular vectors are 5.54×10^{-5}, 3.5×10^{-3}, and 0.43, respectively. The first singular vector satisfies finite duration constraint almost exactly. The second singular vector fairly satisfies this constraint. However, highest order coefficients of last singular vector, possess considerable amount of energy, that seems to cause decomposition error.

Denote the relative error of the decomposition as

$$E^A = \frac{\left\| \mathbf{A}(z) - \mathbf{U}(z)\Sigma(z)\tilde{\mathbf{V}}(z) \right\|_F}{\|\mathbf{A}(z)\|_F}$$

in which $\|\cdot\|_F$ is the extension of Frobenius norm for polynomial matrices and is defined by

$$\|\mathbf{A}(z)\|_F = \sqrt{\sum_n \|A[n]\|_F^2}.$$

Since in our optimization procedure we only seek for finite duration approximation, $\mathbf{U}(z)$ and $\mathbf{V}(z)$ are only approximately paraunitary. Therefore, we also define relative error of paraunitarity as

$$E^U = \frac{\left\| \tilde{\mathbf{U}}(z)\mathbf{U}(z) - \mathbf{I} \right\|_F}{r}.$$

An upper bound for E^U can be obtained as

$$E^U \leq 2\sqrt{\frac{K-M}{K}\sum_{i=1}^{r} E_i^{(u)}(1 - E_i^{(u)})} + \sqrt{\frac{K-M}{K}\sum_{i=1}^{r} (E_i^{(u)})^2},$$

which means as average energy on $K-M$ highest order goes to zero, E^U diminishes.

The relative error of this decomposition is $E^A = 1.18 \times 10^{-2}$ while the error of $\mathbf{U}(z)$ and $\mathbf{V}(z)$ are $E^U = 3.3 \times 10^{-2}$ and $E^V = 3.08 \times 10^{-2}$, respectively. The paraunitarity error is relatively high in contrast with decomposition error. This is due to the difference between the first two singular values and the last singular value.

A plot of relative errors E^A, E^U, and E^V for various amount of M is shown in Figure3. The number of frequency samples is fixed at $K = 2M + 2(N_{max} - N_{min})$.

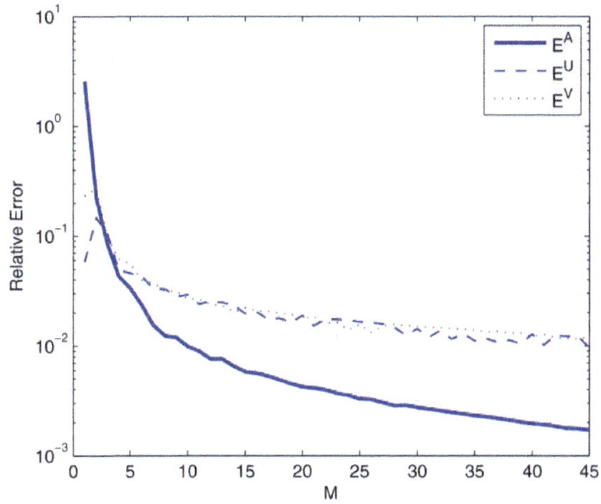

Figure 3: Relative error versus$M$for a decomposition with approximately positive singular values.$K=2M=2$.

The number of frequency samples$K$is an optional choice, however as discussed in Section 4, it should satisfy$K \geq 2M+N_{max}-N_{min}-1$. In order to demonstrate the effect of number of frequency samples on the decomposition error, a plot of relative error versus different amount of$K$is depicted in Figure4. Increasing the number of frequency samples does not lead to reduction of relative error. Moreover, it increases computational burden. Therefore, a value near $2M+(N_{max}-N_{min})-1$ is a reasonable choice for the number of frequency samples.

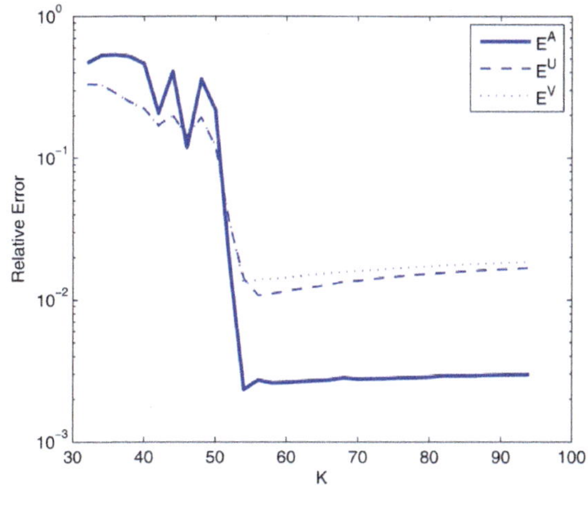

Figure 4: Relative error versus$K$for a decomposition with approximately positive singular values.$M=31$.

Now, lets relax the problem by allowing singular values to be complex and using (22). A plot of E_i^u and E_i^v versus iteration for each pair of singular vectors is depicted in Figure5. The decomposition length is $M=9$ (order is 8) and we use $K=2M+(N_{max}-N_{min})=20$ number of DFT points.

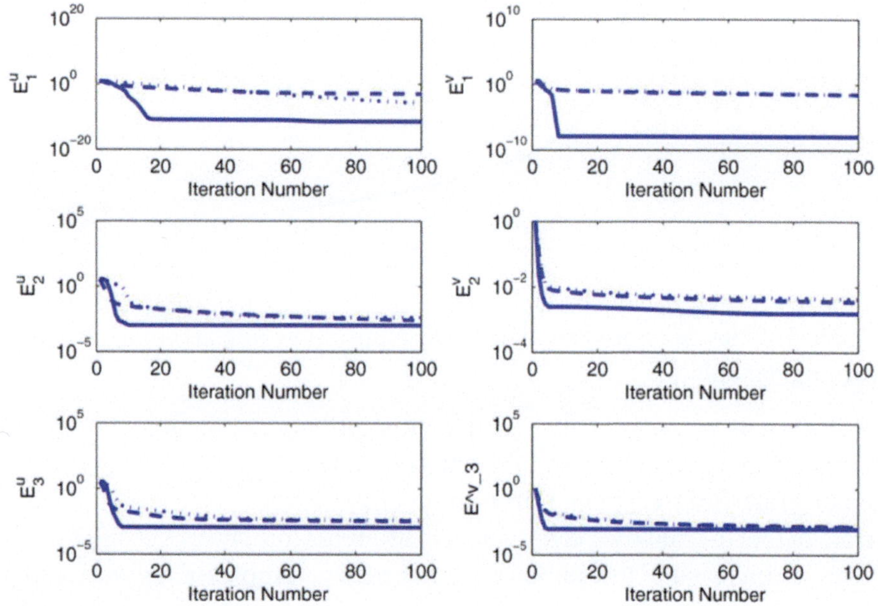

Figure 5: Average highest order coefficients energy E_i versus iteration number for a decomposition with complex singular values.Dotted line: Cauchy points. Dashed line: Alternative minimization. Solid Line: proposed algorithm.

Again Dogleg method converges very rapidly while alternative minimization and Cauchy point converge slowly. The final value of average energy for three left singular vectors are 1.23×10^{-10}, 9.7×10^{-4}, and 10^{-3}, respectively. This is while these values for right singular vectors are 1.12×10^{-10}, 1.4×10^{-3}, and 8.7^{-4}, respectively.

Note that the average energy of highest order coefficients for the third pair of singular vectors alleviate meaningfully. Figure1shows that the third singular value goes to zero and then returns to positive values. If we constrain singular values to be positive, a phase jump of ϖ radian, is imposed to one of third singular vectors near the frequency which singular vector goes to zero. However, by letting singular values to be complex, the zero crossing occur which requires no discontinuity of singular vectors. The relative error of this decomposition is $E^A=4.9\times10^{-3}$ while the error of $\mathbf{U}(z)$

and $\mathbf{V}(z)$ are $E^U = 2.5 \times 10^{-3}$ and $E^V = 3.5 \times 10^{-3}$, respectively. In contrast with constraining singular values to be positive, having complex singular values decrease decomposition and paraunitarity error significantly.

Plots of relative errors E^A, E^U, and E^V for various amount of M and K are shown in Figures 6 and 7, respectively. Letting singular values be complex causes significant reduction of all relative errors. As it was mentioned, Figure 7 shows that increasing K from $2M + N_{max} - N_{min} - 1$ causes no improvement in relative errors while it makes additional computational burden.

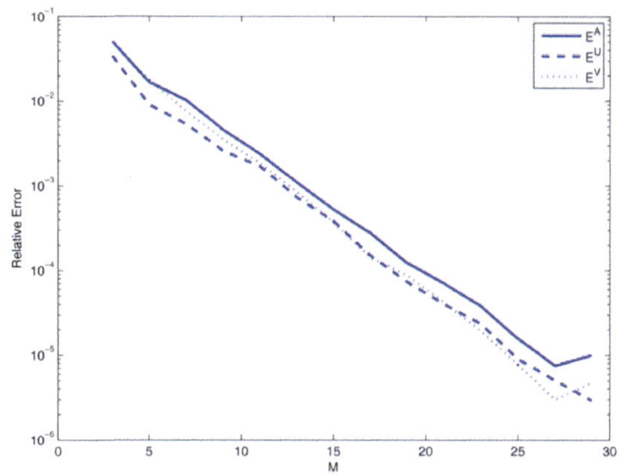

Figure 6: Relative error versus M for a decomposition with complex singular values. $K = 2M + 2$.

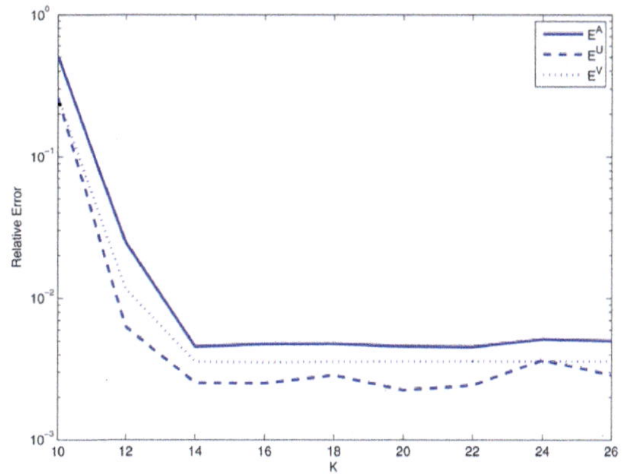

Figure 7: Relative error versus K for a decomposition with complex singular values. $M = 9$.

McWhirter and coauthors [11] have reported the relative error of decomposition. Provided that paraunitary matrices $\mathbf{U}(z)$ and $\mathbf{V}(z)$ are of order 33, the relative error of their algorithm is 0.0469. This is while our algorithm only requires paraunitary matrices of order 3 for relative error of 0.035 with positive singular values and relative error of 2.45×10^{-6} with complex singular values. In addition, in the new approach, exploiting paraunitary matrices of order 33, the relative error is 0.0032 with positive singular values and 4.7×10^{-6} with complex singular values.

This large difference is not caused by iteration numbers because we compare results while all algorithms relatively converges, and with continuation of iterations trivial improvement are obtained. The main reason lies on different constraints of the solution presented in [11] in contrast to our proposed method. While they impose paraunitary constraint on $\tilde{\mathbf{U}}(z)\mathbf{A}(z)\mathbf{V}(z)$ to yield a diagonalized $\Sigma(z)$, we impose the finite duration constraint and obtain approximation of $\mathbf{U}(z)$ and $\mathbf{V}(z)$ with fair fitting to the decomposed matrices at each frequency samples. Therefore, we can consider this method as a finite duration polynomial regression of matrices which is obtained by uniformly sampling $\mathbf{U}(z)$ and $\mathbf{V}(z)$ on the unit circle in z-plane.

As a second example, consider EVD of the following para-Hermitian matrix

$$\mathbf{A}(z) = \begin{bmatrix} .5z^2 + 3 + .5z^{-2} & -.5z^2 + .5z^{-2} \\ .5z^2 - .5z^{-2} & -.5z^2 + 1 - .5z^{-2} \end{bmatrix}$$

The exact smooth EVD of this matrix is of finite order

$$\mathbf{U}(z) = \frac{1}{2}\begin{bmatrix} 1 + z^{-1} & 1 - z^{-1} \\ 1 - z^{-1} & 1 + z^{-1} \end{bmatrix},$$
$$\Lambda(z) = \begin{bmatrix} z^1 + 2 + z^{-1} & 0 \\ 0 & -z^1 + 2 - z^{-1} \end{bmatrix}. \quad (35)$$

Frequency behavior of eigenvalues can be seen in Figure8. Since eigenvalues intersect at two frequencies, smooth decomposition and spectrally majorized decomposition result two distinct solutions.

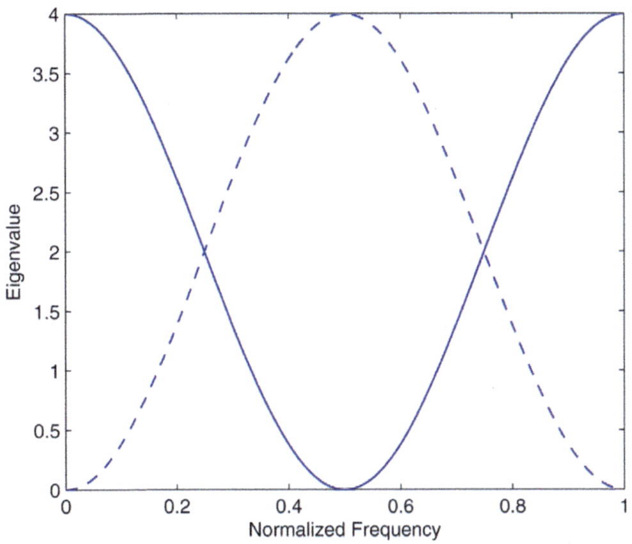

Figure 8: Eigenvalues of smooth decomposition versus frequency.

To perform smooth decomposition, we need to track and rearrange eigenvectors to avoid any discontinuity using Algorithm 2. The resulting $|c_{ij}^{u'}[k]|$are shown in Figure9for$k$=0, 1, ...,K−1. Using these$|c_{ij}^{u'}[k]|$the Algorithm 2 swap first and second eigenvalues and eigenvectors for$k$=12:32 which results in continuity of eigenvalues and eigenvectors.

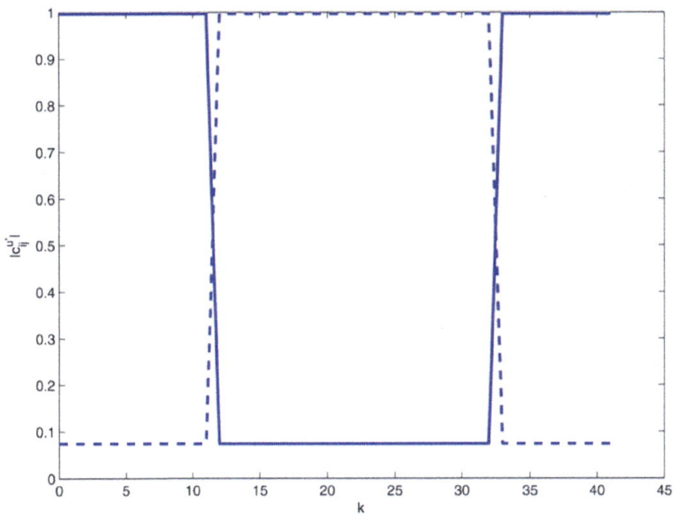

Figure 9: Rearrangement of eigenvalues and eigenvectors.K=42. Dashed Line:

$c_{12}^{u'}[k]$ and $c_{21}^{u'}[k]$. Solid Line: $c_{11}^{u'}[k]$ and $c_{22}^{u'}[k]$.

Now that all eigenvalues and eigenvectors are rearranged in DFT domain, it's time for phase alignment of eigenvectors. A plot of E_i versus iteration for $M=3$ and smooth decomposition is depicted in Figure 10. It is predictable that dogleg algorithm converges rapidly while the alternative minimization and Cauchy point has a long way to converge.

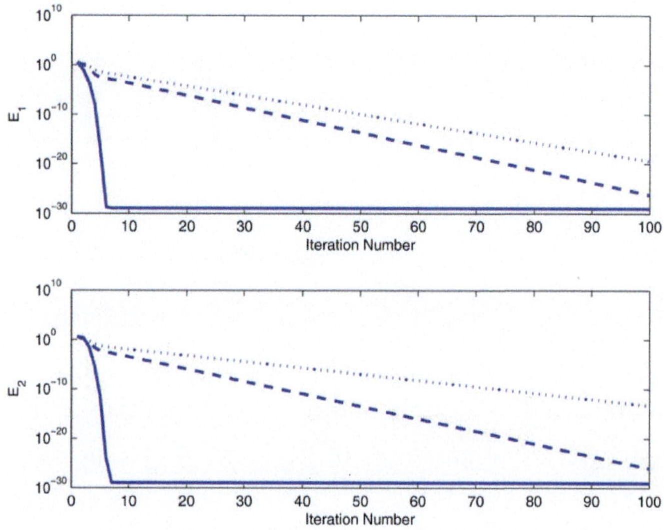

Figure 10: E_i versus iteration number corresponding to smooth decomposition. Dotted line: Cauchy points. Dashed line: Alternative minimization. Solid Line: proposed algorithm.

Since the energy of highest order coefficients of eigenvectors are trifling, using the proposed method for smooth decomposition results in very high accuracy, as seem in the figures. Relative error of smooth decomposition versus M is shown in Figure 11.

While using frequency smooth EVD of (35) leads to relative error below 10^{-5} for $M \geq 3$ with a few number of iterations, Spectrally majorized EVD requires a lot more polynomial order to reach a reasonable relative error.

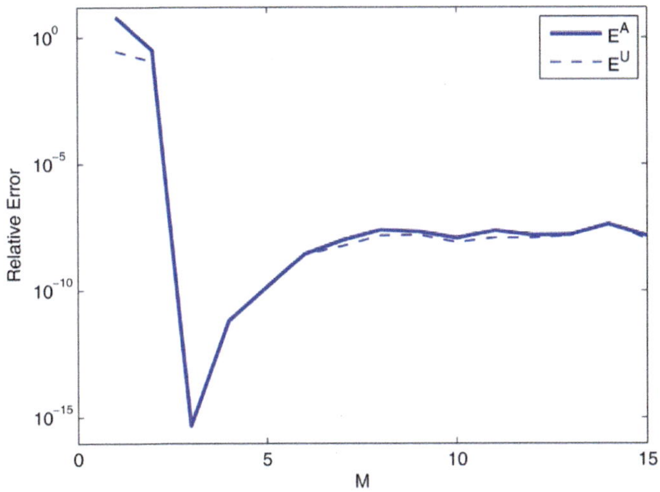

Figure 11: Relative error of smooth decomposition versus M.

Unlike smooth decomposition which requires rearrangement of eigenvalues and eigenvectors, spectral majorization requires only to sort eigenvalues at each frequency sample in decreasing order. Most of conventional EVD algorithm sort eigenvalues in decreasing order, which we should only align eigenvector phases using 3. A plot of E_i versus iteration for M=20 and spectrally majorized decomposition is depicted in Figure 12.

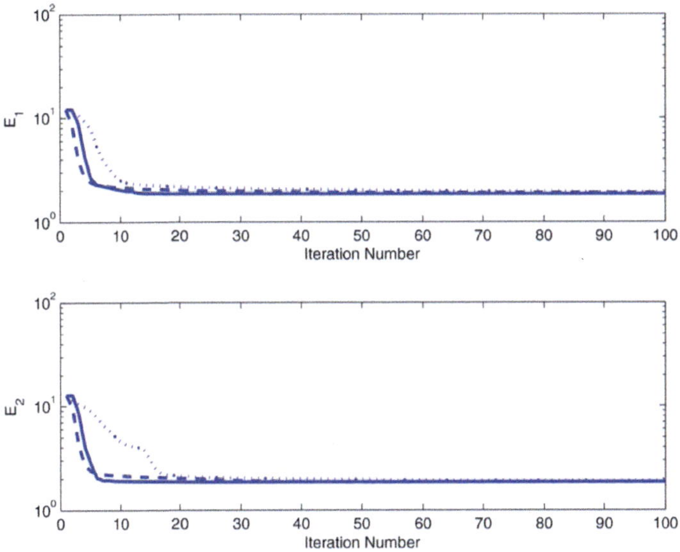

Figure 12: E_i versus iteration number corresponding to spectrally majorized de-

composition.Dotted line: Cauchy points. Dashed line: Alternative minimization. Solid Line: proposed algorithm.

Due to an abrupt change in eigenvectors at the intersection frequency of eigenvalues, increasing the decomposition order leads to a slow decay of relative error. Figure13shows the relative error as a function of$M$.

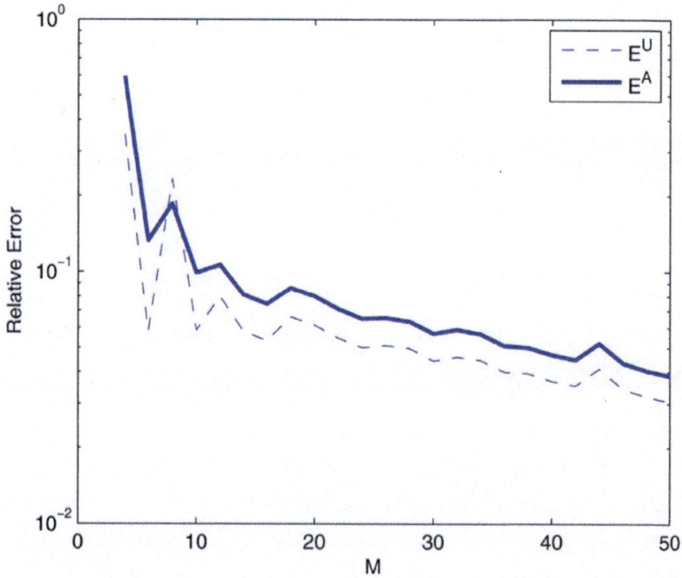

Figure 13: Relative error of spectrally majorized decomposition versus$M$.

To see the difference between smooth and spectrally majorized decomposition results, eigenvalues of spectrally majorized decomposition is shown in Figure14, which is comparable with Figure8which corresponds to eigenvalues of smooth decomposition. Therefore, a low order polynomial is required using smooth decomposition and much higher polynomial order for spectrally majorized decomposition. Even with$M=20$ the decomposition have relatively high error.

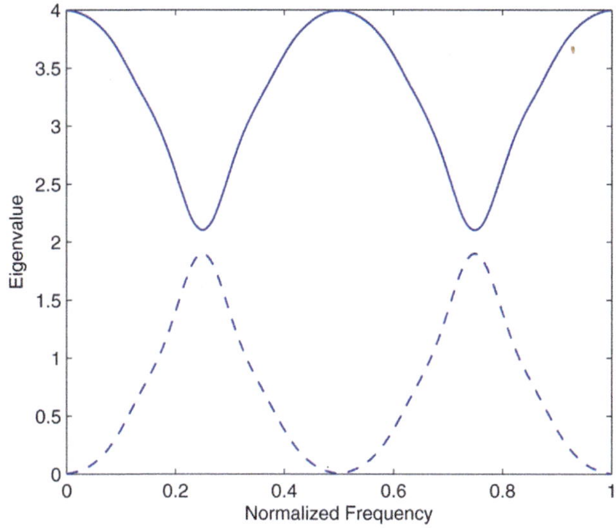

Figure 14: Eigenvalues of spectrally majorized decomposition versus frequency.M=20.

CONCLUSION

An algorithm for polynomial EVD and SVD based on DFT formulation has been presented. One of the advantages of the DFT formulation is that it enables us to control properties of decomposition. Among these properties, we introduce how to setup the decomposition to achieve spectrally majorization and frequency smoothness. We have shown, if singular values (eigenvalues) intersect at some frequency, then simultaneous achievement of spectral majorization and smooth decomposition is not possible. In this situation, setting up the decomposition to possess spectral majorization requires considerably higher order polynomial decomposition and more computational complexity. Highest order polynomial coefficients of singular vectors (eigenvectors) are utilized as square error to obtain a compact decomposition based on phase alignment of frequency samples. The algorithm has the flexibility to compute a decomposition with approximately positive singular values, and a more relaxed decomposition with complex singular values. A solution for this nonlinear quadratic problem is proposed via Newton's method. Since we apply an approximate Hessian matrix to assist the Newton optimization, a fast convergence is achieved. The algorithm capability to control the order of polynomial elements of decomposed matrices and to select properties of decomposition, make the proposed method as a good choice for filterbank and MIMO precoding applications. Finally, performance of the proposed algorithm under different conditions is demonstrated via simulations. Simulation results reveal superior decomposition accuracy in contrast with coefficient domain algorithms due to relaxation of paraunitarity.

REFERENCES

1. Kailath T:*Linear Systems*. NJ: Prentice Hall, Englewood Cliffs; 1980.
2. Tugnait J, Huang B.: Multistep linear predictors-based blind identification and equalization of multiple-input multiple-output channels.*IEEE Trans. Signal Process*2000, 48(1):569-571.
3. Fischer R: Sorted spectral factorization of matrix polynomials in MIMO communications.*IEEE Trans. Commun*2005, 53(6):945-951. 10.1109/TCOMM.2005.849639
4. Zamiri-Jafarian H, Rajabzadeh M: A polynomial matrix SVD approach for time domain broadband beamforming in MIMO-OFDM systems. *IEEE Vehicular Technology Conference, VTC Spring 2008*2008, 802-806.
5. Brandt R: Polynomial matrix decompositions: evaluation of algorithms with an application to wideband MIMO communications. 2010.
6. Palomar D, Lagunas M, Pascual A, Neira A: Practical implementation of jointly designed transmit-receive space-time IIR filters.*International Symposium on Signal Processing and Its Applications, ISSPA*2001, 521-524.
7. Lambert R, Bell A: Blind separation of multiple speakers in a multipath environment.*Proceedings of the International Conference on Acoustics, Speech, and Signal Processing*1997, 423-426.
8. Redif S, McWhirter J, Baxter P, Cooper T: Robust broadband adaptive beamforming via polynomial eigenvalues.*OCEANS 2006*2006, 1-6.
9. Vaidyanathan P:*Multirate Systems and, Filterbanks*. Prentice Hall, Englewood Cliffs; 1993.
10. Foster J, McWhirter J, Chamber J: A novel algorithm for calculating the QR decomposition for a polynomial matrix.*Proceedings of the International Conference on Acoustics, Speech, and Signal Processing*2009, 3177-3180.
11. Foster J, Mcwhirter J, Davies M, Chambers J: An algorithm for calculating the QR and singular value decompositions of polynomial matrices.*IEEE Trans. Signal Process*2010, 58(3):1263-1274.
12. Cescato D, Bolcskei H: QR decomposition of Laurent polynomial matrices sampled on the unit circle.*IEEE Trans. Inf. Theory*2010, 56(9):4754-4761.
13. Mcwhirter J, Baxter P, Cooper T, Redif S: An EVD algorithm for

para-Hermitian polynomial matrices.*IEEE Trans. Signal Process*2007, 55(5):2158-2169.

14. Tkacenko A: Approximate eigenvalue decomposition of para-Hermitian systems through successive FIR paraunitary transformations. In*Proceedings of the International Conference on Acoustics, Speech, and Signal Processing*. (Dallas, Texas, USA; 2010:4074-4077.
15. Lambert R: Multichannel blind deconvolution: FIR matrix algebra and separation of multipath mixtures. 1996.
16. Vaidyanathan P: Theory of optimal orthonormal subband coders.*IEEE Trans. Signal Process*1998, 46(4):1528-1543.
17. Tkacenko A, Vaidyanathan P: On the Spectral Factor Ambiguity of FIR Energy Compaction Filter Banks.*IEEE Trans. Signal Process*2006, 54(1):146-160.
18. Brandt R, Bengtsson M: Wideband MIMO channel diagonalization in the time domain.*International Symposium on Personal, Indoor, and Mobile Radio Communication*2011, 1958-1962.
19. Dieci L, Eirola T: On smooth decomposition of matrices.*SIAM J. Matrix Anal. Appl*1999, 20(3):800-819. 10.1137/S0895479897330182
20. Redif S, McWhirter J, Weiss S: Design of FIR paraunitary filter banks for subband coding using a polynomial eigenvalue decomposition. *IEEE Trans. Signal Process*2011, 59(11):5253-5264.
21. Oppenheim A, Schafer R, Buck J:*Discrete-Time Signal Processing*. Prentice Hall, Englewood Cliffs; 1999.
22. Nocedal J, Wright S:*Numerical Optimization*. New York: Springer; 1999.

Chapter 6

Canonical Polyadic Decomposition of Third-order Semi-nonnegative Semi-symmetric Tensors using LU and QR Matrix Factorizations

Lu Wang[1,2,4], Laurent Albera[1,2,4,5], Amar Kachenoura[1,2,4], Huazhong Shu[3,4] and Lotfi Senhadji[1,2,4]

[1]INSERM, UMR 1099, F-35000 Rennes, France

[2]LTSI, Université de Rennes 1, Bât. 22, Campus de Beaulieu, F-35000 Rennes, France

[3]Laboratory of Image Science and Technology, School of Computer Science and Engineering, Southeast University, Qun Xian Building, No. 2 Sipailou, 210096 Nanjing, China

[4]Centre de Recherche en Information Biomédicale Sino-Français (CRIBs), F-35000 Rennes, France

[5]Centre INRIA Rennes-Bretagne Atlantique, Bât. 12G, Campus de Beaulieu, F-35042 Rennes, France

ABSTRACT

Semi-symmetric three-way arrays are essential tools in blind source separation (BSS) particularly in independent component analysis (ICA).

Citation (APA): Wang, L., Albera, L., Kachenoura, A., Shu, H., & Senhadji, L. (2014). Canonical polyadic decomposition of third-order semi-nonnegative semi-symmetric tensors using LU and QR matrix factorizations. EURASIP Journal on Advances in Signal Processing, 2014(1), 150. (23 pages), DOI: https://doi.org/10.1186/1687-6180-2014-150

These arrays can be built by resorting to higher order statistics of the data. The canonical polyadic (CP) decomposition of such semi-symmetric three-way arrays allows us to identify the so-called mixing matrix, which contains the information about the intensities of some latent source signals present in the observation channels. In addition, in many applications, such as the magnetic resonance spectroscopy (MRS), the columns of the mixing matrix are viewed as relative concentrations of the spectra of the chemical components. Therefore, the two loading matrices of the three-way array, which are equal to the mixing matrix, are nonnegative. Most existing CP algorithms handle the symmetry and the nonnegativity separately. Up to now, very few of them consider both the semi-nonnegativity and the semi-symmetry structure of the three-way array. Nevertheless, like all the methods based on line search, trust region strategies, and alternating optimization, they appear to be dependent on initialization, requiring in practice a multi-initialization procedure. In order to overcome this drawback, we propose two new methods, called JD^{+}_{LU} and JD^{+}_{QR}, to solve the problem of CP decomposition of semi-nonnegative semi-symmetric three-way arrays. Firstly, we rewrite the constrained optimization problem as an unconstrained one. In fact, the nonnegativity constraint of the two symmetric modes is ensured by means of a square change of variable. Secondly, a Jacobi-like optimization procedure is adopted because of its good convergence property. More precisely, the two new methods use LU and QR matrix factorizations, respectively, which consist in formulating high-dimensional optimization problems into several sequential polynomial and rational subproblems. By using both LU and QR matrix factorizations, we aim at studying the influence of the used matrix factorization. Numerical experiments on simulated arrays emphasize the advantages of the proposed methods especially the one based on LU factorization, in the presence of high-variance model error and of degeneracies such as bottlenecks. A BSS application on MRS data confirms the validity and improvement of the proposed methods.

Keywords: Canonical polyadic decomposition, Semi-nonnegative semi-symmetric tensor, Joint diagonalization by congruence, Individual differences in scaling analysis, Blind source separation, Independent component analysis, Magnetic resonance spectroscopy

INTRODUCTION

Higher order (HO) arrays, commonly called tensors, play an important role in numerous applications, such as chemometrics [1], telecommunications [2], and biomedical signal processing [3]. They can be seen as HO extensions of vectors (one-way arrays) and matrices (two-way arrays). In many practical situations, the available data measurements cannot be arranged into a tensor form directly, that is to say, the observation diversity is insufficient either in time or frequency. However, if the latent data satisfies the statistical independence assumption, which is reasonable in many applications, meaningful HO arrays can be built by resorting to HO statistics (HOS) of the data [4]. In this instance, the HO arrays are partially symmetric or Hermitian due to the special algebraic structure of the basic HOS, such as moments and cumulants. In independent component analysis (ICA), the latent physical phenomena which are assumed to be statistically independent can be revealed by decomposing the HO array into factors. There exists several ways to decompose a given HO array, such as the Tucker model [5,6]. Among the existing reliable HO array decomposition models, the canonical polyadic (CP) decomposition model has attracted much attention. Indeed, its uniqueness can be ensured under the sufficient conditions established by Kruskal [7]. In addition, unlike the HO singular value decomposition (HOSVD) [6], the CP model does not impose any orthogonality constraint on its factors.

Theoretically, a polyadic decomposition exactly fits an array by a sum of rank-one terms [8]. A CP decomposition is defined as a polyadic decomposition with a minimal number of rank-one terms which are needed to exactly fit a given HO array. Currently, the CP decomposition is gaining importance in several applications, for example, in exploratory data analysis [9], sensor array processing [10], telecommunications [11,12], ICA [13], and in multiple-input multiple-output radar systems [14]. A multitude of methods were developed to compute the CP decomposition. They include the iterative alternating least squares (ALS) procedure [15], which gains popularity due to its simplicity of implementation and low numerical complexity. Uschmajew proved the local convergence property of ALS under some conditions [16]. However, this convergence can be slow. Therefore, an enhanced line search (ELS) procedure was proposed by Rajih et al. [17] to cope with the slow convergence problem of ALS. Other approaches were also proposed, such as the conjugate gradient algorithm [18] and joint eigenvalue decomposition-based algorithms [19,20], to cite a few. Some HO arrays enjoy certain properties, such as i)symmetryand ii)nonnegativity, which cannot be simply handled by the aforementioned general CP decomposition methods. Therefore, special CP models become more and more important. The first special form of the CP model for three-way arrays that are symmetric in two modes brings the concept of individual differences in scaling (INDSCAL) analysis [21]. On one hand, INDSCAL analysis has been studied as a way of multiple factor analysis [22] with applications to chemometrics, psychology, and marketing research. On the other hand, in the domain of signal processing, and more particularly in blind source separation (BSS), the INDSCAL analysis is widely known as the joint diagonalization of a set of matrices by congruence (JDC). During the past two decades, many successful JDC methods have been proposed, such as Yeredor's alternating columns and diagonal center (ACDC) algorithm [23], the joint approximate diagonalization (JAD) algorithm proposed by Cardoso and Souloumiac [24], the fast Frobenius diagonalization (FFDIAG) algorithm proposed by Ziehe et al. [25], Afsari's LUJ1D algorithm [26], and many others [27,28,29,30,31,32,33]. A recent survey of JDC can be found in [34]. The second special form of CP model is defined when all the factors in the CP decomposition are constrained to be nonnegative, commonly known as nonnegative tensor factorization (NTF). NTF can be regarded as the extension of nonnegative matrix factorization (NMF) [35] to higher orders. In many applications, the physical properties are inherently nonnegative, such as chemistry [1] and fluorescence spectroscopy [36,37]. In those applications, the results are only meaningful if the nonnegativity constraint is satisfied. Various methods for computing NTF and also NMF can be found in [38,39].

So far, the CP model with both the symmetry and nonnegativity constraints has not received much attention. Cologner et al. proposed a family of algorithms based on line search and trust region strategies [40]. Wang et al. developed an alternating minimization scheme [41]. Those methods appear to depend on initialization, and therefore in practice require a multi-initialization procedure, leading to an increase of numerical complexity. In this paper, we propose to fit the CP model of a three-way array by imposing both the semi-nonnegativity and the semi-symmetry constraints. More precisely, we impose a nonnegativity constraint on the two symmetric modes of the INDSCAL model, which leads to the semi-nonnegative INDSCAL model or equivalently the CP decomposition of semi-nonnegative semi-symmetric three-way arrays. Such a model is often encountered in ICA problems where a nonnegative mixing matrix is frequently considered. For example, in magnetic resonance spectroscopy (MRS), the columns of the mixing matrix represent the positive concentrations of the source metabolites. Then, the three-way array built by stacking the matrix slices of a cumulant array is both nonnegative and symmetric in two modes. In such a case, the semi-nonnegative INDSCAL problem is equivalent to the JDC problem subject to a nonnegativity constraint on the joint transformation matrix. We propose two new algorithms to solve the semi-nonnegative INDSCAL problem, called JD^{+}_{LU} and JD^{+}_{QR}. Firstly, we rewrite the constrained optimization problem as an unconstrained one. Actually, the nonnegativity constraint is ensured by means of a square change of variable. Secondly, we propose two Jacobi-like approaches using LU and QR matrix factorizations, respectively, which consist in formulating high-dimensional optimization problems into several sequential polynomial and rational subproblems. By using both LU and QR matrix factorizations, we aim at studying the influence of the used matrix factorization. Numerical experiments highlight the advantages of the proposed methods especially JD^{+}_{LU}, in the case of dealing with high-variance model error and with degeneracies such as bottlenecks. A BSS application on MRS signals confirms the validity and improvement of the proposed methods. A part of this work has been recently presented at the 8th IEEE Sensor Array and Multichannel Signal Processing Workshop [42].

The rest of the paper is organized as follows. After the presentation of some notations, the 'Multilinear algebra prerequisites and problem statement' section introduces some basic definitions of the multilinear algebra then gives the semi-nonnegative INDSCAL problem formulation. In the 'Methods' section, we describe the proposed algorithms in detail and also provide an analysis of the numerical complexities. The 'Simulation results' section shows the computer simulation results. Finally, we conclude the paper.

MULTILINEAR ALGEBRA PREREQUISITES AND PROBLEM STATEMENT

Notations

The following notations are used throughout this paper. $\mathbb{R}^{N_1 \times N_2 \times \cdots \times N_i}$ and $\mathbb{R}_{+}^{N_1 \times N_2 \times \cdots \times N_i}$ denote the set of real-valued $(N_1 \times N_2 \times \cdots \times N_i)$ arrays and the set of nonnegative real-valued $(N_1 \times N_2 \times \cdots \times N_i)$ arrays, respectively. Vectors, matrices, and HO arrays are denoted by bold lowercase letters $(\boldsymbol{a}, \boldsymbol{b}, \cdots)$, bold uppercase letters $(\boldsymbol{A}, \boldsymbol{B}, \cdots)$ and bold calligraphic letters $(\boldsymbol{\mathcal{A}}, \boldsymbol{\mathcal{B}}, \cdots)$, respectively. The (i,j)-th entry of a matrix$\boldsymbol{A}$is symbolized by$A_{i,j}$. Sometimes, the MATLAB®; column/row notation is adopted to indicate submatrices of a given matrix or subarrays of a HO array. Also, $\boldsymbol{a}_i$ denotes the i-th column vector of matrix $\boldsymbol{A}$. $\boxdot$ denotes the Hadamard product (element-wise product), and $A^{\boxdot 2} = A \boxdot A$. $\odot$ denotes the Khatri-Rao product. $\boldsymbol{A}^{\square}$ denotes the pseudo inverse of $\boldsymbol{A}$. The superscripts $^{-1}$, ^{T}, and $^{-T}$ stand for the inverse, the transpose, and the inverse after transpose operators, respectively. The (N×N) identity matrix is denoted by $\boldsymbol{I}_N$. $\boldsymbol{0}_N$ stands for N-dimensional vectors of zeros. $|a|$ denotes the absolute value of a. $\|\boldsymbol{A}\|_F$ and det(A) stand for the Frobenius norm and determinant of matrix $\boldsymbol{A}$, respectively. diag($\boldsymbol{A}$) returns a matrix comprising only the diagonal elements of $\boldsymbol{A}$. Diag($\boldsymbol{b}$) is the diagonal matrix whose diagonal elements are given by the vector $\boldsymbol{b}$. off($\boldsymbol{A}$) vanishes the diagonal components of the input matrix $\boldsymbol{A}$. vec($\boldsymbol{A}$) reshapes a matrix $\boldsymbol{A}$ into a column vector by stacking its columns vertically.

Definitions and Problem Formulation

Now we introduce some basic definitions in multilinear algebra which are necessary for the problem formulation.

Definition 1

The outer product $C = u^{(1)} \circ u^{(2)} \circ u^{(3)}$ of three vectors $\boldsymbol{u}^{(i)} \in \mathbb{R}^{N_i}$ $(1 \leq i \leq 3)$ is a three-way array of $\mathbb{R}^{N_1 \times N_2 \times N_3}$ whose elements are defined by $c_{i_1,i_2,i_3} = u_{i_1}^{(1)} u_{i_2}^{(2)} u_{i_3}^{(3)}$.

Definition 2

Each three-way array C expressed as the outer product of three vectors is a rank-1 three-way array.

More generally, the rank of a three-way array is defined as follows:

Definition 3

The rank of an array $\mathcal{C} \in \mathbb{R}^{N_1 \times N_2 \times N_3}$, denoted by rk(C), is the minimal number of rank-1 arrays belonging to $\mathbb{R}^{N_1 \times N_2 \times N_3}$ that yield $\mathcal{C}$ in a linear combination.

Despite the similarity between the definition of the tensor rank and its matrix counterpart, the rank of a three-way array may exceed its dimensions [4].

Definition 4

A three-way array slice is a two-dimensional section (fragment) of a three-way array, obtained by fixing one of the three indices [38].

For example, the k-th frontal slice of a three-way array $\mathcal{C}$ can be denoted by $C_{:,:,k}$ using MATLAB notation, and sometimes it is also denoted by $\boldsymbol{C}^{(k)}$.

The low-rank INDSCAL model of a three-way array is defined as follows:

Definition 5

For a givenP, corresponding to the number of rank-1 terms, the INDSCAL model of a three-way array $\mathcal{C} \in \mathbb{R}^{N\times N\times K}$ can be expressed as:

$$\mathcal{C} = \sum_{p=1}^{P} \boldsymbol{a}_p \circ \boldsymbol{a}_p \circ \boldsymbol{d}_p + \mathcal{V} \qquad (1)$$

where the three-way array $\mathcal{V}$ represents the model residual.

The notation $\mathcal{C} = [\![\boldsymbol{A}, \boldsymbol{A}, \boldsymbol{D}]\!] + \mathcal{V}$ refers to the INDSCAL decomposition (1) of $\mathcal{C}$ with the associated loading matrices $\boldsymbol{A} = [\boldsymbol{a}_1, \cdots, \boldsymbol{a}_P] \in \mathbb{R}^{N\times P}$ andD=[d_1,⋯,d_p]∈ $\mathbb{R}^{K\times P}$. If and only if the residual $\mathcal{V}$ is a null tensor, we have an exact INDSCAL decomposition.

An exact INDSCAL decomposition is considered to be essentially unique when it is only subject to scale and permutation indeterminacies. It means that an INDSCAL decomposition is insensitive to a scaling of the three vectors$\boldsymbol{a}_p$,$\boldsymbol{a}_p$, and$\boldsymbol{d}_p$provided that the product of the three scale numbers is equal to 1, and an arbitrary permutation of the rank-1 terms. A necessary and sufficient uniqueness condition for the INDSCAL model was established by Afsari [43].

The INDSCAL model can also be described by using the frontal slices of $\mathcal{C}$:

$$\forall k \in \{1, 2, \cdots, K\}, \ \boldsymbol{C}^{(k)} = \mathcal{C}_{:,:,k} = \boldsymbol{A}\boldsymbol{D}^{(k)}\boldsymbol{A}^{\mathsf{T}} + \boldsymbol{V}^{(k)} \qquad (2)$$

where$\boldsymbol{D}^{(k)}$is a diagonal matrix whose diagonal contains the elements of the$k$-th row of$\boldsymbol{D}$, andV(k)=V:,:,k.

In this paper, we propose to fit the INDSCAL model of three-way arrays while imposing nonnegativity constraints on both equal loading matrices$\boldsymbol{A}$. It will be referred to as the semi-nonnegative INDSCAL model, as follows:

Problem 1

Given $\mathcal{C} \in \mathbb{R}^{N\times N\times K}$ and an integer P, find a semi-nonnegative INDSCAL model of $\mathcal{C} = [\![\boldsymbol{A}, \boldsymbol{A}, \boldsymbol{D}]\!]$, subject to the (N × P) matrix$\boldsymbol{A}$having nonnegative components.

The semi-nonnegative INDSCAL problem is equivalent to the JDC problem subject to the nonnegativity constraint on the joint transformation matrix. In this paper, we mainly focus on the case of square nonnegative joint transformation matrix, for whichN=P. The case ofN>Pwill be discussed briefly in the next section. Therefore, the problem that we tackle in this paper is defined as follows:

Problem 2

Given a three-way array $\mathcal{C} \in \mathbb{R}^{N\times N\times K}$ with K symmetric frontal slices $\boldsymbol{C}^{(k)} \in \mathbb{R}^{N\times N}$, find a (N × N) joint transformation matrix$\boldsymbol{A}$and K diagonal matrices$\boldsymbol{D}^{(k)}$of dimension (N × N) such that:

$$\forall k \in \{1,2,\cdots,K\},\ \boldsymbol{C}^{(k)} = \boldsymbol{A}\boldsymbol{D}^{(k)}\boldsymbol{A}^T + \boldsymbol{V}^{(k)} \tag{3}$$

by minimizing the residual term $\boldsymbol{V}^{(k)}$ in a least-squares sense, subject to $\boldsymbol{A}$ having nonnegative components.

JDC cost functions

If the residual array $\mathcal{V}$ is a realization of a Gaussian random array, it is logical to fit the INDSCAL model by the following direct least square (DLS) criterion [23,44]:

$$J_{\mathrm{DLS}}(\boldsymbol{A},\boldsymbol{D}) = \sum_{k=1}^{K} \left\| \boldsymbol{C}^{(k)} - \boldsymbol{A}\boldsymbol{D}^{(k)}\boldsymbol{A}^{\mathrm{T}} \right\|_F^2 \tag{4}$$

and to minimize (4) with respect to $\boldsymbol{A}$ and $\boldsymbol{D}$. Note that, in the field of ICA, only the loading matrix $\boldsymbol{A}$ is of interest since it corresponds to the mixing matrix of several latent source signals. The minimization of (4) with respect to $\boldsymbol{D}$, when $\boldsymbol{A}$ is fixed, was given by Yeredor in [23]:

$$\boldsymbol{D}^{(k)} = \mathrm{Diag}\left\{\left[(\boldsymbol{A}^{\mathrm{T}}\boldsymbol{A}) \boxdot (\boldsymbol{A}^{\mathrm{T}}\boldsymbol{A})\right]^{-1} (\boldsymbol{A} \odot \boldsymbol{A})^{\mathrm{T}}\, \mathrm{vec}\left(\boldsymbol{C}^{(k)}\right)\right\} \tag{5}$$

When $\boldsymbol{A}$ is orthogonal, we can replace $\boldsymbol{D}^{(k)}$ by $\mathrm{Diag}\{(\mathrm{A}\odot\mathrm{A})^{\mathrm{T}}\mathrm{vec}(\mathrm{C}^{(k)})\}$ in (4). Then, the extra parameter $\boldsymbol{D}$ can be eliminated and the minimization of (4) is equivalent to minimizing the following indirect least square (IDLS) criterion [45,46]:

$$J_{\mathrm{IDLS\text{-}O}}(\boldsymbol{A}) = \sum_{k=1}^{K} \left\| \mathrm{off}\left(\boldsymbol{A}^{\mathrm{T}}\boldsymbol{C}^{(k)}\boldsymbol{A}\right) \right\|_F^2 \tag{6}$$

In some cases such as in ICA, the orthogonality assumption of $\boldsymbol{A}$ can be satisfied by using a spatial whitening procedure [47]. However, it is known that the whitening procedure may introduce additional errors. Therefore, many algorithms propose to relax the orthogonality constraint by introducing the following cost function [25,31]:

$$J_{\mathrm{IDLS}}(\boldsymbol{A}) = \sum_{k=1}^{K} \left\| \mathrm{off}\left(\boldsymbol{A}^{-1}\boldsymbol{C}^{(k)}\boldsymbol{A}^{-\mathrm{T}}\right) \right\|_F^2 \tag{7}$$

Frequently, the minimization of criterion (7) is performed on a matrix $\boldsymbol{Z} \overset{\mathrm{def}}{=} \boldsymbol{A}^{-1}$ instead of $\boldsymbol{A}$ for simplicity, and $\boldsymbol{Z}$ is called the joint diagonalizer. To use this criterion, the matrix $\boldsymbol{A}$ (or $\boldsymbol{Z}$) should be properly constrained in order to avoid the trivial zero solution and/or degenerate solutions [34].

Besides the criterions (4) and (7), Afsari [26] presented a new cost function, which is invariant to column scaling of $\boldsymbol{A}$. Pham proposed an information theoretic criterion [48], which requires each matrix $\boldsymbol{C}^{(k)}$ to be positive definite. Tichavský and Yeredor gave a special weighted least square criterion [49].

METHODS

Problem Reformulation

Existing semi-nonnegative INDSCAL algorithms are based on the minimization of the cost function (4) [40,41]. They are able to achieve a better estimation of $\boldsymbol{A}$ than ACDC when the data satisfies the semi-nonnegative INDSCAL model at the cost of a higher computational complexity. We propose to use criterion (7) based on elementary factorizations of $\boldsymbol{A}$ due to the fast convergence property of this kind of procedures. Generally, it is quite difficult to impose the nonnegativity constraint on $\boldsymbol{A}$ while computing its inverse $\boldsymbol{A}^{-1}$ by minimizing (7). Let us consider the structure of $\boldsymbol{\mathcal{C}} = [\![\boldsymbol{A}, \boldsymbol{A}, \boldsymbol{D}]\!]$ with the following assumptions:

- $\boldsymbol{A} \in \mathbb{R}_{+}^{N \times N}$ is nonsingular;
- $\boldsymbol{D} \in \mathbb{R}^{K \times N}$ does not contain zero entries.

Then, each frontal slice of $\boldsymbol{\mathcal{C}}$ is nonsingular and its inverse can be expressed as follows:

$$\left(\boldsymbol{C}^{(k)}\right)^{-1} = \boldsymbol{A}^{-\mathsf{T}} \left(\boldsymbol{D}^{(k)}\right)^{-1} \boldsymbol{A}^{-1} \quad (8)$$

We use $\boldsymbol{C}^{(k,-1)}$ to denote $(\mathrm{C}^{(k)})^{-1}$ for simplicity. Eq.8 shows that $\boldsymbol{C}^{(k,-1)}$ also preserves the jointly diagonalizable structure. Furthermore, instead of $\boldsymbol{A}^{-1}$, $\boldsymbol{A}$ serves as the joint diagonalizer. Then, $\boldsymbol{A}$ can be estimated by minimizing the following modified criterion based on (7):

$$J(A) = \sum_{k=1}^{K} \left\| \operatorname{off}\left(A^{\mathsf{T}} C^{(k,-1)} A\right) \right\|_F^2 \quad (9)$$

By such a manipulation, most algorithms based on criterion (7) can now estimate $\boldsymbol{A}$ directly. However, none of them can guarantee the nonnegativity of $\boldsymbol{A}$. In order to impose the nonnegativity constraint on $\boldsymbol{A}$, we resort to use a square change of variable which was introduced by Chu et al. [50] for NMF, next adopted by Royer et al. for NTF [37] and by Coloigner et al. for semi-nonnegative INDSCAL [40]:

$$\boldsymbol{A} = \boldsymbol{B} \boxdot \boldsymbol{B} = \boldsymbol{B}^{\boxdot 2} \quad (10)$$

where $\boldsymbol{B} \in \mathbb{R}^{N\times N}$. Then, problem 2 can be reformulated as follows:

Problem 3

Given $\mathcal{C} = [\![\boldsymbol{A}, \boldsymbol{A}, \boldsymbol{D}]\!] \in \mathbb{R}^{N\times N\times K}$, find the square nonnegative loading matrix $A=B^{\boxdot 2}$ such that $\boldsymbol{B}$ minimizes the following cost function:

$$J(\boldsymbol{B}) = \sum_{k=1}^{K} \left\| \text{off}\left(\left(\boldsymbol{B}^{\boxdot 2}\right)^T \boldsymbol{C}^{(k,-1)} \boldsymbol{B}^{\boxdot 2} \right) \right\|_F^2 \tag{11}$$

LU and QR parameterizations of B

In order to minimize (11), one may consider a gradient-like approach. However, the performance of this kind of method is sensitive to the initial guess and to the search step size. In addition, the calculation of gradient of (11) with respect to $\boldsymbol{B}$ is computationally expensive due to the existence of the Hadamard product. Other algorithms, using Jacobi-like procedures [25,26,31], parameterize $\boldsymbol{A}$ as a product of several special elementary matrices and estimate each elementary matrix successively. We propose to follow such a minimization scheme.

Now let us recall the following definitions and lemmas:

Definition 6

A unit upper (or lower) triangular matrix is an upper (or lower, respectively) triangular matrix whose main diagonal elements are equal to 1.

Definition 7

An elementary upper (or lower) triangular matrix with parameters $\{i,j,u_{i,j}\}$ and $i<j$ is a unit upper (or lower, respectively) triangular matrix whose non-diagonal elements are zeros except the (i,j)-th entry, which is equal to $u_{i,j}$.

$\boldsymbol{U}^{(i,j)}(u_{i,j})$ with $1 \le i<j\le N$ denotes an elementary upper triangular matrix:

$$\boldsymbol{U}^{(i,j)}(u_{i,j}) = \begin{pmatrix} \boldsymbol{I}_{i-1} & \vdots & \boldsymbol{0} & \vdots & \boldsymbol{0} \\ \cdots & 1 & \cdots & u_{i,j} & \cdots \\ \boldsymbol{0} & \vdots & \boldsymbol{I}_{j-i-1} & \vdots & \boldsymbol{0} \\ \cdots & 0 & \cdots & 1 & \cdots \\ \boldsymbol{0} & \vdots & \boldsymbol{0} & \vdots & \boldsymbol{I}_{N-j} \end{pmatrix} \tag{12}$$

Similarly, $\boldsymbol{L}^{(i,j)}(\ell_{i,j})$ with $1 \le j<i\le N$ corresponds to an elementary lower triangular matrix.

Definition 8

A Givens rotation matrix with parameters $\{i,j,\theta_{i,j}\}$ and $i<j$ is equal to an identity matrix except for the (i,i)-th, (j,j)-th, (i,j)-th, and (j,i)-th entries, which are equal to $\cos(\theta_{i,j})$, $\cos(\theta_{i,j})$, $-\sin(\theta_{i,j})$, and $\sin(\theta_{i,j})$, respectively.

$\boldsymbol{Q}^{(i,j)}(\theta_{i,j})$ with $1 \le i<j\le N$ indicates the corresponding Givens rotation matrix:

$$\boldsymbol{Q}^{(i,j)}(\theta_{i,j}) = \begin{pmatrix} \boldsymbol{I}_{i-1} & \vdots & \boldsymbol{0} & \vdots & \boldsymbol{0} \\ \ldots & \cos(\theta_{i,j}) & \ldots & -\sin(\theta_{i,j}) & \ldots \\ \boldsymbol{0} & \vdots & \boldsymbol{I}_{j-i-1} & \vdots & \boldsymbol{0} \\ \ldots & \sin(\theta_{i,j}) & \ldots & \cos(\theta_{i,j}) & \ldots \\ \boldsymbol{0} & \vdots & \boldsymbol{0} & \vdots & \boldsymbol{I}_{N-j} \end{pmatrix} \quad (13)$$

Lemma 1

Any (N × N) unit lower triangular matrix$\boldsymbol{L}$whose (i,j)-th component is $\ell_{i,j}$(i > j) can be factorized as the following product of N (N-1)/2elementary lower triangular matrices [51,Chapter 3]:

$$\boldsymbol{L} = \prod_{j \in \mathscr{J}_1} \prod_{i \in \mathscr{I}_1(j)} \boldsymbol{L}^{(i,j)}(\ell_{i,j}) \quad (14)$$

where the two sets of indices$\mathscr{J}_1$ *and* $\mathscr{I}_1(j)$are defined by $\mathscr{J}_1 = \{1, 2, \ldots, N\}$ *and* $\mathscr{I}_1(j) = \{j+1, j+2, \ldots, N\}$ for the sake of convenience. Similarly, any (N × N) unit upper triangular matrix$\boldsymbol{U}$whose (i,j)-th component is equal to $u_{i,j}$(i < j) can be factorized as a product of elementary upper triangular matrices as follows:

$$\boldsymbol{U} = \prod_{i \in \mathscr{I}_2} \prod_{j \in \mathscr{J}_2(i)} \boldsymbol{U}^{(i,j)}(u_{i,j}) \quad (15)$$

where$\mathscr{I}_2$ *and* $\mathscr{J}_2(i)$are two sets of indices, defined by $\mathscr{I}_2 = \{N-1, N-2, \ldots, 1\}$ *and* $\mathscr{J}_2(i) = \{N, N-1, \ldots, i+1\}$.

Lemma 2

Any (N × N) orthonormal matrix$\boldsymbol{Q}$can be factorized as the following product of N (N-1)/2Givens rotation matrices [52, Chapter 14]:

$$\boldsymbol{Q} = \prod_{i \in \mathscr{I}_2} \prod_{j \in \mathscr{J}_2(i)} \boldsymbol{Q}^{(i,j)}(\theta_{i,j}) \quad (16)$$

where$\mathscr{I}_2$ *and* $\mathscr{J}_2(i)$are defined in Lemma 1.

For any nonsingular matrix$\boldsymbol{B} \in \mathbb{R}^{N \times N}$, the LU matrix factorization decomposes it as$\boldsymbol{B}=\boldsymbol{LU\Lambda\Pi}$, where$\boldsymbol{L} \in \mathbb{R}^{N \times N}$is a unit lower triangular matrix,$\boldsymbol{U} \in \mathbb{R}^{N \times N}$is a unit upper triangular matrix, $\boldsymbol{\Lambda} \in \mathbb{R}^{N \times N}$is a diagonal matrix, and$\boldsymbol{\Pi} \in \mathbb{R}^{N \times N}$is a permutation matrix.$\boldsymbol{B}$also admits the QR matrix factorization as$\boldsymbol{B}=\boldsymbol{QR\Lambda}$, where$\boldsymbol{Q} \in \mathbb{R}^{N \times N}$is an orthonormal matrix,$\boldsymbol{R} \in \mathbb{R}^{N \times N}$is a unit upper triangular matrix, and$\boldsymbol{\Lambda} \in \mathbb{R}^{N \times N}$is a diagonal matrix. Due to the indeterminacies of the JDC problem, the global minimum of (11), say $\boldsymbol{B}$, can be expressed as $\boldsymbol{B}=\boldsymbol{L}\boldsymbol{U}$ and $\boldsymbol{B}=\boldsymbol{Q}\boldsymbol{R}$ without loss of generality. Moreover, by incorporating Lemma 1 and Lemma 2, we obtain the two following elementary factorizations of $\boldsymbol{B}$:

$$\boldsymbol{B} = \prod_{j\in\mathcal{J}_1}\prod_{i\in\mathcal{I}_1(j)} \boldsymbol{L}^{(i,j)}(\ell_{i,j}) \prod_{i\in\mathcal{I}_2}\prod_{j\in\mathcal{J}_2(i)} \boldsymbol{U}^{(i,j)}(u_{i,j}) \tag{17}$$

$$\boldsymbol{B} = \prod_{i\in\mathcal{I}_2}\prod_{j\in\mathcal{J}_2(i)} \boldsymbol{Q}^{(i,j)}(\theta_{i,j}) \prod_{i\in\mathcal{I}_2}\prod_{j\in\mathcal{J}_2(i)} \boldsymbol{U}^{(i,j)}(u_{i,j}) \tag{18}$$

As a consequence, the minimization of (11) with respect to $\boldsymbol{B}$ is converted to the estimate of $N(N-1)$ parameters: $\ell_{i,j}$ and $u_{i,j}$ for the LU decomposition (17), or $\theta_{i,j}$ and $u_{i,j}$ for the QR decomposition (18). Instead of simultaneously computing the $N(N-1)$ parameters, we propose two Jacobi-like procedures which perform $N(N-1)$ sequential optimizations. This yields two new algorithms: i) the first algorithm based on (17), named JD^{+}_{LU}, estimates each $\ell_{i,j}$ and $u_{i,j}$ successively, and ii) the second one based on (18), called JD^{+}_{QR}, estimates each $\theta_{i,j}$ and $u_{i,j}$ sequentially.

Now, the difficulty is how to estimate four kinds of parameter, namely $\boldsymbol{L}^{(i,j)}(\ell_{i,j})$ and $\boldsymbol{U}^{(i,j)}(u_{i,j})$ for JD^{+}_{LU}, and $\boldsymbol{Q}^{(i,j)}(\theta_{i,j})$ and $\boldsymbol{U}^{(i,j)}(u_{i,j})$ for JD^{+}_{QR}. Two points should be noted here: i) $\boldsymbol{L}^{(i,j)}(\ell_{i,j})$ and $\boldsymbol{U}^{(i,j)}(u_{i,j})$ belong to the same category of matrices; therefore, they can be estimated by the same algorithmic procedure just with an emphasis on the relation between the i and j indices ($i<j$ for $\boldsymbol{U}^{(i,j)}(u_{i,j})$ and $j<i$ for $\boldsymbol{L}^{(i,j)}(\ell_{i,j})$); ii) for both JD^{+}_{LU} and JD^{+}_{QR} algorithms, the procedure of estimating $\boldsymbol{U}^{(i,j)}(u_{i,j})$ is identical. Consequently, the principal problem is reduced to estimating two kinds of parameters, namely $\boldsymbol{U}^{(i,j)}(u_{i,j})$ and $\boldsymbol{Q}^{(i,j)}(\theta_{i,j})$.

Minimization with respect to the elementary upper triangular matrix $U^{(i,j)}(u_{i,j})$

In this section, we minimize (11) with respect to $\boldsymbol{U}^{(i,j)}(u_{i,j})$ with $1 \leq i<j\leq N$. Let $\tilde{\boldsymbol{A}}$ and $\tilde{\boldsymbol{B}}$ denote the current estimate of $\boldsymbol{A}$ and $\boldsymbol{B}$ before estimating the parameter $u_{i,j}$, respectively. Let $\tilde{\boldsymbol{A}}^{(\text{new})}$ and $\tilde{\boldsymbol{B}}^{(\text{new})}$ stand for $\tilde{\boldsymbol{A}}$ and $\tilde{\boldsymbol{B}}$ updated by $\boldsymbol{U}^{(i,j)}(u_{i,j})$, respectively. Furthermore, the update of $\tilde{\boldsymbol{B}}$ is defined as follows:

$$\tilde{\boldsymbol{B}}^{(\text{new})} = \tilde{\boldsymbol{B}}\boldsymbol{U}^{(i,j)}(u_{i,j}) \tag{19}$$

In order to compute the parameter $u_{i,j}$, a typical way is to minimize the criterion (11) with respect to $u_{i,j}$ by replacing matrix $\tilde{\boldsymbol{B}}$ by $\tilde{\boldsymbol{B}}^{(\text{new})}$. For the sake of convenience, we denote $J(u_{i,j})$ instead of $J(\tilde{\boldsymbol{B}}^{(\text{new})})$. Then, $J(u_{i,j})$ can be expressed as follows:

$$J(u_{i,j}) = \sum_{k=1}^{K} \left\| \text{off}\left\{ \left[\left(\tilde{\boldsymbol{B}}^{(\text{new})} \right)^{\boxdot 2} \right]^{\mathsf{T}} \boldsymbol{C}^{(k,-1)} \left[\left(\tilde{\boldsymbol{B}}^{(\text{new})} \right)^{\boxdot 2} \right] \right\} \right\|_F^2 \tag{20}$$

The expression of the Hadamard square of the update $\tilde{\boldsymbol{B}}^{(\text{new})}$ is shown in the following proposition:

Proposition 1

$\tilde{\boldsymbol{A}}^{(new)} = \left(\tilde{\boldsymbol{B}}^{(new)}\right)^{\boxdot 2} = \left(\tilde{\boldsymbol{B}}\boldsymbol{U}^{(i,j)}(u_{i,j})\right)^{\boxdot 2}$ can be expressed as a function of $u_{i,j}$ as follows:

$$\tilde{\boldsymbol{A}}^{(new)} = \left(\tilde{\boldsymbol{B}}^{(new)}\right)^{\boxdot 2} = \tilde{\boldsymbol{B}}^{\boxdot 2}\, \boldsymbol{U}^{(i,j)}\left(u_{i,j}^2\right) + 2\, u_{i,j} \left(\tilde{\boldsymbol{b}}_i \boxdot \tilde{\boldsymbol{b}}_j\right) \boldsymbol{e}_j^T \tag{21}$$

where $\tilde{\boldsymbol{b}}_i$ *and* $\tilde{\boldsymbol{b}}_j$ denote the i-th and j-th columns of $\tilde{\boldsymbol{B}}$, respectively, and $\boldsymbol{e}_j$ is the j-th column of the identity matrix $\boldsymbol{I}_N$.

Inserting (21) into the cost function (20), we have:

$$J(u_{i,j}) = \sum_{k=1}^{K} \left\| \text{off}\left(\tilde{\boldsymbol{C}}^{(k,\text{new})}\right) \right\|_F^2 = \sum_{k=1}^{K} \left\| \text{off} \left(\underbrace{\boldsymbol{U}^{(i,j)}\left(u_{i,j}^2\right)^{\mathsf{T}} \tilde{\boldsymbol{C}}^{(k)} \boldsymbol{U}^{(i,j)}\left(u_{i,j}^2\right)}_{①} + \underbrace{u_{i,j}\boldsymbol{U}^{(i,j)}\left(u_{i,j}^2\right)^{\mathsf{T}} \tilde{\boldsymbol{c}}^{(k,1)} \boldsymbol{e}_j^{\mathsf{T}}}_{②} + \underbrace{u_{i,j}\boldsymbol{e}_j \tilde{\boldsymbol{c}}^{(k,2)} \boldsymbol{U}^{(i,j)}\left(u_{i,j}^2\right)}_{③} + \underbrace{u_{i,j}^2 \tilde{c}^{(k,3)} \boldsymbol{e}_j \boldsymbol{e}_j^{\mathsf{T}}}_{④} \right) \right\|_F^2 \tag{22}$$

where $\tilde{\boldsymbol{C}}^{(k)} = \tilde{\boldsymbol{A}}^{\mathsf{T}} \boldsymbol{C}^{(k,-1)} \tilde{\boldsymbol{A}}$, $\tilde{\boldsymbol{c}}^{(k,1)} = 2\tilde{\boldsymbol{A}}^{\mathsf{T}} \boldsymbol{C}^{(k,-1)} \left(\tilde{\boldsymbol{b}}_i \boxdot \tilde{\boldsymbol{b}}_j\right)$, $\tilde{\boldsymbol{c}}^{(k,2)} = 2\left(\tilde{\boldsymbol{b}}_i \boxdot \tilde{\boldsymbol{b}}_j\right)^{\mathsf{T}} \times \boldsymbol{C}^{(k,-1)} \tilde{\boldsymbol{A}}$, and $\tilde{c}^{(k,3)} = 4\left(\tilde{\boldsymbol{b}}_i \boxdot \tilde{\boldsymbol{b}}_j\right)^{\mathsf{T}} \boldsymbol{C}^{(k,-1)} \left(\tilde{\boldsymbol{b}}_i \boxdot \tilde{\boldsymbol{b}}_j\right)$ are a (N×N) constant matrix, a (N× 1) constant column vector, a (1 ×N) constant row vector, and a constant scalar, respectively. The term ① in (22) transforms the j-th column and the j-th row of $\tilde{\boldsymbol{C}}^{(k)}$. The term ② in (22) is a zero matrix except its j-th column containing non-zero elements, while the term ③ contains non-zero entries only on its j-th row. The term ④ is a zero matrix except its (j,j)-th component being non-zero. In addition, $\tilde{\boldsymbol{C}}^{(k,\text{new})}$=①+②+③+④ is a (N×N) symmetric matrix. Hence, (22) shows that only the j-th column and j-th row of $\tilde{\boldsymbol{C}}^{(k,\text{new})}$ involve the parameter $u_{i,j}$, while the other elements remain constant. Therefore, the minimization of the cost function (20) is equivalent to minimizing the sum of the squares of the j-th columns of $\tilde{\boldsymbol{C}}^{(k,\text{new})}$ except their (j,j)-th elements with $k\in\{1,\cdots,K\}$. The required elements of $\tilde{\boldsymbol{C}}^{(k,\text{new})}$ can be expressed by the following proposition.

Proposition 2

The elements of the j-th column except the (j,j)-th entry of $\tilde{C}^{(k,\text{new})}$ is a second-degree polynomial function in $u_{i,j}$ as follows, for every value n different of j:

$$\tilde{C}_{n,j}^{(k,new)} = \tilde{C}_{n,i}^{(k)} u_{i,j}^2 + \tilde{c}_n^{(k,1)} u_{i,j} + \tilde{C}_{n,j}^{(k)} \tag{23}$$

where $\tilde{C}_{n,i}^{(k)}$ and $\tilde{C}_{n,j}^{(k)}$ are the (n,i)-th and (n,j)-th components of matrix $\tilde{C}^{(k)}$, respectively, and $\tilde{c}_n^{(k,1)}$ is the n-th element of vector $\tilde{\boldsymbol{c}}^{(k,1)}$.

The proof of this proposition is straightforward. Indeed, we can show that the elements of the j-th column except the (j,j)-th entry of the term ① in (22) can be expressed by $\tilde{C}_{n,i}^{(k)} u_{i,j}^2 + \tilde{C}_{n,j}^{(k)}$ with $1 \leq n \leq N$ and $n \neq j$, and those elements of the term ② in (22) are equal to $\tilde{c}_n^{(k,1)} u_{i,j}$ with $1 \leq n \leq N$ and $n \neq j$. The sum of these elements directly leads to (23). The terms ③ and ④ do not need to be considered, since they do not affect the off-diagonal elements in the j-th column. Proposition 2 shows that the minimization of the cost function (20) can be expressed in the following compact matrix form:

$$J(u_{i,j}) = \sum_{k=1}^{K} \left\| \boldsymbol{E}^{(k)} \boldsymbol{u}_{i,j} \right\|_F^2 = \boldsymbol{u}_{i,j}^{\mathsf{T}} \boldsymbol{Q}_E \boldsymbol{u}_{i,j} \tag{24}$$

where $\boldsymbol{Q}_E = \sum_{k=1}^{K} (\boldsymbol{E}^{(k)})^{\mathsf{T}} \boldsymbol{E}^{(k)}$ is a (3×3) symmetric coefficient matrix. $\boldsymbol{E}^{(k)}$ is a $((N-1) \times 3)$ matrix defined as follows: the first column contains the i-th column of $\tilde{C}^{(k)}$ without the j-th element, the second column contains vector $\tilde{\boldsymbol{c}}^{(k,1)}$ without the j-th entry, and the third column contains the j-th column of $\tilde{C}^{(k)}$ without the j-th component. $\boldsymbol{u}_{i,j} = \left[u_{i,j}^2, u_{i,j}, 1\right]^{\mathsf{T}}$ is a three-dimensional parameter vector.

Equation (24) shows that $J(u_{i,j})$ is a fourth-degree polynomial function. The global minimum $u_{i,j}$ can be obtained by computing the roots of its derivative and selecting the one yielding the smallest value of (24). Once the optimal $u_{i,j}$ is computed, $\tilde{\boldsymbol{B}}^{(\text{new})}$ is updated by (19) and the joint diagonalizer $\tilde{A}^{(\text{new})}$ is updated by computing $(\tilde{\boldsymbol{B}}^{(\text{new})})$□2. Then, the same procedure is repeated to compute the next $u_{i,j}$ with another (i,j) index.

The minimization of (11) with respect to the elementary lower triangular matrix $\boldsymbol{L}^{(i,j)}(\ell_{i,j})$ with $1 \leq j < i \leq N$ can be computed in the same way. Proposition 2 is also valid for the parameter $\ell_{i,j}$ when $1 \leq j < i \leq N$. The detailed derivation is omitted here. The processing of all the $N(N-1)$ parameters $u_{i,j}$ and $\ell_{i,j}$ is called a LU sweep. In addition, for estimating $\boldsymbol{L}^{(i,j)}(\ell_{i,j})$, the (i,j) index obeys the following order:

$$(2,1),(3,1),\ldots,(N,1),(3,2),(4,2),\ldots,(N,2),\ldots,$$
$$(N-1,N-2),(N,N-2),(N,N-1) \quad (25)$$

Regarding $\boldsymbol{U}^{(i,j)}(u_{i,j})$, the (i,j) index varies according to the following sequence:

$$(N-1,N),(N-2,N),(N-2,N-1),\ldots,$$
$$(2,N),(2,N-1),\ldots,(2,3),(1,N),(1,N-1),\ldots,(1,2) \quad (26)$$

The proposed $\mathrm{JD}_{\mathrm{LU}}^{+}$ algorithm is comprised of several LU sweeps.

Minimization with respect to the Givens rotation matrix $Q^{(i,j)}(\theta_{i,j})$

Now we minimize (11) with respect to $\boldsymbol{Q}^{(i,j)}(\theta_{i,j})$ with $1 \leq i<j \leq N$. By abuse of notation, in this section, we continue to use $\tilde{\boldsymbol{A}}$ and $\tilde{\boldsymbol{B}}$ to denote the current estimate of $\boldsymbol{A}$ and $\boldsymbol{B}$, respectively, before estimating the parameter $\theta_{i,j}$. Also, let $\tilde{\boldsymbol{A}}^{(\text{new})}$ and $\tilde{\boldsymbol{B}}^{(\text{new})}$ stand for $\tilde{\boldsymbol{A}}$ and $\tilde{\boldsymbol{B}}$ updated by $\boldsymbol{Q}^{(i,j)}(\theta_{i,j})$, respectively. The update of $\tilde{\boldsymbol{B}}$ is defined as follows:

$$\tilde{\boldsymbol{B}}^{(\text{new})} = \tilde{\boldsymbol{B}}\boldsymbol{Q}^{(i,j)}(\theta_{i,j}) \quad (27)$$

Similarly, for computing the parameter $\theta_{i,j}$, we can minimize the criterion (11) with respect to $\theta_{i,j}$ by replacing matrix $\tilde{\boldsymbol{B}}$ by $\tilde{\boldsymbol{B}}^{(\text{new})}$. We denote $J(\theta_{i,j})$ instead of $J(\tilde{\boldsymbol{B}}^{(\text{new})})$ for convenience purpose. Then, $J(\theta_{i,j})$ can be expressed as follows:

$$J(\theta_{i,j}) = \sum_{k=1}^{K} \left\| \text{off}\left\{ \left[\left(\tilde{\boldsymbol{B}}^{(\text{new})} \right)^{\boxdot 2} \right]^{\mathsf{T}} \boldsymbol{C}^{(k,-1)} \left[\left(\tilde{\boldsymbol{B}}^{(\text{new})} \right)^{\boxdot 2} \right] \right\} \right\|_F^2 \quad (28)$$

The Hadamard square of the update $\tilde{\boldsymbol{B}}^{(\text{new})}$ now can be rewritten as shown in the following proposition.

Proposition 3

$$\tilde{\boldsymbol{A}}^{(new)} = \left(\tilde{\boldsymbol{B}}^{(new)}\right)^{\boxdot 2} = \left(\tilde{\boldsymbol{B}}\boldsymbol{Q}^{(i,j)}(\theta_{i,j})\right)^{\boxdot 2}$$

can be written as a function of $\theta_{i,j}$ as follows:

$$\tilde{\boldsymbol{A}}^{(new)} = \left(\tilde{\boldsymbol{B}}^{(new)}\right)^{\boxdot 2} = \tilde{\boldsymbol{B}}^{\boxdot 2} \left(\boldsymbol{Q}^{(i,j)}(\theta_{i,j})\right)^{\boxdot 2}$$
$$+ \sin(2\theta_{i,j}) \left(\tilde{\boldsymbol{b}}_i \boxdot \tilde{\boldsymbol{b}}_j\right) \left(\boldsymbol{e}_i^T - \boldsymbol{e}_j^T\right) \quad (29)$$

where $\tilde{\boldsymbol{b}}_i$ and $\tilde{\boldsymbol{b}}_j$ denote the i-th and j-th columns of $\tilde{\boldsymbol{B}}$, respectively, and $\boldsymbol{e}_i$ and $\boldsymbol{e}_j$ are the i-th and j-th columns of the identity matrix $\boldsymbol{I}_N$, respectively.

Inserting (29) into the cost function (28), we obtain:

$$J(\theta_{i,j}) = \sum_{k=1}^{K} \left\| \text{off}\left(\tilde{C}^{(k,\text{new})}\right) \right\|_F^2 = \sum_{k=1}^{K} \left\| \text{off}\left(\underbrace{\left[\left(Q^{(i,j)}(\theta_{i,j})\right)^{\boxdot 2}\right]^{\mathsf{T}} \tilde{C}^{(k)} \left(Q^{(i,j)}(\theta_{i,j})\right)^{\boxdot 2}}_{①} \right.\right.$$

$$+ \underbrace{\sin(2\theta_{i,j}) \left[\left(Q^{(i,j)}(\theta_{i,j})\right)^{\boxdot 2}\right]^{\mathsf{T}} \tilde{c}^{(k,1)} \left(e_i^{\mathsf{T}} - e_j^{\mathsf{T}}\right)}_{②}$$

$$\left.\left. + \underbrace{\sin(2\theta_{i,j})(e_i - e_j)\tilde{c}^{(k,2)}\left(Q^{(i,j)}(\theta_{i,j})\right)^{\boxdot 2}}_{③} + \underbrace{\sin^2(2\theta_{i,j})\tilde{c}^{(k,3)}(e_i - e_j)\left(e_i^{\mathsf{T}} - e_j^{\mathsf{T}}\right)}_{④} \right) \right\|_F^2 \tag{30}$$

where

$\tilde{C}^{(k)} = \tilde{A}^{\mathsf{T}} C^{(k,-1)} \tilde{A}$, $\tilde{c}^{(k,1)} = \tilde{A}^{\mathsf{T}} C^{(k,-1)} \left(\tilde{b}_i \boxdot \tilde{b}_j\right)$,

$\tilde{c}^{(k,2)} = \left(\tilde{b}_i \boxdot \tilde{b}_j\right)^{\mathsf{T}} C^{(k,-1)} \tilde{A}$, and $\tilde{c}^{(k,3)} = \left(\tilde{b}_i \boxdot \tilde{b}_j\right)^{\mathsf{T}} C^{(k,-1)} \left(\tilde{b}_i \boxdot \tilde{b}_j\right)$ are a (N×N) constant matrix, a (N× 1) constant column vector, a (1 ×N) constant row vector, and a constant scalar, respectively. The term①in (30) transforms thei-th andj-th columns and thei-th andj-th rows of $\tilde{c}^{(k)}$. The term②in (30) is a zero matrix except itsi-th andj-th columns containing non-zero elements, while the term③contains non-zero entries only on itsi-th andj-th rows. The term④is a zero matrix except its (i,i)-th, (j,j)-th, (i,j)-th, and (j,i)-th components being non-zero. $\tilde{C}^{(k,\text{new})}$=①+②+③+④is a (N×N) symmetric matrix. Hence, (30) shows that only thei-th andj-th columns and thei-th andj-th rows of $\tilde{C}^{(k,\text{new})}$ involve the parameter $\theta_{i,j}$, while the other components remain constant. It is noteworthy that the (i,j)-th and (j,i)-th components are twice affected by the transformation. Considering the symmetry of $\tilde{C}^{(k,\text{new})}$, we propose to minimize the sum of the squares of the (i,j)-th entries of theKmatrices $\tilde{C}^{(k,\text{new})}$, instead of minimizing all the off-diagonal entries. Although minimizing this quantity is not equivalent to minimizing the global cost function (28), such a simplified minimization scheme is commonly adopted in many algorithms, such as [20,31]. We denote this local minimization by $\tilde{J}(\theta_{i,j})$. The (i,j)-th component of $\tilde{C}^{(k,\text{new})}$ is expressed in the following proposition.

Proposition 4

The (i,j)-th entry of $\tilde{C}^{(k,\text{new})}$ can be expressed as a function of $\theta_{i,j}$ as follows:

$$\tilde{C}_{i,j}^{(k,new)} = -\sin^2(2\theta_{i,j})\tilde{c}^{(k,3)} + \sin^2(\theta_{i,j})\left(\tilde{C}_{i,i}^{(k)}\cos^2(\theta_{i,j}) + \tilde{C}_{j,i}^{(k)}\sin^2(\theta_{i,j})\right) + \cos^2(\theta_{i,j})\left(\tilde{C}_{i,j}^{(k)}\cos^2(\theta_{i,j}) + \tilde{C}_{j,j}^{(k)}\sin^2(\theta_{i,j})\right) + \sin(2\theta_{i,j})\left(\tilde{c}_i^{(k,1)}\cos^2(\theta_{i,j}) + \tilde{c}_j^{(k,1)}\sin^2(\theta_{i,j})\right) - \sin(2\theta_{i,j})\left(\tilde{c}_j^{(k,2)}\cos^2(\theta_{i,j}) + \tilde{c}_i^{(k,2)}\sin^2(\theta_{i,j})\right) \quad (31)$$

where $\tilde{C}_{i,i}^{(k)}$, $\tilde{C}_{j,j}^{(k)}$, $\tilde{C}_{i,j}^{(k)}$, *and* $\tilde{C}_{j,i}^{(k)}$ are the (i,i)-th, (j,j)-th, (i,j)-th, and (j,i)-th components of matrix $\tilde{c}^{(k)}$, respectively. $\tilde{c}_i^{(k,q)}$ *and* $\tilde{c}_j^{(k,q)}$ are thei-th andj-th elements of vector $\tilde{c}^{(k,q)}$ withq∈{1,2}, respectively.

It is straightforward to show that the (i,j)-th entry of the term①in (30) can be expressed by $\sin^2(\theta_{i,j})\cos^2(\theta_{i,j})\left(\tilde{C}_{i,i}^{(k)} + \tilde{C}_{j,j}^{(k)}\right) + \sin^4(\theta_{i,j})C_{j,i}^{(k)} + \cos^4(\theta_{i,j})\tilde{C}_{i,j}^{(k)}$, the (i,j)-th element of the term②issin(2θ$_{i,j}$) $\left(\cos^2(\theta_{i,j})\tilde{c}_i^{(k,1)} + \sin^2(\theta_{i,j})\tilde{c}_j^{(k,1)}\right)$, the (i,j)-th component of the term③is equal to−sin(2θi,j) $\left(\sin^2(\theta_{i,j})\,\tilde{c}_i^{(k,2)} + \cos^2(\theta_{i,j})\tilde{c}_j^{(k,2)}\right)$, and that of the term④is−sin²(2θ$_{i,j}$) $\tilde{c}^{(k,3)}$. Then, Proposition 4 can be proved.

In order to simplify the notation of (31), we resort to the Weierstrass change of variable:t$_{i,j}$=tan(θ$_{i,j}$). Then, we obtain:

$$\sin(2\theta_{i,j}) = \frac{2t_{i,j}}{1+t_{i,j}^2},\ \cos(2\theta_{i,j}) = \frac{1-t_{i,j}^2}{1+t_{i,j}^2},\ \sin^2(\theta_{i,j}) = \frac{t_{i,j}^2}{1+t_{i,j}^2},\ \cos^2(\theta_{i,j}) = \frac{1}{1+t_{i,j}^2} \quad (32)$$

By substituting (32) into (31), we obtain an alternative expression of the (i,j)-th entry of $\tilde{C}^{(k,\mathrm{new})}$ which is described in the following proposition. Then, the minimization of $\tilde{J}(\theta_{i,j})$ transforms to $\tilde{J}(t_{i,j})$.

Proposition 5

The (i,j)-th entry of $\tilde{C}^{(k,\mathrm{new})}$ can be expressed by a rational function of t$_{i,j}$ as follows:

$$\tilde{C}_{i,j}^{(k,new)} = \frac{f_4^{(k)}t_{i,j}^4 + f_3^{(k)}t_{i,j}^3 + f_2^{(k)}t_{i,j}^2 + f_1^{(k)}t_{i,j} + f_0^{(k)}}{\left(1+t_{i,j}^2\right)^2} \quad (33)$$

where

$f_4^{(k)} = \tilde{C}_{j,i}^{(k)}, f_3^{(k)} = -2\tilde{c}_i^{(k,1)}, f_2^{(k)} = \tilde{C}_{i,i}^{(k)} + \tilde{C}_{j,j}^{(k)} + 2\tilde{c}_j^{(k,2)} - 4\tilde{c}^{(k,3)}, f_1^{(k)} = 2\tilde{c}_i^{(k,2)} - \tilde{c}_j^{(k,1)}$, *and* $f_0^{(k)} = \tilde{C}_{j,j}^{(k)}$.

Eq. 33 easily shows that the sum of the squares of the (i,j)-th entries of theK-matrices $\tilde{C}^{(k,\text{new})}$, is a rational function in$t_{i,j}$, namely $\tilde{J}(t_{i,j})$, where the degrees of the numerator and the denominator are 8 and 8, respectively. $\tilde{J}(t_{i,j})$can be expressed in the following compact matrix form:

$$\tilde{J}(t_{i,j}) = \sum_{k=1}^{K} \left\| \left(f^{(k)}\right)^{\mathsf{T}} \tau_{i,j} \right\|_F^2 = \tau_{i,j}^{\mathsf{T}} Q_F \, \tau_{i,j} \tag{34}$$

where $Q_F = \sum_{k=1}^{K} f^{(k)} \left(f^{(k)}\right)^{\mathsf{T}}$ is a (5 × 5) symmetric coefficient matrix, $f^{(k)} = \left[f_4^{(k)}, f_3^{(k)}, f_2^{(k)}, f_1^{(k)}, f_0^{(k)}\right]^{\mathsf{T}}$ is a five-dimensional vector, and$\tau_{i,j}$is a five-dimensional parameter vector defined as follows:

$$\tau_{i,j} = \frac{1}{\left(1+t_{i,j}^2\right)^2} \left[t_{i,j}^4,\ t_{i,j}^3,\ t_{i,j}^2,\ t_{i,j},\ 1\right]^{\mathsf{T}} \tag{35}$$

The global minimum$t_{i,j}$can be obtained by computing the roots of its derivative and selecting the one yielding the smallest value of $\tilde{J}(t_{i,j})$. Once$t_{i,j}$is obtained,$\theta_{i,j}$can be computed from the inverse tangent function$\theta_{i,j}$=arctan($t_{i,j}$). It is noteworthy that the found$\theta_{i,j}$cannot guarantee to decrease the actual cost function (28). If$\theta_{i,j}$leads to an increase of (28), we reset$\theta_{i,j}$= 0. Otherwise, $\tilde{B}^{(\text{new})}$ is updated as described in (27) and the joint diagonalizer$\tilde{A}^{(\text{new})}$ is updated by computing($\tilde{B}^{(\text{new})}$)□2. The same procedure will be repeated to compute$\theta_{i,j}$with the next (i,j) index. The order of the (i,j) indices is defined in Eq.26. The processing of all theN(N- 1)/2 parameters$\theta_{i,j}$and also the otherN(N- 1)/2 parameters$u_{i,j}$is called a QR sweep. Several QR sweeps yield the proposedJD_{QR}^{+}algorithm.

Both of theJD_{LU}^{+}andJD_{QR}^{+}algorithms can be stopped when the value of cost function (11) or its relative change between two successive sweeps fall below a fixed small positive threshold. Such a stopping criterion is guaranteed to be met since the cost function is non-increasing in each Jacobi-like sweep.

Practical Issues

In practice, we observe that if each frontal slice of the three-way array$\mathcal{C}$ is almost exactly jointly diagonalizable due to a high signal-to-noise ratio (SNR), the classical non-constrained JDC methods can also give a nonnegative$A$with high probability. In this situation, the explicit nonnegativity constraint could be unnecessary and could increase the computational burden. Therefore, we propose to relax the nonnegativity constraint by directly decomposing $\boldsymbol{A}$ into elementary LU and QR forms, respectively, instead of using the decompositions of $\boldsymbol{B}$ as follows:

$$\boldsymbol{A} = \prod_{j\in\mathcal{J}_1}\prod_{i\in\mathcal{I}_1(j)} \boldsymbol{L}^{(i,j)}(\ell_{i,j}) \prod_{i\in\mathcal{I}_2}\prod_{j\in\mathcal{J}_2(i)} \boldsymbol{U}^{(i,j)}(u_{i,j}) \tag{36}$$

$$\boldsymbol{A} = \prod_{i\in\mathcal{I}_2}\prod_{j\in\mathcal{J}_2(i)} \boldsymbol{Q}^{(i,j)}(\theta_{i,j}) \prod_{i\in\mathcal{I}_2}\prod_{j\in\mathcal{J}_2(i)} \boldsymbol{U}^{(i,j)}(u_{i,j}) \tag{37}$$

where the index sets $\mathcal{I}_1(j)$, $\mathcal{J}_1$, $\mathcal{I}_2$, and $\mathcal{J}_2(i)$ are defined in Lemma 1. By inserting (36) and (37) into the cost function (9), the ways of estimating the two sets of parameters $\{\ell_{i,j}, u_{i,j}\}$ and $\{\theta_{i,j}, u_{i,j}\}$ are identical to those of Afsari's LUJ1D and QRJ1D methods [26], respectively. Therefore, in practice, in order to give an automatically SNR-adaptive method, for $\mathrm{JD}^{+}_{\mathrm{LU}}$, in each Jacobi-like iteration, we suggest to compute $u_{i,j}$ by LUJ1D first. If all the elements in the j-th column of $\tilde{\boldsymbol{A}}\boldsymbol{U}^{(i,j)}(u_{i,j})$ have the same sign ε, the update $\tilde{\boldsymbol{A}}^{(\mathrm{new})} = \varepsilon\tilde{\boldsymbol{A}}\boldsymbol{U}^{(i,j)}(u_{i,j})$ is adopted. Otherwise, $u_{i,j}$ is computed by minimizing (20) and $\tilde{\boldsymbol{A}}^{(\mathrm{new})}$ is updated by computing (21). Each $\ell_{i,j}$ is computed similarly. Furthermore, the proposed $\mathrm{JD}^{+}_{\mathrm{QR}}$ and QRJ1D are combined in the same manner.

Afsari reported in [26] that if the rows of matrices $\tilde{\boldsymbol{C}}^{(k)}$ ($k\in\{1,\cdots,K\}$) are not balanced in their norms, the computation of the parameter could be inaccurate. In order to cope with this effect, we apply Afsari's row balancing scheme every few sweeps. Such a scheme updates each $\tilde{\boldsymbol{C}}^{(k,\mathrm{new})}$ by $\tilde{\boldsymbol{C}}^{(k,\mathrm{new})} = \boldsymbol{\Lambda}\tilde{\boldsymbol{C}}^{(k)}\boldsymbol{\Lambda}$ and $\tilde{\boldsymbol{A}}^{(\mathrm{new})}$ by $\tilde{\boldsymbol{A}}^{(\mathrm{new})} = \tilde{\boldsymbol{A}}\boldsymbol{\Lambda}$ using a diagonal matrix $\boldsymbol{\Lambda} \in \mathbb{R}_{+}^{N\times N}$, whose diagonal elements are defined as follows:

$$\Lambda_{n,n} = \frac{1}{\sqrt{\sum_{k=1}^{K}\left\|\tilde{\boldsymbol{C}}^{(k)}_{n,:}\right\|^2}}, \quad n\in\{1,2,\cdots,N\} \tag{38}$$

where $\tilde{\boldsymbol{C}}^{(k)}_{n,:}$ denotes the n-th row of $\tilde{\boldsymbol{C}}^{(k)}$.

In ICA, when a non-square matrix $\boldsymbol{A} \in \mathbb{R}_{+}^{N\times P}$ with $N>P$ is encountered, the invertibility assumption of the frontal slices $\boldsymbol{C}^{(k)}$ does not hold. In this situation, we can compress $\boldsymbol{A}$ by means of a nonnegative matrix $\boldsymbol{W}_{+} \in \mathbb{R}_{+}^{P\times N}$ such that the resulting matrix $\bar{A}=W_{+}A$ is a nonnegative square matrix. Then, the $\mathrm{JD}^{+}_{\mathrm{LU}}$ and $\mathrm{JD}^{+}_{\mathrm{QR}}$ algorithms can be used to compute the compressed loading matrix $\bar{A}$. $\boldsymbol{W}_{+}$ can be computed by using the nonnegative compression algorithm (NN-

COMP) that we proposed in [53]. More precisely, given a realization of an observation vector, we obtain the square root of the covariance matrix, denoted by $\boldsymbol{\Upsilon} \in \mathbb{R}^{N\times P}$. The classical prewhitening matrix is computed by $\boldsymbol{W} = \boldsymbol{\Upsilon}^{\sharp} \in \mathbb{R}^{P\times N}$ where # denotes the pseudo inverse operator [47]. Then, the NN-COMP algorithm computes a linear transformation matrix $\boldsymbol{\Psi} \in \mathbb{R}^{P\times P}$ such that $\boldsymbol{W}_{+}=\boldsymbol{\Psi W}$ has nonnegative components. Once $\bar{A}$ is estimated, the original matrix $\boldsymbol{A}$ is obtained as follows:

$$\boldsymbol{A} = \boldsymbol{W}^{\sharp}\boldsymbol{\Psi}^{-1}\bar{\boldsymbol{A}} = \boldsymbol{\Upsilon}\boldsymbol{\Psi}^{-1}\bar{\boldsymbol{A}} \quad (39)$$

It should be noted that generally $\boldsymbol{A}$ does not need to be computed in such an ICA problem, since the sources can be estimated directly by means of $\bar{A}$.

Numerical Complexity

The numerical complexities of $\mathrm{JD}^{+}_{\mathrm{LU}}$ and $\mathrm{JD}^{+}_{\mathrm{QR}}$ are analyzed in terms of the number of floating point operations (flops). A flop is defined as a multiplication followed by an addition. In practice, only the number of multiplications, required to identify the loading matrix $\boldsymbol{A} \in \mathbb{R}_{+}^{N\times N}$ from a three-way array $\boldsymbol{C} \in \mathbb{R}^{N\times N\times K}$, is considered, which does not affect the order of magnitude of the numerical complexity.

For both algorithms, the inverses $\boldsymbol{C}^{(k,-1)}$ ($k\in\{1,\cdots,K\}$) of the frontal slices of $\boldsymbol{C}$ cost N^3K flops, the initialization of $\tilde{\boldsymbol{C}}^{(k)}_{\mathrm{ini}} = \tilde{\boldsymbol{A}}^{\mathsf{T}}_{\mathrm{ini}}\boldsymbol{C}^{(k,-1)}\tilde{\boldsymbol{A}}_{\mathrm{ini}}$ requires $2N^3K$ flops, and at each sweep, the calculation of parameters $u_{i,j}$ needs $N(N-1)(5N^2+12N-8)K/2$ flops. In addition, in the case of the $\mathrm{JD}^{+}_{\mathrm{LU}}$ algorithm, the calculation cost of $\tilde{\boldsymbol{A}}^{(\mathrm{new})}$, $\tilde{\boldsymbol{B}}^{(\mathrm{new})}$, and $\tilde{\boldsymbol{C}}^{(k,\mathrm{new})}$, with $k\in\{1,\cdots,K\}$, is $N(N-1)(4N+(4N+1)K)$ flops, and the numerical complexity of computing the parameters $\ell_{i,j}$ is equal to that of $u_{i,j}$. Regarding the $\mathrm{JD}^{+}_{\mathrm{QR}}$ algorithm, for each sweep, the complexity of calculating the parameters $\theta_{i,j}$ is equal to $N(N-1)(5N^2+3N+29)K/2$ flops, and the estimation of $\tilde{\boldsymbol{A}}^{(\mathrm{new})}$, $\tilde{\boldsymbol{B}}^{(\mathrm{new})}$, and $\tilde{\boldsymbol{C}}^{(k,\mathrm{new})}$, with $k\in\{1,\cdots,K\}$, costs $N(N-1)(5N+(12N+20)K/2)$ flops. In practice, the proposed $\mathrm{JD}^{+}_{\mathrm{LU}}$ and $\mathrm{JD}^{+}_{\mathrm{QR}}$ techniques are combined with LUJ1D and QRJ1D [26], respectively, leading to the magnitude of global numerical complexities of $\mathrm{JD}^{+}_{\mathrm{LU}}$ and $\mathrm{JD}^{+}_{\mathrm{QR}}$ being between $O(N^3K)$ and $O(N^4K)$. A recent nonnegative JDC method called $\mathrm{ACDC}^{+}_{\mathrm{LU}}$ [41] is also based on a square change of variable and LU matrix factorization. It minimizes the cost function (4) with respect to $\boldsymbol{A}$ and $\boldsymbol{D}$ alternately, leading to a higher numerical complexity. By means of the reformulation of the cost function, the proposed methods avoid the estimation of $\boldsymbol{D}$, therefore achieving a lower complexity compared to $ACDC^{+}_{LU}$. The explicit expressions of the overall complexity of JD^{+}_{LU}, JD^{+}_{QR}, and $ACDC^{+}_{LU}$ [41], as well as those of four classical JDC algorithms, namely ACDC [23], FFDIAG [25], LUJ1D [26], and QRJ1D [26], are listed in Table1. One can notice that numerical complexities of the proposed JD^{+}_{LU} and JD^{+}_{QR} methods are at most one order of magnitude higher than those of the four JDC algorithms and still lower than that of $ACDC^{+}_{LU}$. Moreover, JD^{+}_{LU} is less computationally expensive than JD^{+}_{QR}.

Table 1: Numerical complexities of seven JDC algorithms in terms of flops

	Numerical complexity
ACDC	$(13/3N^3K+3N^4+2N^2K+N^3+N^2)N_s$
FFDIAG	$(2N^3K+N^3+2N^2K+4N(N-1))N_s$
LUJ1D	$(4NK+N-2K)N(N-1)N_s$
QRJ1D	$(6NK+2.5N+1.5K)N(N-1)N_s$
$ACDC^{+}_{LU}$	$((15N^2+4N)KN(N-1)+4/3N^2K+N^3+N^2)\ N_s^1$
	$+((33N^2+7N)KN(N-1)+4/3N^2K+N^3+N^2)\ N_s^2$
JD^{+}_{LU}	$3N^3K+(4NK+N-2K)N(N-1)\ N_s^1$
	$+((5N^2+16N-7)K+4N)N(N-1)\ N_s^2$
JD^{+}_{QR}	$3N^3K+(6NK+2.5N+1.5K)N(N-1)\ N_s^1$
	$+((5N^2+15.5N+21)K+7N)N(N-1)\ N_s^2$

(N,N,K): the dimensions of the three-way array $\boldsymbol{\mathcal{C}}$. For ACDC, FFDIAG, LUJ1D, and QRJ1D, N_s is the number of total sweeps. For $ACDC^{+}_{LU}$, JD^{+}_{LU}, and JD^{+}_{QR}, N_s^1 is the number of sweeps without nonnegativity constraint; N_s^2 is the number of sweeps with explicit nonnegativity constraint.

SIMULATION RESULTS

This section is twofold. In the first part, the performance of the proposed JD^{+}_{LU} and JD^{+}_{QR} algorithms is evaluated with simulated semi-nonnegative semi-symmetric three-way arrays $\boldsymbol{\mathcal{C}}$. Several experiments are designed to study the convergence property, the influence of SNR, the impact of the third dimension K of $\boldsymbol{\mathcal{C}}$, the effect of the coherence of the loading matrix $\boldsymbol{D}$, and the influence of the condition number of the diagonal matrices $\boldsymbol{D}^{(k)}$. We also evaluate the proposed methods for estimating a non-square matrix $\boldsymbol{A}$. The proposed algorithms are compared with four classical nonorthogonal JDC methods, namely ACDC [23], FFDIAG [25], LUJ1D [26], QRJ1D [26], and the nonnegative JDC method $\text{ACDC}^{+}_{\text{LU}}$ [41]. In the second part, the source separation ability of the proposed algorithms is studied through a BSS application. In this context, the $\text{JD}^{+}_{\text{LU}}$ and $\text{JD}^{+}_{\text{QR}}$ are used to jointly diagonalize several matrix slices of the fourth-order cumulant array [40] of the observations and compared with several classical ICA [47,54,55] and NMF [56] methods.

Simulated Semi-nonnegative INDSCAL Model

The synthetic semi-nonnegative semi-symmetric three-way array $\mathcal{C} = [\![\boldsymbol{A}, \boldsymbol{A}, \boldsymbol{D}]\!] \in \mathbb{R}^{N\times N\times K}$ of rank N is generated randomly according to the semi-nonnegative INDSCAL model (3). When used without further specification, all the algorithms are manipulated under the following conditions:

i) Model generation: The loading matrix $\boldsymbol{A} \in \mathbb{R}_{+}^{N\times N}$ is randomly drawn from a uniform distribution on the interval [0,1]. The loading matrix $\boldsymbol{D} \in \mathbb{R}^{K\times N}$ is drawn from a Gaussian distribution with a mean of 1 and a deviation of 0.5. The pure array $\mathcal{C}$ is perturbed by a residual INDSCAL noise array $\mathcal{V}$. The loading matrices of $\mathcal{V}$ are drawn from a zero-mean unit-variance Gaussian distribution. The resulting noisy three-way array can be written as follows:

$$\mathcal{C}_N = \frac{\mathcal{C}}{\|\mathcal{C}\|_F} + \sigma_N \frac{\mathcal{V}}{\|\mathcal{V}\|_F} \tag{40}$$

where σ_N is a scalar controlling the noise level. Then, the SNR is defined by $\text{SNR} = -20\log_{10}(\sigma_N)$.

ii) Initialization: In each Monte Carlo trial, all the algorithms are initialized by a same random matrix whose components obey the uniform distribution over [0,1].

iii) Afsari's row balancing scheme: The LUJ1D, QRJ1D, $\text{JD}^{+}_{\text{LU}}$, and $\text{JD}^{+}_{\text{QR}}$ algorithms perform the row balancing scheme once per run of five sweeps.

iv) Stopping criterion: All the algorithms stop either when the relative error of the corresponding criterion between two successive sweeps is less than 10^{-5} or when the number of sweeps exceeds 200. A sweep of ACDC includes a full AC phase and a DC phase.

v) Performance measurement: The performance is measured by means of the error between the true loading matrix A and the estimate $\tilde{A}$, the numerical complexity, and the CPU time. We define the following scale-invariant and permutation-invariant distance [40]:

$$\alpha(A,\tilde{A}) = \frac{1}{N}\sum_{n=1}^{N} \min_{(n,n')\in I_n^2} d(a_n,\tilde{a}_{n'}) \tag{41}$$

where a_n and $\tilde{a}_{n'}$ are the n-th column of A and the n'-th column of $\tilde{A}$, respectively. I_n^2 is defined recursively by $I_1^2=\{1,\cdots,N\}\times\{1,\cdots,N\}$, and $I_{n+1}^2 = I_n^2 - J_n^2$ where $J_n^2 = \mathrm{argmin}_{(n,n')\in I_n^2} d(a_n,\tilde{a}_{n'})$. In addition, $d(a_n,\tilde{a}_{n'})$ is defined as the pseudo-distance between two vectors [13]:

$$d(a_n,\tilde{a}_{n'}) = 1 - \frac{\left\| a_n^{\top}\tilde{a}_{n'} \right\|^2}{\|a_n\|^2\|\tilde{a}_{n'}\|^2} \tag{42}$$

The criterion (41) is an upper bound of the optimal permutation-invariant criterion. It avoids the burdensome computation of all the permutations. A small value of (41) means a good performance in the sense that $\tilde{A}$ is close to A.

vi) Test environment: The simulations are carried out in Matlab v7.14 on Mac OS X and run on Intel Quad-Core CPU 2.8 GHz with 32 GB memory. Moreover, we repeat all the experiments with 500 Monte Carlo trials.

Convergence

In this experiment, the convergences of the $\mathrm{JD}_{\mathrm{LU}}^{+}$ and $\mathrm{JD}_{\mathrm{QR}}^{+}$ algorithms are compared to those of ACDC, FFDIAG, LUJ1D, QRJ1D, and $\mathrm{ACDC}_{\mathrm{LU}}^{+}$. The dimensions of the three-way array $\mathcal{C}_N \in \mathbb{R}^{N\times N\times K}$ are set to N= 5 and K= 15. The performance is assessed under three SNR conditions: SNR = - 5, 10, and 25 dB, respectively. Figure 1 shows the convergence curves measured in terms of the cost function as a function of sweeps. It shows that FFDIAG, LUJ1D, and QRJ1D exhibit fast convergence behavior. They converge in less than 20 sweeps. $\mathrm{ACDC}_{\mathrm{LU}}^{+}$ decreases the cost function (4) quasi-linearly. ACDC and $\mathrm{ACDC}_{\mathrm{LU}}^{+}$ do not converge in a maximum of 200 sweeps. The proposed JD+LU algorithm converges in about 100 sweeps when SNR =25 dB and SNR =10 dB, and it converges in about 40 sweeps when SNR =-5 dB. Regarding $\mathrm{JD}_{\mathrm{QR}}^{+}$, it reduces the cost function (11) to the values relatively higher than those achieved by $\mathrm{JD}_{\mathrm{LU}}^{+}$ and converges in about 50 sweeps whatever the SNR is. It seems that FFDIAG, LUJ1D, and QRJ1D achieve the fastest convergence rate. It should be noted that while an algorithm may converge to a point in which the value of the cost function is close to zero, such a point could be a local minimum far from the desired matrix$\boldsymbol{A}$as shown in Figure2. The top picture in Figure2shows the convergence curves measured in terms of the estimating error$\alpha(A, \tilde{\boldsymbol{A}})$as a function of sweeps when SNR=25 dB. It shows that the solutions of FFDIAG, LUJ1D, and QRJ1D are still far from optimum.

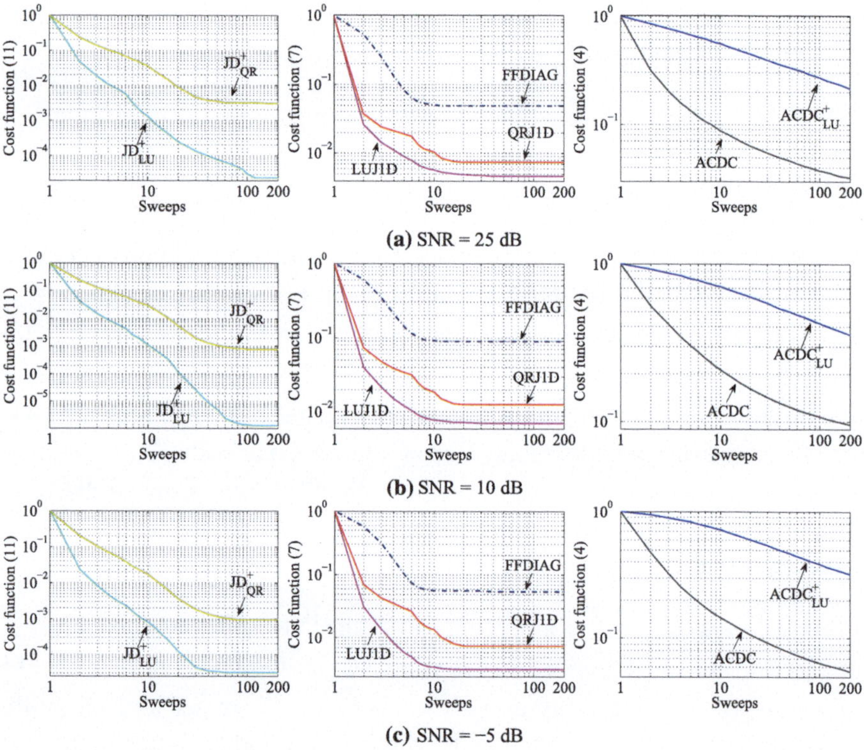

Figure 1: JDC performance versus sweeps.The average value of the cost function evolution of all the algorithms as a function of the number of sweeps with various SNR levels. The dimensions of$C$N are set to$N$= 5 and$K$= 15. The SNR values are set to 25 dB(a), 10 dB(b), and -5 dB(c), respectively.

ACDC and$\text{ACDC}^{+}_{\text{LU}}$give better estimations of$\boldsymbol{A}$than the previous three methods. The best results are achieved by the proposed$\text{JD}^{+}_{\text{LU}}and\text{JD}^{+}_{\text{QR}}$ methods. The middle picture in Figure2displays the convergence curves when SNR=10 dB. It can be observed that ACDC converges to a local minimum which is not the global one and that the performance of the proposed methods is still better than that of the five other algorithms. For a low SNR =-5 dB, as shown in the bottom picture in Figure2, both the methods based on alternating optimization, namely ACDC and $ACDC^{+}_{LU}$, converge to local minima which are less desirable. The proposed algorithms are always able to converge to better results than the classical methods. The average numerical complexities and CPU time of all the algorithms over Monte Carlo trials are shown in Table2. It is observed that FFDIAG, LUJ1D, and QRJ1D require a small amount of calculations, whereas $ACDC^{+}_{LU}$requires a large amount of calculations. The proposed JD^{+}_{LU}just costs a bit more flops and CPU time than ACDC, but it is still much more efficient. Concerning the JD^{+}_{QR}algorithm, it is more costly than JD^{+}_{LU}, with a comparable performance. We can then conclude that JD^{+}_{LU}offers the best performance/complexity compromise in these experiments.

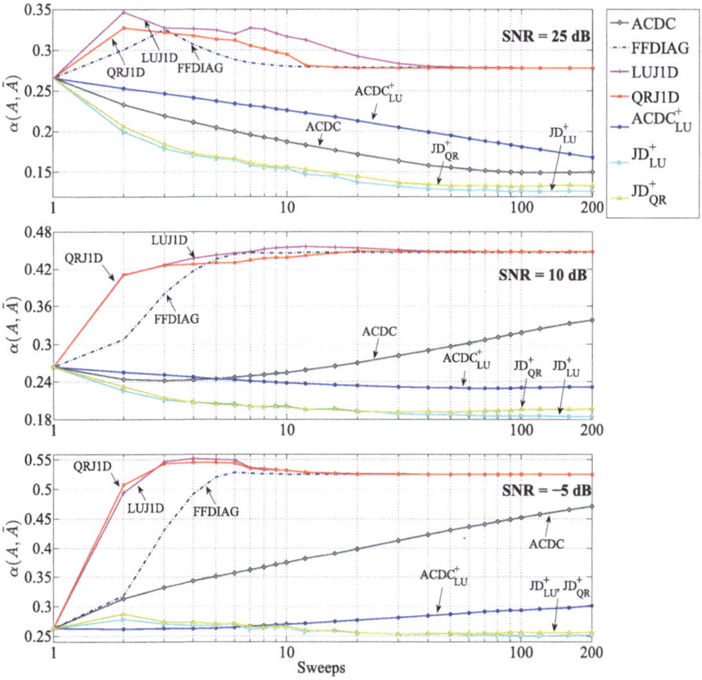

Figure 2: JDC performance versus sweeps.The average errorα(A, $\tilde{A}$)evolution of all the algorithms as a function of the number of sweeps with various SNR levels. The dimensions ofCNare set toN= 5 andK= 15. The SNR values are set to 25 dB (top), 10 dB (middle), and -5 dB (bottom), respectively.

Table 2: Average numerical complexities (in flops) and computation time (in seconds) of the convergence experiment

	SNR = 25 dB		SNR = 10 dB		SNR =-5 dB	
	Complexity	Time	Complexity	Time	Complexity	Time
ACDC	2.1708×10^6	1.1357	2.0338×10^6	1.0535	1.6800×10^6	0.8761
FFDIAG	1.1001×10^5	0.0331	1.0878×10^5	0.0327	9.4380×10^4	0.0287
LUJ1D	2.8903×10^5	0.0660	2.3126×10^5	0.0526	1.4199×10^5	0.0325
QRJ1D	2.2158×10^5	0.0383	2.4445×10^5	0.0421	2.6989×10^5	0.0468
ACDC+LU	2.4462×10^7	2.6735	2.7498×10^7	2.8034	2.9646×10^7	2.9119
JD+LU	2.8487×10^6	0.8107	4.9684×10^6	1.1098	7.1434×10^6	1.2938
JD+QR	3.0766×10^6	1.0455	5.0554×10^6	1.1932	8.2185×10^6	1.3026

Effect of SNR

In this section, we study the behaviors of the seven algorithms as a function of SNR. The dimensions of the three-way arrayCNare set toN= 5 andK= 15. We repeat the experiments with SNR ranging from -30 to 50 dB with a step of 2 dB. The top picture in Figure3depicts the average curves ofα(A, $\tilde{A}$)of the seven algorithms as a function of SNR. The obtained results show that the performance of all the methods increases as SNR grows. For the unconstrained methods, generally, ACDC performs better than FFDIAG, LUJ1D, and QRJ1D. The nonnegativity constraint obviously helps$\text{ACDC}_{\text{LU}}^{+}$, $\text{JD}_{\text{LU}}^{+}$, and$\text{JD}_{\text{QR}}^{+}$to improve the results for lower SNR values. The performance of ACDC and$\text{ACDC}_{\text{LU}}^{+}$remains stable for higher SNR values due to the small number of available sweeps and the lack of good initializations. Generally, the proposed$\text{JD}_{\text{LU}}^{+}and\text{JD}_{\text{QR}}^{+}$algorithms outperform the others when SNR is between -20 and 30 dB and perform similar to FFDIAG, LUJ1D, and QRJ1D when SNR is above 45 dB. The average numerical complexity and CPU time at each SNR level of all the methods in this experiment are shown in the bottom of Figure3. It shows that the proposed methods achieve better estimations of$A$and cost less flops and CPU time than$\text{ACDC}_{\text{LU}}^{+}$. The$\text{JD}_{\text{LU}}^{+}$ gives the best performance/complexity trade-off for all the considered SNR values.

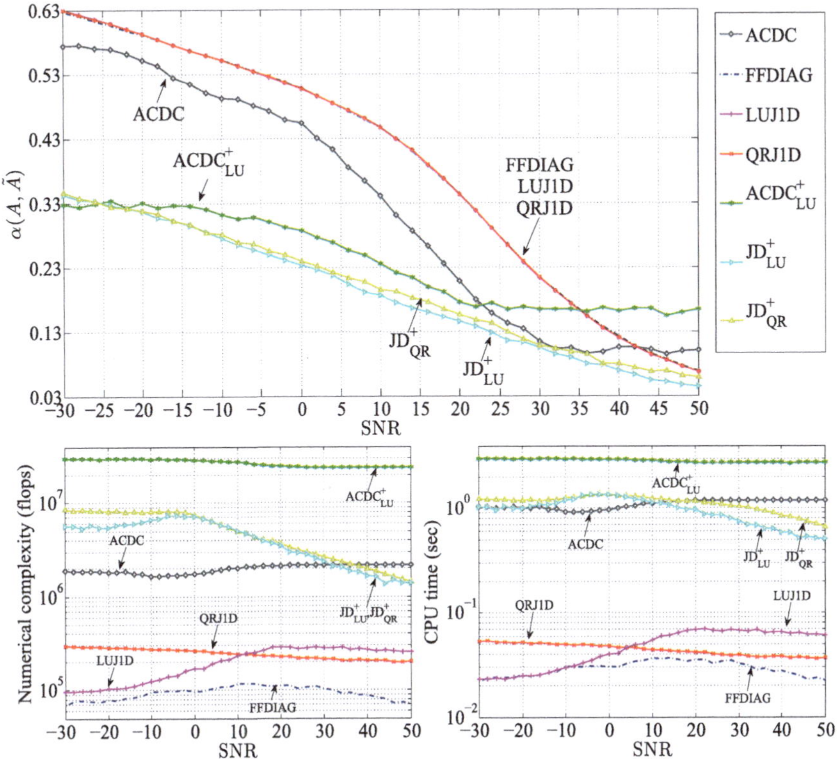

Figure 3: JDC performance versus SNR.The dimensions ofCNare set toN= 5 andK= 15. Top: the average errorα(A, $\tilde{\boldsymbol{A}}$)evolution of all the algorithms as a function of SNR. Bottom: the average numerical complexities (left) and the CPU time (right) of all the algorithms, respectively.

Effect of Dimension K

In ICA, the third dimensionKof the three-way array $\mathcal{C}_N \in \mathbb{R}^{N \times N \times K}$ corresponds to the number of covariance matrices at different lags, or the number of matrix slices derived from a cumulant array. In this section, we study the influence ofKon the performance of the seven algorithms. The first and second dimensions ofCNare set toN= 5. The SNR value is fixed to 10 dB. We repeat the experiment withKranging from 3 to 55. The top picture in Figure4shows the average curves ofα(A, $\tilde{\boldsymbol{A}}$)of all the algorithms as a function ofK. For the five existing methods, ACDC,ACDC+LU, FFDIAG, LUJ1D, and QRJ1D, their performance is quite stable with respect toK. The performance of the proposed methods progresses asKincreases and then practically stabilizes for high values ofK. It indicates that after some point (e.g.,K≥ 20), the additional information brought by an increase ofKdoes not further improve the results. The proposed JD^{+}_{LU}andJD^{+}_{QR}algorithms maintain competitive advantages through all theKvalues. The two images in the bottom of Figure4present the average numerical complexity and CPU time of all the algorithms in this experiment, respectively. It shows that the numerical complexity of$JD^{+}_{LU}$and$JD^{+}_{QR}$is between that of ACDC and $ACDC^{+}_{LU}$. TheJD^{+}_{LU}andJD^{+}_{QR}methods seem to be the most effective algorithms compared to the other methods.

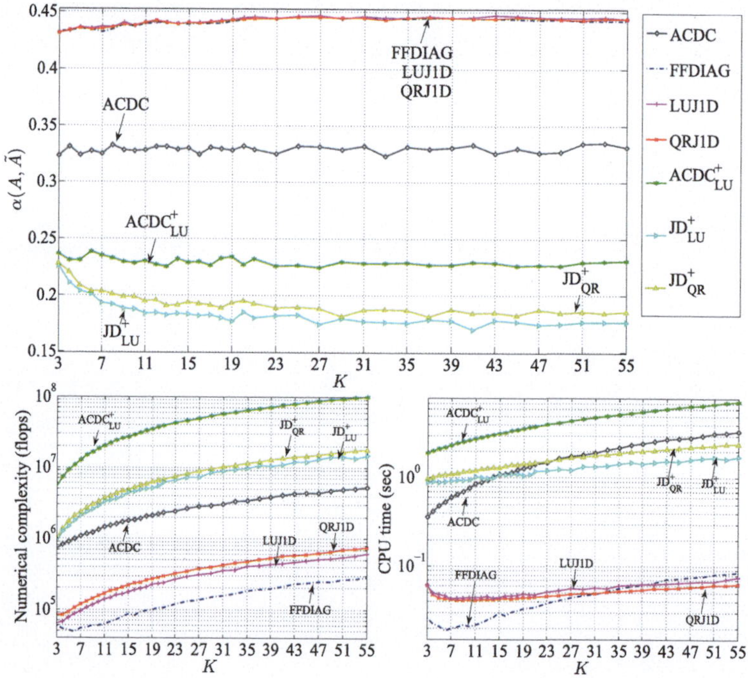

Figure 4: JDC performance versus dimensionK.The first and second dimensions ofCNand the SNR value are set toN=5 and SNR=10 dB, respectively. Top: the average errorα(A, $\tilde{A}$)evolution of all the algorithms as a function of dimensionK. Bottom: the average numerical complexities (left) and the CPU time (right) of all the algorithms, respectively.

Effect of Coherence of D

In this experiment, the effect of the coherence of the third loading matrix***D***of the three-way array$C = [\![A, A, D]\!]$is evaluated. Let$d_n$and$d_m$denote then-th andm-th columns of***D***, respectively. The angle$\psi_{n,m}$between$d_n$and$d_m$can be derived by using the following Euclidean dot product formula $\boldsymbol{d}_n^{\mathsf{T}}\boldsymbol{d}_m = \|\boldsymbol{d}_n\|\|\boldsymbol{d}_m\|\cos(\psi_{n,m})$. Then, the coherence ρ of **D** is defined as the maximum absolute cosine of angle $\psi_{n,m}$ between the columns of **D** as follows:

$$\rho = \max_{\substack{n,m \\ n \neq m}} |\cos(\psi_{n,m})| \text{ with } \cos(\psi_{n,m}) = \frac{\boldsymbol{d}_n^{\mathsf{T}}\boldsymbol{d}_m}{\|\boldsymbol{d}_n\|\|\boldsymbol{d}_m\|} \tag{43}$$

The quantity ρ is also known as the modulus of uniqueness of JDC [43]. By its definition (43), ρ falls in the range of [0,1]. The JDC problem is considered to be ill-conditioned when ρ is close to 1. Such an ill-conditioned problem can be met in ICA when **A** has nearly collinear column vectors. For example, in order to perform ICA, provided that all the sources are non-Gaussian, which is often the case in practice, we can build a three-way array $\boldsymbol{C}$ by stacking the matrix slices of the fourth-order cumulant array of the observation data. Then, the loading matrix **D** can be expressed as follows:

$$\boldsymbol{D} = (\boldsymbol{A} \odot \boldsymbol{A})\boldsymbol{C}_{4,\{s\}} \tag{44}$$

where $\boldsymbol{C}_{4,\{s\}} = \text{diag}[\mathcal{C}_{1,1,1,1,\{s\}}, \cdots, \mathcal{C}_{N,N,N,N,\{s\}}]$ is a (N×N) diagonal matrix with $C_{n,n,n,n,\{s\}}$ being the fourth-order cumulant of the n-th source, n∈{1,⋯,N}, and where ⊙ denotes the Khatri-Rao product. It can be observed that the coherence of the columns of **A** will influence the coherence of the matrix **D**. In the following test, the dimensions of the three-way array CN are set to N= 5 and K= 15. The SNR value is fixed to 10 dB. In order to control ρ, firstly, we randomly generate an orthogonal matrix $\boldsymbol{D} \in \mathbb{R}^{15\times 5}$ so that ρ= 0 by orthogonalizing a (15×5) random matrix. Secondly, we rotate its five columns such that all the internal angles between any columns are equal to a predefined value ψ. Therefore, ρ is only controlled by the angle ψ and equals to |cos(ψ)|. We repeat the experiment with the angle ψ ranging from 0 to π/2 with a step of π/60. A small ψ value means a large ρ value. The top picture in Figure 5 displays the average curves of α(A, $\tilde{\boldsymbol{A}}$) of all the algorithms as a function of ψ. It shows that the nonnegativity constrained methods $\text{ACDC}_{\text{LU}}^{+}$, $\text{JD}_{\text{LU}}^{+}$, and $\text{JD}_{\text{QR}}^{+}$, outperform the unconstrained ones ACDC, FFDIAG, LUJ1D, and QRJ1D. The proposed algorithms are more efficient, particularly when the coherence level is high. The average numerical complexity and CPU time displayed in the bottom of Figure 5 indicate that the $\text{JD}_{\text{LU}}^{+}$ algorithm provides the best performance/complexity compromise, while the $\text{JD}_{\text{QR}}^{+}$ algorithm is also competitive with regard to $\text{ACDC}_{\text{LU}}^{+}$.

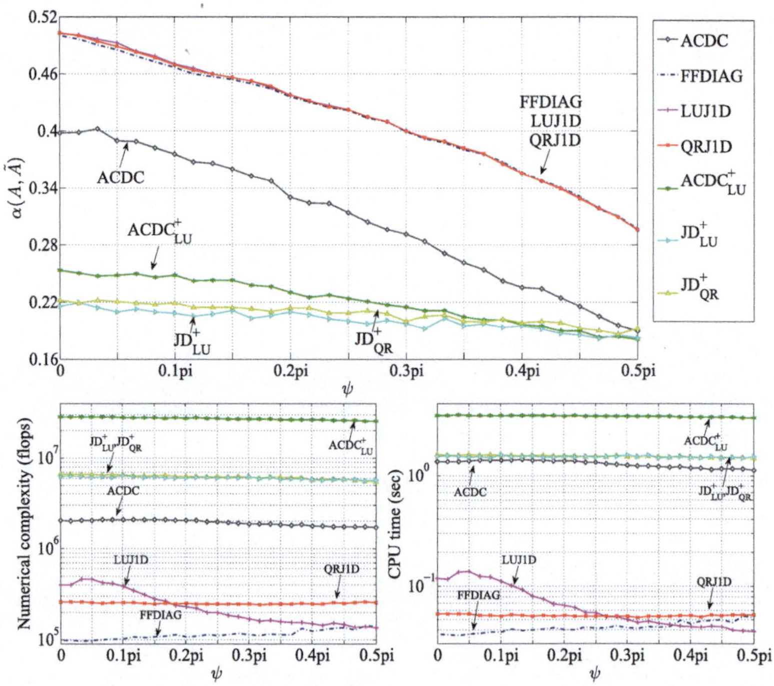

Figure 5: JDC performance versus coherence.The dimensions ofCNand the SNR value are set toN= 5,K= 15, and SNR=10 dB, respectively. Top: the average errorα(A, $\tilde{A}$)evolution of all the algorithms as a function of internal angleψbetween any two columns ofD. Bottom: the average numerical complexities (left) and the CPU time (right) of all the algorithms, respectively.

Effect of Condition Number of $D^{(k)}$

When the JDC problem is considered, a diagonal matrix$\boldsymbol{D}^{(k)}$could contain some diagonal elements which, despite being non-zero, are many orders of magnitude lower than some other elements, leading to an ill-conditioned matrix$\boldsymbol{C}^{(k)}$. For the proposed methods, the inverse of such a matrix$\boldsymbol{C}^{(k)}$would contain numerical errors. In this experiment, we study the performance of the seven algorithms as a function of the condition number of one of the diagonal matrices$\boldsymbol{D}^{(k)}$. The dimensions of the three-way arrayCNare set toN= 5 andK= 15. The SNR value is set to 10 dB. We vary the condition number of the first diagonal matrix$\boldsymbol{D}^{(1)}$from 1 to 1,000 by fixing the ratio of its largest diagonal element to its smallest diagonal element. The top picture in Figure6displays the average curves of the estimating errorα(A, $\tilde{A}$)of the seven algorithms as a function of the condition number of$\boldsymbol{D}^{(1)}$. The results reveal that a highly ill-conditioned diagonal matrix$\boldsymbol{D}^{(1)}$has a clear negative effect on the estimation accuracy of all the algorithms. The nonnegativity constrained methodsACDC+LU, JD^{+}_{LU}, and JD^{+}_{QR}outperform the classical algorithms ACDC, FFDIAG, LUJ1D, and QRJ1D whatever the condition number is. The proposed $JD^{+}_{LU}$and JD^{+}_{QR}algorithms maintain advantages when the condition number is less than 100. Regarding the cases of larger condition numbers,$ACDC^{+}_{LU}$is more superior since it does not need to invert the highly ill-conditioned matrix.

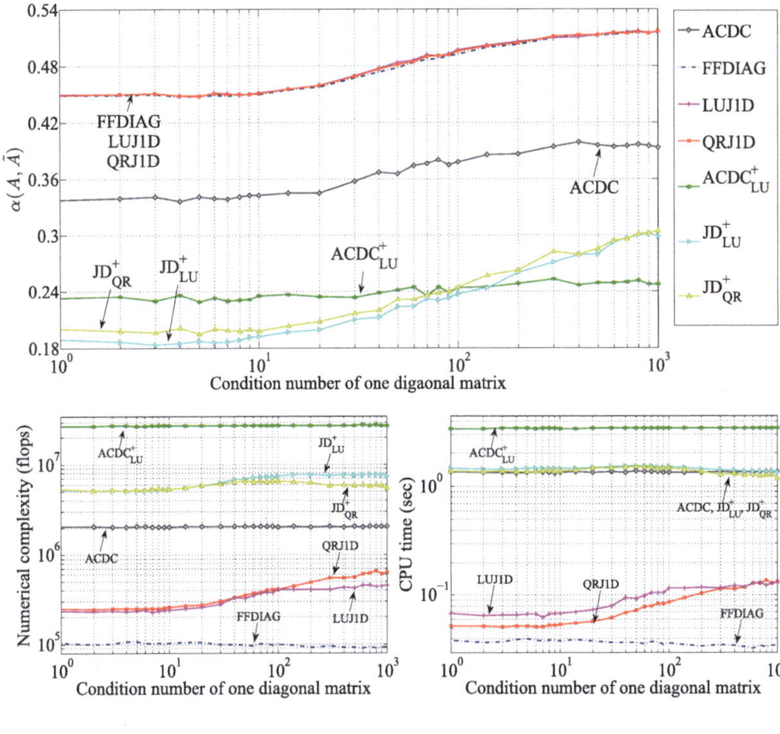

Figure 6: JDC performance versus condition number.The dimensions ofCNand the SNR value are set toN= 5,K= 15, and SNR=10 dB, respectively. Top: the average error$\alpha(A, \tilde{\boldsymbol{A}})$evolution of all the algorithms as a function of the condition number of one of the diagonal matrices$\boldsymbol{D}^{(k)}$. Bottom: the average numerical complexities (left) and the CPU time (right) of all the algorithms, respectively.

It is worthy pointing out that in practice, we can choose these sufficiently well-conditioned matrices$\boldsymbol{C}^{(k)}$for the proposed methods, whose condition numbers are below a predefined threshold. In addition, a weighted cost function for which weights would depend on the condition number of each matrix can be considered. On the other hand, the performance of the classical methods can also be improved by choosing a particular subset of available matrices [57] and by properly weighting the cost functions [49]. In order to give a fair comparison, all the algorithms operate on the same set of matrices in all the experiments of this paper. In addition, the average numerical complexity and CPU time at each condition number of all the methods in this experiment are shown in the bottom of Figure6. It shows that the proposed methods give the best performance/complexity trade-off compared to $\text{ACDC}_{\text{LU}}^{+}$ whatever the condition number is.

Test with a Non-square Matrix A

As described in the section of practical issues, when a non-square matrix $\boldsymbol{A} \in \mathbb{R}_{+}^{N\times P}$ with N>P is met in ICA, we propose to compress it by a nonnegative compression matrix $\boldsymbol{W}_{+} \in \mathbb{R}_{+}^{P\times N}$ [53], such that the resulting matrix $\bar{A} = W_{+}A$ is a (P×P) nonnegative square matrix. Then, the proposed methods can be applied to estimate $\bar{A}$. Similar to the classical prewhitening, the nonnegative compression step could introduce numerical errors. In this experiment, we compare our methods to ACDC and $\text{ACDC}_{\text{LU}}^{+}$ through a simulated ICA model. The latter algorithms can directly estimate a non-square matrix $\boldsymbol{A}$ from the fourth-order cumulant matrix slices. The ICA model is established as follows:

$$\boldsymbol{x}[f] = \boldsymbol{A}\boldsymbol{s}[f] + \boldsymbol{v}[f] \tag{45}$$

where $\boldsymbol{x}[f] = [x_1[f], \cdots, x_N[f]]^{\mathrm{T}}$ observation vector, $\boldsymbol{s}[f] = [s_1[f], s_2[f], s_3[f]]^{\mathrm{T}}$ is the (3 × 1) zero-mean unit-variance source vector whose elements are independently drawn from a uniform distribution over $[-\sqrt{3}, \sqrt{3}]$, $\boldsymbol{v} = [v_1[f], \cdots, v_N[f]]^{\mathrm{T}}$ is the (N× 1) zero-mean unit-variance Gaussian noise vector, and $\boldsymbol{A}$ is the (N× 3) nonnegative mixing matrix whose components are independently drawn from a uniform distribution over [0,1]. In this context, the SNR is defined by:

$$\text{SNR} = 20\log_{10}(\|\{\boldsymbol{A}\boldsymbol{s}[f]\}\|_F / \|\{\boldsymbol{v}[f]\}\|_F) \tag{46}$$

For the proposed $\text{JD}_{\text{LU}}^{+}$ and $\text{JD}_{\text{QR}}^{+}$ algorithms, the given realization of $\{\boldsymbol{x}[f]\}$ is compressed by means of a matrix $\boldsymbol{W}_{+} \in \mathbb{R}_{+}^{3\times N}$ computed using the method proposed in [53], leading to a three-dimensional vector $\{\bar{x}[f]\}$. We compute the fourth-order cumulant array of $\{\bar{x}[f]\}$ and choose the first three matrix slices in order to build a three-way array. Hence, $\text{JD}_{\text{LU}}^{+}$ and $\text{JD}_{\text{QR}}^{+}$ decompose a (3×3×3) array. Once the compressed mixing matrix $\bar{A}$ is estimated, the original mixing matrix is obtained by Eq.39. Regarding ACDC and $ACDC^{+}_{LU}$, the fourth-order cumulant array of $\{\boldsymbol{x}[f]\}$ is directly computed without compression. We apply ACDC and $ACDC^{+}_{LU}$ on two three-way arrays with different third dimensions. The first array of dimension (N×N× 3) is built by choosing the first three matrix slices from the fourth-order cumulant array, while the second array of dimension (N×N×N) is built using the first N matrix slices. We study the impact of the number of observations N on the performance of the JDC algorithms, by varying N from 4 to 24. The SNR value is fixed to 5 dB. The number of samples used to estimate the cumulants is set to 10^3. Figure 7 shows the average curves of the estimating error $\alpha(A, \tilde{A})$ of all the algorithms as a function of N. As it can be seen, when $N \leq 15$, the larger the value of N, the more accurate estimation of $\boldsymbol{A}$ is achieved. When $N > 15$, the further increase of N does not bring significant improvement in terms of the estimation accuracy. ACDC and $ACDC^{+}_{LU}$ give better results when the array with a larger third dimension is considered. Their results on (N×N×N) arrays outperform the proposed methods when $N = 4$. $ACDC^{+}_{LU}$ also gives the best estimation on (N×N×N) arrays with $N = 5$. It suggests that the numerical errors introduced by the compression step limit the performance of the proposed methods when only a small number of observation is available. Such a negative effect can be partially compensated by using a large number of observations, since the proposed JD^{+}_{LU} and JD^{+}_{QR} methods maintain the highest performance in terms of estimation accuracy when $N \geq 6$. The performance ACDC and $ACDC^{+}_{LU}$ can be further improved by using a $(N \times N \times N^2)$ array, which contains all the N^2 fourth-order cumulant matrix slices. However, it leads to a higher numerical complexity especially for a large value of N. Regarding the proposed JD^{+}_{LU} and JD^{+}_{QR} methods, their performance can also be improved by using all the nine matrix slices of the fourth-order cumulant array of the compressed observation vector. Nevertheless, the experimental result has already shown that by using only a small number of matrix slices, JD^{+}_{LU} and JD^{+}_{QR} can maintain lower numerical complexities than ACDC and $ACDC^{+}_{LU}$, while achieving better estimation results, when a large value of N is considered. Therefore, despite the negative influence of the nonnegative compression, the proposed methods still offer a good performance/complexity compromise to estimate a non-square matrix $\boldsymbol{A}$.

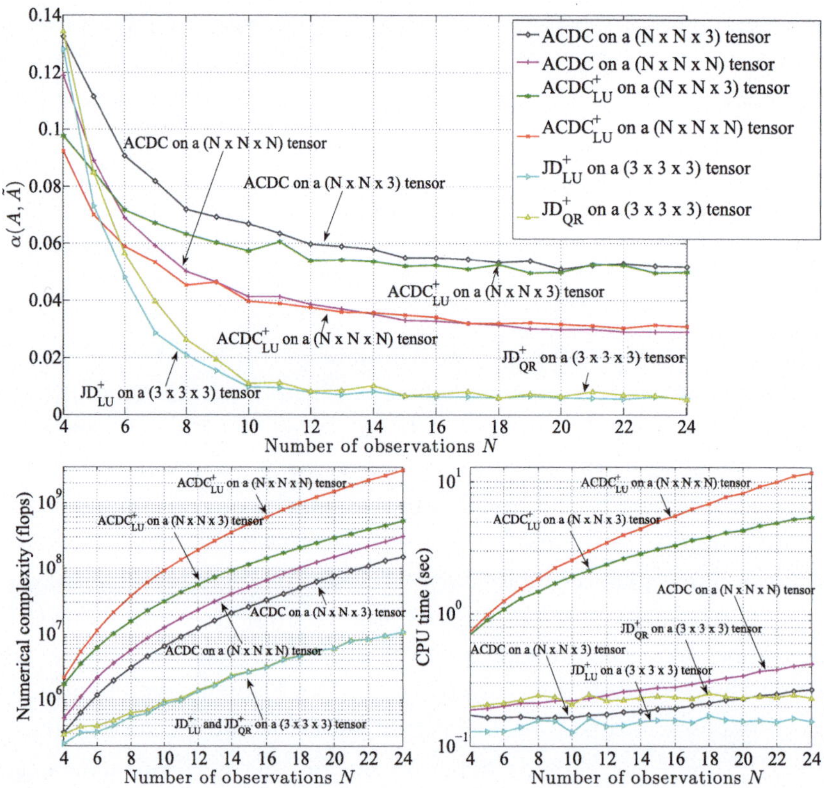

Figure 7: JDC performance on an ICA model versus number of observations. The number of sources$P$and the SNR value are set to$P$= 3 and SNR = 5 dB, respectively. Top: the average error$\alpha(A, \tilde{A})$evolution of all the algorithms as a function of the number of observations$N$. Bottom: the average numerical complexities (left) and the CPU time (right) of all the algorithms, respectively.

BSS Application on MRS Data

In this section, we aim to illustrate the potential capability of the proposed $JD^{+}_{LU}$and JD^{+}_{QR}algorithms for solving a real-life BSS problem by an application carried on simulated MRS data.

MRS is a powerful non-invasive analytical technique for analyzing the chemical content of MR-visible nuclei and therefore enjoys particular advantages for assessing metabolism. The chemical property of each nucleus determines the frequency at which it appears in the MR spectrum, giving rise to peaks corresponding to specific metabolites [58]. Therefore, the MRS observation spectra can be modeled as the mixture of the spectrum of each constitutional source metabolite. More specifically, it can be written as the noisy linear instantaneous mixing model described in Eq.45, where $\boldsymbol{x}[f] \in \mathbb{R}^N$ is the MRS observation vector, $\boldsymbol{s}[f] \in \mathbb{R}^P$ is the source vector representing the statistically quasi-independent source metabolites, $\boldsymbol{\nu} \in \mathbb{R}^N$ is the instrumental noise vector, and $\boldsymbol{A} \in \mathbb{R}_+^{N \times P}$ is the nonnegative mixing matrix containing the positive concentrations of the source metabolites. SNR is defined as in Eq.46. In this experiment, two simulated MRS source metabolites $\{s_1[f]\}$ and $\{s_2[f]\}$, namely the Choline (Cho) and Myo-inositol (Ins) (see Figure8b), are generated by Lorentzian and Gaussian functions [59]. Each of the sources contains 10^3samples. The observation vector$\boldsymbol{x}$[f] is generated according to (45). The components of the (N×2) mixing matrix$\boldsymbol{A}$are randomly drawn from a uniform distribution. The additive noise$\boldsymbol{v}$[f] is modeled as a zero-mean unit-variance Gaussian vector. The ICA methods based on the proposed JD_{LU}^{+}andJD_{QR}^{+}algorithms, namely JD_{LU}^{+}-ICA and JD_{QR}^{+}-ICA, consist of four steps: i) compressing $\{\boldsymbol{x}[f]\}$ by means of a matrix $W_+ \in \mathbb{R}_+^{2 \times N}$ [53], ii) estimating the fourth-order cumulant array of the compressed observations and stacking all the cumulant matrix slices in a three-way array, iii) decomposing the resulting three-way array by means of JD_{LU}^{+}andJD_{QR}^{+}, respectively, and iv) reconstructing the sources. The JD_{LU}^{+}-ICA and JD_{QR}^{+}-ICA are compared to four state-of-the-art BSS algorithms, namely two efficient ICA methods CoM_2[54] and SOBI [47], the nonnegative ICA (NNICA) method with a line search along the geodesic [55], and the NMF method [56] based on alternating nonnegativity least squares. The performance is assessed by means of the error $\alpha(\{\boldsymbol{s}[f]\}^{\mathsf{T}}, \{\tilde{\boldsymbol{s}}[f]\}^{\mathsf{T}})$ between the true sources$\boldsymbol{s}$[f] and its estimate $\tilde{\boldsymbol{s}}[f]$, the numerical complexity, and the CPU time. To find out the detailed analysis of the numerical complexity of the classical ICA algorithms, the reader can refer to the book chapter [60]. Figure8shows an example of the separation results of all the methods withN= 32 observations and a SNR of 10 dB. Regarding CoM_2, SOBI, NNICA, and NMF, there are some obvious disturbances presented in the estimated metabolites. As far as JD_{LU}^{+}-ICA and JD_{QR}^{+}-ICA are concerned, the estimated source metabolites are quasi-perfect. Furthermore, the comprehensive performance of all the methods will be studied by the following experiments with 200 independent Monte Carlo trials.

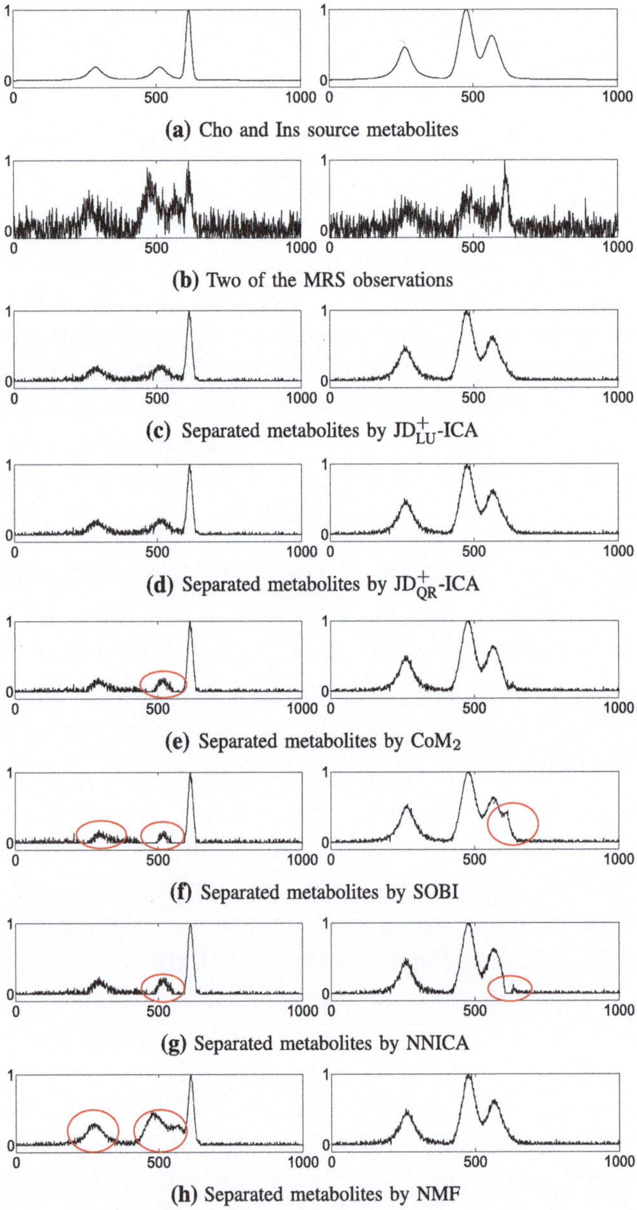

Figure 8: BSS results on MRS data.An example of the results of blind separation of two simulated MRS metabolites. The number of observationsNis set to 32, and the SNR value is 10 dB.(a)Cho and Ins source metabolites.(b)Two of the observations.**(c-h)**Separated metabolites by JD^{+}_{LU}-ICA, JD^{+}_{QR}-ICA, CoM_2, SOBI, NNICA, and NMF, respectively.

In the first experiment, the effect of the number of observationsNis evaluated. The SNR is fixed to 10 dB. The six methods are compared withNranging from 4 to 116 with a step of 4. The average curves of error $\alpha(\{s[f]\}^{T}, \{\tilde{s}[f]\}^{T})$as a function ofNare shown in the left image of Figure9. It can be seen that the estimating errors of all the methods improve asNincreases. It suggests that in noisy BSS contexts, using more sensors often yields better results. The proposed JD^{+}_{LU}-ICA and JD^{+}_{QR}-ICA methods maintain the competitive advantages. The average curves of the numerical complexities of this experiment are shown in the bottom left picture of Figure9. We can notice that the numerical complexities of all the methods increase withN. The complexities of JD^{+}_{LU}-ICA and JD^{+}_{QR}-ICA seem identical in the logarithmic scaled plot, which is because theoretically their complexities are mainly dominated by the computation of the nonnegative compression step and of the cumulants. Indeed, JD^{+}_{LU}-ICA is more computationally efficient than JD^{+}_{QR}-ICA in the step of CP decomposition of the cumulant array. This can be verified by the average CPU time of those methods, shown in the bottom right image of Figure9. We can observe that JD^{+}_{LU}-ICA is slower than CoM_2, but it is faster than NNICA, SOBI, and NMF.

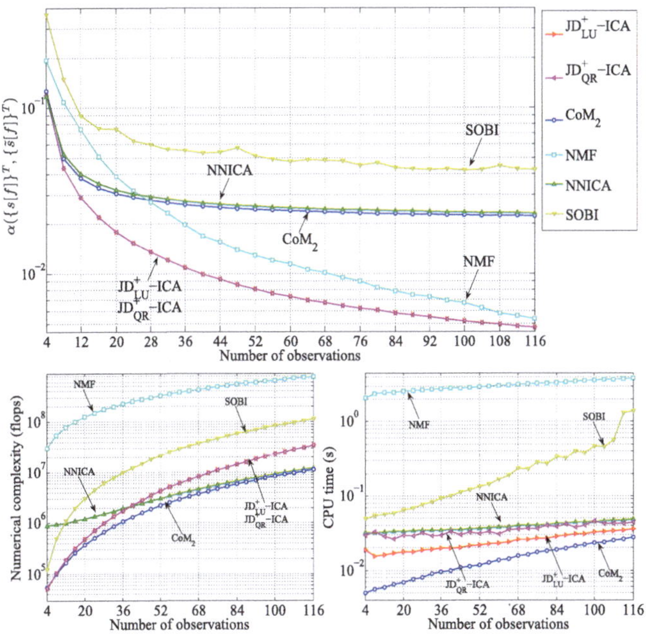

Figure 9: BSS performance on MRS data versus the number of observations.

Average results of blind separation of two simulated MRS metabolites. The SNR value is set to 10 dB. Left: the average error $\alpha(\{s[f]\}^{\mathrm{T}}, \{\tilde{s}[f]\}^{\mathrm{T}})$ evolution of all the algorithms as a function of the number of observations. Right: the average numerical complexities (top) and the CPU time (bottom) of all the algorithms, respectively.

In the second experiment, we study the influence of SNR on the performance of the six methods. The number of observationsNis set to 32. SNR is varied from 0 to 50 dB with a step of 2 dB. The average curves of the estimating error $\alpha(\{s[f]\}^{\mathrm{T}}, \{\tilde{s}[f]\}^{\mathrm{T}})$ as well as those of the numerical complexities and CPU time as a function of SNR of all the six methods are shown in Figure10. The proposed $\mathrm{JD}^{+}_{\mathrm{LU}}$-ICA and $\mathrm{JD}^{+}_{\mathrm{QR}}$-ICA methods provide the best estimation results with moderate computational complexities and CPU time. Generally speaking, the $\mathrm{JD}^{+}_{\mathrm{LU}}$-ICA algorithm offers the best performance/ complexity trade-off in this BSS experimental context.

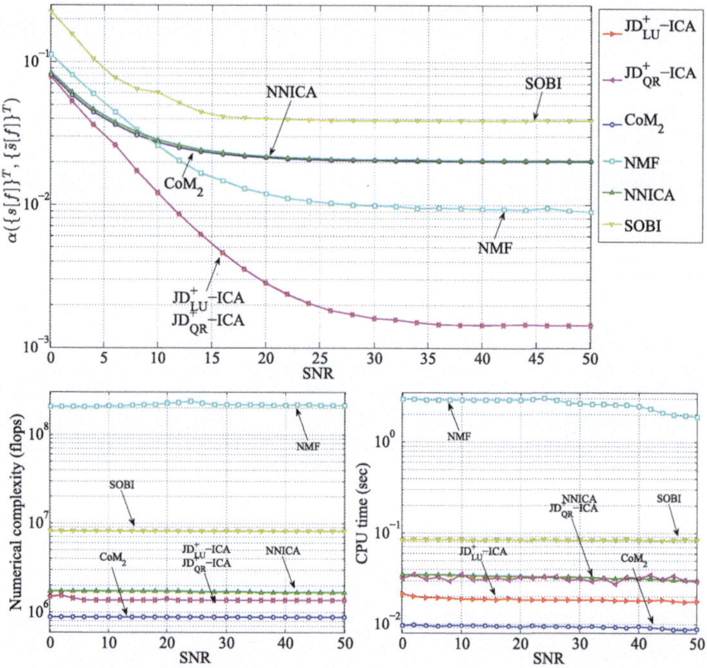

Figure 10: BSS performance on MRS data versus SNR.Average results of blind separation of two simulated MRS metabolites. The number of observations is set toN=32. Left: the average error $\alpha(\{s[f]\}^{\mathrm{T}}, \{\tilde{s}[f]\}^{\mathrm{T}})$ evolution of all the algorithms as a function of SNR. Right: the average numerical complexities (top) and the CPU time (bottom) of all the algorithms, respectively.

CONCLUSIONS

We have proposed two methods, called JD^{+}_{LU} and JD^{+}_{QR}, in order to achieve the CP decomposition of semi-nonnegative semi-symmetric three-way arrays. The nonnegativity constraint is imposed on the two symmetric modes of three-way arrays by means of a square change of variable, giving rise to an unconstrained joint diagonalization by congruence problem. Therefore, the nonnegative loading matrix can be estimated by computing the joint diagonalizer. We consider the elementary LU and QR parameterizations of the Hadamard square root of the nonnegative joint diagonalizer, leading to two Jacobilike optimization procedures. In each Jacobi-like iteration, the optimization is formulated into a minimization of a polynomial or rational function with respect to only one parameter. In addition, the numerical complexity for each algorithm has been analyzed.

The performance of the proposed JD^{+}_{LU} and JD^{+}_{QR}algorithms is evaluated with simulated semi-nonnegative semi-symmetric three-way arrays. Four classical nonorthogonal JDC methods without nonnegativity constraints including ACDC [23], FFDIAG [25], LUJ1D [26], and QRJ1D [26] and one nonnegative JDC method $ACDC^{+}_{LU}$ [41] are tested as reference methods. The performance is assessed in terms of the matrix estimation accuracy, the numerical complexity, and the CPU time. The convergence property, the influence of SNR, the impact of dimension, the effect of coherence, and the influence of condition number are extensively studied by Monte Carlo experiments. The obtained results show that the proposed algorithms offer better estimation accuracy by means of exploiting the nonnegativitya priori. The JD^{+}_{LU} algorithm provides the best performance/complexity compromise.

The proposed algorithms are suitable tools for solving the ICA problems where a nonnegative mixing matrix is considered, such as in MRS. In this case, the three-way array built by stacking the matrix slices of a HO cumulant array fulfills the semi-nonnegative semi-symmetric structure. We proposed two ICA methods, namely JD^{+}_{LU}-ICA and JD^{+}_{QR}-ICA, based on CP decomposition of the fourth-order cumulant array using JD^{+}_{LU} and JD^{+}_{QR}, respectively. The source separation ability of the proposed algorithms is verified through a BSS application carried out on simulated MRS data. The JD^{+}_{LU}-ICA and JD^{+}_{QR}-ICA are compared to one NMF method [56], one nonnegative ICA method [55], and two classical ICA methods, namely CoM_2[54] and SOBI [47]. The performance is comprehensively studied as a function of the number of observations and of SNR. The experimental results demonstrate the improvement of the proposed methods in terms of the source estimation accuracy and also show that exploiting the twoa prioriof the data, namely the nonnegativity of the mixing matrix and the statistical independence of the sources, allows us to achieve better estimation results. The JD^{+}_{LU}-ICA algorithm provides the best performance/complexity trade-off.

REFERENCES

1. Smilde A, Bro R, Geladi P:Multi-way Analysis: Applications in the Chemical Sciences. Wiley, West Sussex; 2004.
2. de Almeida ALF, Favier G, Ximenes LR: Space-time-frequency (STF) MIMO communication systems with blind receiver based on a generalized PARATUCK2 model.IEEE Trans. Signal Process2013, 61(8):1895-1909.
3. De Vos M, Vergult A, De Lathauwer L, De Clercq W, Van Huffel S, Dupont P, Palmini A, Van Paesschen W: Canonical decomposition of ictal scalp EEG reliably detects the seizure onset zone.Neuroimage2007, 37(3):844-854. 10.1016/j.neuroimage.2007.04.041
4. Comon P, Luciani X, de Almeida ALF: Tensor decompositions, alternating least squares and other tales.J. Chemometr2009, 23: 393-405. 10.1002/cem.1236
5. Tucker LR: Some mathematical notes on three-mode factor analysis. Psychometrika1966, 31(3):279-311. 10.1007/BF02289464
6. De Lathauwer L, De Moor B, Vandewalle J: A multilinear singular value decomposition.SIAM J. Matrix Anal. Appl2000, 21(4):1253-1278. 10.1137/S0895479896305696
7. Kruskal JB: Three-way arrays: rank and uniqueness of trilinear decompositions, with application to arithmetic complexity and statistics.Lin. Algebra Appl1977, 18(2):98-138.
8. Hitchcock FL: The expression of a tensor or a polyadic as a sum of products.J. Math. Phys1927, 6(1):164-189.
9. Kroonenberg PM:Applied Multiway Data Analysis. Wiley, Hoboken; 2008.
10. Sidiropoulos ND, Bro R, Giannakis GB: Parallel factor analysis in sensor array processing.IEEE Trans. Signal Process2000, 48(8):2377-2388. 10.1109/78.852018
11. de Almeida ALF, Favier G, Motab JCM: PARAFAC-based unified tensor modeling for wireless communication systems with application to blind multiuser equalization.Signal Process2007, 87(2):337-351. 10.1016/j.sigpro.2005.12.014
12. de Almeida ALF, Favier G: Double Khatri-Rao space-time-frequency coding using semi-blind PARAFAC based receiver.IEEE Signal Process. Lett2013, 20(5):471-474.

13. Albera L, Ferréol A, Comon P, Chevalier P: Blind identification of overcomplete mixtures of sources (BIOME).Lin. Algebra Appl2004, 391: 3-30.

14. Röemer F, Haardt M: Tensor-based channel estimation and iterative refinements for two-way relaying with multiple antennas and spatial reuse.IEEE Trans. Signal Process2010, 58(11):5720-5735.

15. Harshman RA, Lundy ME: PARAFAC: parallel factor analysis.Comput. Stat. Data Anal1994, 18(1):39-72. 10.1016/0167-9473(94)90132-5

16. Uschmajew A: Local convergence of the alternating least squares algorithm for canonical tensor approximation.SIAM. J. Matrix Anal. Appl2012, 33(2):639-652. 10.1137/110843587

17. Rajih M, Comon P, Harshman RA: Enhanced line search: a novel method to accelerate PARAFAC.SIAM J. Matrix Anal. Appl2008, 30(3):1128-1147. 10.1137/06065577

18. Acar E, Dunlavy DM, Kolda TG: A scalable optimization approach for fitting canonical tensor decompositions.J. Chemometr2011, 25(2):67-86. 10.1002/cem.1335

19. Röemer F, Haardt M: A semi-algebraic framework for approximate CP decompositions via simultaneous matrix diagonalizations (SECSI). Signal Process2013, 93(9):2722-2738. 10.1016/j.sigpro.2013.02.016

20. Luciani X, Albera L: Canonical polyadic decomposition based on joint eigenvalue decomposition.Chemometr. Intell. Lab2014, 132: 152-167.

21. Carroll JD, Chang J-J: Analysis of individual differences in multidimensional scaling via an n-way generalization of Eckart-Young decomposition.Psychometrika1970, 35(3):283-319. 10.1007/BF02310791

22. Husson F, Pagés J: INDSCAL model: geometrical interpretation and methodology.Comput. Stat. Data Anal2006, 50(2):358-378. 10.1016/j.csda.2004.08.005

23. Yeredor A: Non-orthogonal joint diagonalization in the least-squares sense with application in blind source separation.IEEE Trans. Signal Process2002, 50(7):1545-1553. 10.1109/TSP.2002.1011195

24. Cardoso JF, Souloumiac A: Jacobi angles for simultaneous diagonalization.SIAM J. Matrix Anal. Appl1996, 17: 161-164. 10.1137/S0895479893259546

25. Ziehe A, Laskov P, Nolte G, Muller K-R: A fast algorithm for joint diagonalization with non-orthogonal transformations and its application

to blind source separation.J. Mach. Learn. Res2004, 5: 777-800.

26. Afsari B: Simple LU and QR based non-orthogonal matrix joint diagonalization. InICA 2006, Springer LNCS 3889. Charleston, SC, USA; 5–8 March 2006.
27. Van der Veen AJ: Joint diagonalization via subspace fitting techniques. InProc. ICASSP '01. Salt Lake, City, UT; 7–11 May 2001:2773-2776.
28. Yeredor A: On using exact joint diagonalization for noniterative approximate joint diagonalization.IEEE Signal Process. Lett2005, 12(9):645-648.
29. Vollgraf R, Obermayer K: Quadratic optimization for simultaneous matrix diagonalization.IEEE Trans. Signal Process2006, 54(9):3270-3278.
30. Li XL, Zhang XD: Nonorthogonal joint diagonalization free of degenerate solution.IEEE Trans. Signal Process2007, 55(5):1803-1814.
31. Souloumiac A: Nonorthogonal joint diagonalization by combining Givens and hyperbolic rotations.IEEE Trans. Signal Process2009, 57(6):2222-2231.
32. Xu XF, Feng DZ, Zheng WX: A fast algorithm for nonunitary joint diagonalization and its application to blind source separation.IEEE Trans. Signal Process2011, 59(7):3457-3463.
33. Chabriel G, Barrère J: A direct algorithm for nonorthogonal approximate joint diagonalization.IEEE Trans. Signal Process2012, 60(1):39-47.
34. Chabriel G, Kleinsteuber M, Moreau E, Shen H, Tichavský P, Yeredor A: Joint matrices decompositions and blind source separation: A survey of methods, identification, and applications.IEEE Signal Process. Mag2014, 31(3):34-43.
35. Lee DD, Seung HS: Learning the parts of objects by non-negative matrix factorization.Nature1999, 401(6755):788-791. 10.1038/44565
36. Zhang Q, Wang H, Plemmons RJ, Pauca VP: Tensor methods for hyperspectral data analysis: a space object material identification study.J. Opt. Soc. Am. A. Opt. Image Sci. Vis2008, 25(12):3001-3012. 10.1364/JOSAA.25.003001
37. Royer J-P, Thirion-Moreau N, Comon P: Computing the polyadic decomposition of nonnegative third order tensors.Signal Process2011, 91(9):2159-2171. 10.1016/j.sigpro.2011.03.006
38. Cichocki A, Zdunek R, Phan AH, Amari S:Nonnegative Matrix and

Tensor Factorizations: Applications to Exploratory Multi-way Data Analysis and Blind Source Separation. Wiley, West Sussex; 2009.

39. Zhou GX, Cichocki A, Zhao Q, Xie SL: Nonnegative matrix and tensor factorizations : an algorithmic perspective.IEEE Signal Process. Mag2014, 31(3):54-65.

40. Coloigner J, Karfoul A, Albera L, Comon P: Line search and trust region strategies for canonical decomposition of semi-nonnegative semi-symmetric 3rd order tensors.Lin. Algebra Appl2014, 450(1):334-374.

41. Wang L, Albera L, Kachenoura A, Shu HZ, Senhadji L: Nonnegative joint diagonalization by congruence based on LU matrix factorization. IEEE Signal Process. Lett2013, 20(8):807-810.

42. Wang L, Albera L, Kachenoura A, Shu HZ, Senhadji L: CP decomposition of semi-nonnegative semi-symmetric tensors based on QR matrix factorization. InSAM'14, Proceedings of the Eighth IEEE Sensor Array and Multichannel Signal Processing Workshop. A Coruna, Spain; 22–25 June 2014:449-452.

43. Afsari B: Sensitivity analysis for the problem of matrix joint diagonalization.SIAM J. Matrix Anal. Appl2008, 30(3):1148-1171. 10.1137/060655997

44. Wax M, Sheinvald J: A least-squares approach to joint diagonalization. IEEE Signal Process. Lett1997, 4(2):52-53.

45. Dégerine S, Kane E: A comparative study of approximate joint diagonalization algorithms for blind source separation in presence of additive noise.IEEE Trans. Signal Process2007, 55(6):3022-3031.

46. Fadaili EM, Thirion-Moreau N, Moreau E: Nonorthogonal joint diagonalization/zero diagonalization for source separation based on time-frequency distributions.IEEE Trans. Signal Process2007, 55(5):1673-1687.

47. Belouchrani A, Abed-Meraim K, Cardoso JF, Moulines E: A blind source separation technique using second-order statistics.IEEE Trans. Signal Process1997, 45(2):434-444. 10.1109/78.554307

48. Pham DT: Joint approximate diagonalization of positive definite Hermitian matrices.SIAM J. Matrix Anal. Appl2001, 22: 1837-1848.

49. Tichavský P, Yeredor A: Fast approximate joint diagonalization incorporating weight matrices.IEEE Trans. Signal Process2009, 57(3):878-891.

50. Chu M, Diele F, Plemmons R, Ragni S: Optimality computation and interpretation of nonnegative matrix factorizations. Technical report, Wake Forest University 2004
51. Meyer CD:Matrix Analysis and Applied Linear Algebra. SIAM, Philadelphia; 2000.
52. Vaidyanathan PP:Multirate Systems and Filter Banks. PTR Prentice Hall, United States; 1993.
53. Wang L, Kachenoura A, Albera L, Karfoul A, Shu HZ, Senhadji L: Nonnegative compression for semi-nonnegative independent component analysis. InSAM'14, Proceedings of the Eighth IEEE Sensor Array and Multichannel Signal Processing Workshop. A Coruna, Spain; 22–25 June 2014:81-84.
54. Comon P: Independent component analysis, a new concept?Signal Process1994, 36(3):287-314. 10.1016/0165-1684(94)90029-9
55. Plumbley MD: Algorithms for nonnegative independent component analysis.IEEE Trans. Neural Netw2003, 14(3):534-543. 10.1109/TNN.2003.810616
56. Kim H, Park H: Nonnegative matrix factorization based on alternating nonnegativity constrained least squares and active set method.SIAM J. Matrix Anal. Appl2008, 30(2):713-730. 10.1137/07069239X
57. De Lathauwer: Algebraic methods after prewhitening. InHandbook of Blind Source Separation, ed. by P Comon, C Jutten,. Elsevier, Oxford; 2010:155-177. Chap. 5
58. Befroy DE, Shulman GI: Magnetic resonance spectroscopy studies of human metabolism.Diabetes2011, 60(5):1361-1369. 10.2337/db09-0916
59. Moussaoui S: Séparation de sources non-négatives: application au traitement des signaux de spectroscopie. PhD thesis, Université Henri Poincaré, (2005)
60. Albera L, Comon P, Parra LC, Karfoul A, Kachenoura A, Senhadji L: Biomedical applications. InHandbook of Blind Source Separation ed. by P Comon, C Jutten,. Elsevier, Oxford; 2010:737-777. Chap. 18

Chapter 7

Sparse Signal Subspace Decomposition based on Adaptive Over-complete Dictionary

Hong Sun[1,2] , Cheng-wei Sang[1] and Didier Le Ruyet[3]

[1]School of Electronic Information, Wuhan University, Luojia Hill, Wuhan 430072, China

[2]Signal and Image Processing Department, Telecom ParisTech, 46 rue Barrault, 75013 Paris, France

[3]CEDRIC Laboratory, CNAM, 292 rue Saint Martin, 75003 Paris, France

ABSTRACT

This paper proposes a subspace decomposition method based on an over-complete dictionary in sparse representation, called "sparse signal subspace decomposition" (or 3SD) method. This method makes use of a novel criterion based on the occurrence frequency of atoms of the dictionary over the data set. This criterion, well adapted to subspace decomposition over a dependent basis set, adequately reflects the intrinsic characteristic of

Citation (APA): Sun, H., Sang, C. W., & Le Ruyet, D. (2017). Sparse signal subspace decomposition based on adaptive over-complete dictionary. EURASIP Journal on Image and Video Processing, 2017(1), 50. (10 pages), DOI: https://doi.org/10.1186/s13640-017-0200-7

regularity of the signal. The 3SD method combines variance, sparsity, and component frequency criteria into a unified framework. It takes benefits from using an over-complete dictionary which preserves details and from subspace decomposition which rejects strong noise. The 3SD method is very simple with a linear retrieval operation. It does not require any prior knowledge on distributions or parameters. When applied to image denoising, it demonstrates high performances both at preserving fine details and suppressing strong noise.

Keywords: Subspace decomposition, Sparse representation, Frequency of components, PCA, K-SVD, Image denoising

INTRODUCTION

Signal subspace methods (SSMs) are efficient techniques to reduce the dimensionality of data and to filter out noise [1]. The fundamental idea under SSM is to project the data on a basis made of two subspaces, one mostly containing the signal and the other the noise. The two subspaces are separated by a thresholding criterion associated with some measures of information.

The two most popular methods of signal subspace decomposition are wavelet shrinkage [2] and principal component analysis (PCA) [3]. Both techniques have proved to be quite efficient. However, wavelet decomposition depending on signal statistics is not equally adapted to different data and requires some knowledge on prior distributions or parameters of signals to efficiently choose the thresholds for shrinkage. A significant advantage of the PCA is its adaptability to data. The separation criterion is based on energy which may be seen as a limitation in some cases as illustrated in the next section.

In recent years, sparse coding has attracted significant interest in the field of signal denoising [4]. A sparse representation is a decomposition of a signal on a very small set of components of an over-complete basis (called dictionary) which is adapted to the processed data. A difficult aspect for signal subspace decomposition based on such a sparse representation is to define the most appropriate criterion to identify the principal components (called atoms) from the learned dictionary to build the principal subspace. The non-orthogonal property of the dictionary does not allow to use the energy criterion for this purpose, as done with PCA.

To solve this problem, we introduce a new criterion to measure the importance of atoms and propose a SSM under the criterion of the occurrence frequency of atoms. We thus make benefit both from the richness of over-complete dictionaries which preserves details of information and from signal subspace decomposition which rejects strong noise.

The remainder of this paper is organized as follows: Section 2 presents two related works to signal decomposition. Section 3 introduces the proposed sparse signal subspace decomposition based on the adaptive over-complete dictionary. Some experimental results and analysis are presented in Section 4. Finally, we draw the conclusion in Section 5.

REVIEW OF PCA AND SPARSE CODING METHODS

We start with a brief description of two well-established approaches to signal decomposition that are relevant and related to the approach proposed in the next section.

PCA-based Subspace Decomposition

The basic tool of SSM is principal component analysis (PCA). PCA makes use of an orthonormal basis to capture on a small set of vectors (the signal subspace) as much energy as possible from the observed data. The other basis vectors are expected to contain noise only, and the signal projection on these vectors is rejected.

Consider a data set $\{\mathbf{x}_m \in \mathbb{R}^{N\times 1}\}_{m=1}^{M}$ grouped in a matrix form $\mathbf{X}$ of size N×M: $\mathbf{X} = \{\mathbf{x}_m\}_{m=1}^{M}$. The PCA is based on singular value decomposition (SVD) with singular values σ_i in descending order obtained from:

$$\mathbf{X} = \mathbf{U}\mathbf{A} = \mathbf{U}\Sigma\mathbf{V}^T \quad (1)$$

where $\mathbf{U}$ and $\mathbf{V}$ are unitary matrices of sizes N×N and M×M, respectively ($\mathbf{U}^T\mathbf{U}=\mathbf{I}_N$, $\mathbf{V}^T\mathbf{V}=\mathbf{I}_M$), and $\Sigma=\begin{bmatrix}\operatorname{diag}[\sigma_1,\cdots,\sigma_r],\mathbf{0}\\ \mathbf{0}\end{bmatrix}$ of size N×M with $\sigma_1\geq\sigma_2\geq\cdots\geq\sigma_r>0$, $\sigma_r > 0$, $\{\sigma_i\}_{i=1}^{r}$ are positive real known as the singular values of $\mathbf{X}$ with rank r ($r\leq N$).

Equation (1) can be rewritten in a vector form as:

$$[\mathbf{x}_1 \quad \mathbf{x}_2\cdots\mathbf{x}_m\cdots\mathbf{x}_M] = [\mathbf{u}_1 \quad \mathbf{u}_2\cdots\mathbf{u}_n\cdots\mathbf{u}_N].[\boldsymbol{\alpha}_1 \quad \boldsymbol{\alpha}_2\cdots\boldsymbol{\alpha}_m\cdots\boldsymbol{\alpha}_M] \quad (2)$$

where $\mathbf{U} = \{\mathbf{u}_n \in \mathbb{R}^{N\times 1}\}_{n=1}^{N}$ and $\mathbf{A} = \{\alpha_m \in \mathbb{R}^{N\times 1}\}_{m=1}^{M}$. Equation (2) means that the data set $\{\mathbf{x}_m\}_{m=1}^{M}$ is expressed on the orthonormal basis $\{\mathbf{u}_n\}_{n=1}^{N}$ as $\{\alpha_m\}_{m=1}^{M}$.

In the SVD decomposition given in Eq. (1), the standard deviation σ_i is used as the measurement for identifying the meaningful basis vector $\mathbf{u}_i$. PCA takes the first P(P<r) components $\{\mathbf{u}_n\}_{n=1}^{P}$ to span the signal subspace, and the remainders $\{\mathbf{u}_n\}_{n=P+1}^{r}$ are considered in a noise subspace orthogonal to the signal subspace. Therefore, projection on the signal subspace will hopefully filter out noise and reveal hidden structures. The reconstructed signal $\hat{\mathbf{S}}_{\mathrm{PCA}}$ of size N×M is obtained by projecting in the signal subspace as:

$$\hat{\mathbf{S}}_{\mathrm{PCA}} = [\mathbf{u}_1 \cdots \mathbf{u}_P \quad \mathbf{0}_{P+1} \cdots \mathbf{0}_N] \cdot [\boldsymbol{\alpha}_1 \quad \boldsymbol{\alpha}_2 \cdots \boldsymbol{\alpha}_m \cdots \boldsymbol{\alpha}_M] \tag{3}$$

The underlying assumption is that information in the data set is almost completely contained in a small linear subspace of the overall space of possible data vectors, whereas additive noise is typically distributed through the larger space isotropically. PCA, using the standard deviation as a criterion, implies that the components of the signal of interest in the data set have a maximum variance and the other components are mainly due to the noise. However, in many practical cases, some components with low variances might actually be important because they carry information relative to the signal details. On the contrary, when dealing with noise with non-Gaussian statistics, it may happen that some noise components may actually have higher variances. At last, note that it is often difficult to provide a physical meaning to the orthonormal basis $\{\mathbf{u}_i\}_{i=1}^{r}$ of the SVD decomposition (Eq. (2)) although they have a very clear definition in the mathematical sense as orthogonal, independent, and normal. It is therefore difficult to impose known constraints on the signal features when they exist after the principal component decomposition.

Sparse Decomposition

Recent years have shown a growing interest in research on the sparse decomposition of M observations $\{\mathbf{x}_m \in \mathbb{R}^N\}_{m=1}^{M}$ based on a dictionary $\mathbf{D} = \{\mathbf{d}_k\}_{k=1}^{K} \in \mathbb{R}^{N\times K}$. When K>N, the dictionary is said to be over-complete. $\mathbf{d}_k \in \mathbb{R}^N$ is a basis vector, also called an atom since it is not necessarily independent. By learning from data set $\{\mathbf{x}_m\}_{m=1}^{M}$, the sparse decomposition is the solution of Eq. (4) [4]:

$$\{\mathbf{D}, \boldsymbol{\alpha}_m\} = \underset{\mathbf{D}, \boldsymbol{\alpha}_m}{\operatorname{argmin}} \parallel \boldsymbol{\alpha}_m \parallel_0 + \parallel \mathbf{D}\boldsymbol{\alpha}_m - \mathbf{x}_m \parallel_2^2 \leq \varepsilon, \quad 1 \leq m \leq M \tag{4}$$

where $\alpha_m = [\alpha_m(1)\ \alpha_m(2)\ \ldots\ \alpha_m(K)]^T \in \mathbb{R}^{K\times 1}$ is the sparse code of the observation $\mathbf{x}_m$. The allowed error tolerance ε can be chosen according to the standard deviation of the noise. An estimate of the underlying signal $\{\mathbf{s}_m\}_{m=1}^M$ embedded in the observed data set $\{\mathbf{x}_m\}_{m=1}^M$ would be:

$$\begin{aligned} &[\hat{\mathbf{s}}_1 \quad \hat{\mathbf{s}}_2 \cdots \hat{\mathbf{s}}_m \cdots \hat{\mathbf{s}}_M] \\ &= [\mathbf{d}_1 \quad \mathbf{d}_2 \cdots \mathbf{d}_k \cdots \mathbf{d}_K] . [\boldsymbol{\alpha}_1 \quad \boldsymbol{\alpha}_2 \cdots \boldsymbol{\alpha}_m \cdots \boldsymbol{\alpha}_M] \\ &\text{or equivalently} \quad \hat{\mathbf{S}} = \mathbf{DA} \end{aligned} \tag{5}$$

where the matrix **A** of size K×M is composed of M sparse column vectors $\boldsymbol{\alpha}_m$.

The first term on the right side of Eq. (4) is a sparsity-inducing regularization that constrains the solution with the fewest number of nonzero coefficients in each of the sparse code vectors $\boldsymbol{\alpha}_m$ (1≤m≤M). The underlying assumption is that a meaningful signal could be represented by combining few atoms. This learned dictionary adapted to sparse signal descriptions has proved to be more effective in signal reconstruction and classification tasks than the PCA method, which is demonstrated in the next section. The second term in Eq. (4) is the residual of the reconstruction, based on the mean-square reconstruction error estimate in the same way as in the PCA method.

On the other hand, we note that the dictionary **D**, a basis in sparse decomposition, is produced by learning noisy data set $\{\mathbf{x}_m\}_{m=1}^M$, so the basis vectors $\{\mathbf{d}_k\}_{k=1}^K$ should be decomposed into a principal subspace and a residual subspace. However, it is impossible to exploit an energy-constrained subspace since $\{\mathbf{d}_k\}_{k=1}^K$ are not necessarily orthogonal or independent.

THE PROPOSED SPARSE SUBSPACE DECOMPOSITION

In this section, we introduce a novel criterion to the subspace decomposition over a learned dictionary and a corresponding index of significance of the atoms. Then we propose a signal sparse subspace decomposition (3SD) method under this new criterion.

Weight Vectors of Learned Atoms

At first, we intend to find out the weight of the atoms. In the sparse representation given in (5), coefficient matrix **A** is composed by M sparse column vectors α_m, each α_m representing the weight of the observation $\mathbf{x}_m$, a local parameter for the m-th observation. Let us consider the row vectors $\{\beta_k\}_{k=1}^{K}$ of coefficient matrix **A** :

$$\mathbf{A} = [\boldsymbol{\alpha}_1\ \boldsymbol{\alpha}_2\ \cdots\ \boldsymbol{\alpha}_M]$$
$$= \begin{bmatrix} \alpha_1(1) & \alpha_2(1) & \cdots & \alpha_M(1) \\ \alpha_1(2) & \alpha_2(2) & \cdots & \alpha_M(2) \\ \vdots & \vdots & \ddots & \vdots \\ \alpha_1(K) & \alpha_2(K) & \cdots & \alpha_M(K) \end{bmatrix} = \begin{bmatrix} \beta_1 \\ \beta_2 \\ \vdots \\ \beta_K \end{bmatrix}$$
$$\text{where}\quad \beta_k = [\alpha_1(k)\ \alpha_2(k)\ \ldots\ \alpha_M(k)] \in \mathbb{R}^{1\times M} \tag{6}$$

Note that the row vector $\boldsymbol{\beta}_k$ is not necessarily sparse. Then Eq. (5) can be rewritten as:

$$\hat{\mathbf{S}} = \mathbf{DA}$$
$$= [\mathbf{d}_1 \cdots \mathbf{d}_k \cdots \mathbf{d}_K] . \left[\beta_1^T \cdots \beta_k^T \cdots \beta_K^T\right]^T \tag{7}$$

Equation (7) means that the row vector $\boldsymbol{\beta}_k$ is the weight of the atom $\mathbf{d}_k$, which is a global parameter over the data set **X**. Denoting $\|\boldsymbol{\beta}_k\|_0$ the ℓ_0 zero pseudo-norm of $\boldsymbol{\beta}_k$. $\|\boldsymbol{\beta}_k\|_0$ is the number of occurrences of atom $\mathbf{d}_k$ over the data set $\{\mathbf{x}_m\}_{m=1}^{M}$. We call it the frequency of the atom $\mathbf{d}_k$ denoted by f_k:

$$f_k \triangleq \text{Frequency}(\mathbf{d}_k|\mathbf{X}) = \|\boldsymbol{\beta}_k\|_0 \tag{8}$$

In the sparse decomposition, basis vectors $\{\mathbf{d}_k\}_{k=1}^{K}$ are prototypes of signal segments. That allows us to take them as a signal patterns. Thereupon, some important features of this signal pattern could be considered as a criterion to identify significant atoms. It is demonstrated [5] that f_k is a good description of the signal texture. Intuitively, a signal pattern must occur in meaningful signals with higher frequency even with a lower energy. On the contrary, a noise pattern would hardly be reproduced in observed data even with a higher energy.

It is reasonable to take this frequency f_k as a relevance criterion to decompose the over-complete dictionary into a principal signal subspace and a remained noise subspace. Here, we use the word "subspace," but in fact, these two subspaces are not necessarily independent.

Subspace Decomposition based on Over-complete Dictionary

Taking vectors $\{\beta_k\}_{k=1}^{K}$, we calculate their ℓ^0-norms $\{\|\beta_k\|_0\}_{k=1}^{K}$ and rank them in descending order as follows. The index k of vectors $\{\beta_k\}_{k=1}^{K}$ are belonging to the set $\mathbf{C}=\{1,2,\cdots,k,\cdots,K\}$. A one-to-one index mapping function π is defined as:

$$\pi(\mathbf{C} \to \mathbf{C}) : k = \pi(\tilde{k}), \qquad k, \tilde{k} \in \mathbf{C}$$
$$s.t. \ \|\boldsymbol{\beta}_{\pi(1)}\|_0 \geq \|\boldsymbol{\beta}_{\pi(2)}\|_0 \geq \cdots \geq \|\boldsymbol{\beta}_{\pi(\tilde{k})}\|_0 \geq \cdots \geq \|\boldsymbol{\beta}_{\pi(K)}\|_0 \tag{9}$$

By the permutation π of the row index k of matrix $A=\left[\beta_1^T \cdots \beta_k^T \cdots \beta_K^T\right]^T$, the reordered coefficient matrix $\tilde{\mathbf{A}}$ becomes

$$\tilde{\mathbf{A}} = \left[\boldsymbol{\beta}_{\pi(1)}^T \quad \boldsymbol{\beta}_{\pi(2)}^T \cdots \boldsymbol{\beta}_{\pi(k)}^T \cdots \boldsymbol{\beta}_{\pi(K)}^T\right]^T \tag{10}$$

With corresponding reordered dictionary $\tilde{\mathbf{D}} = \{\mathbf{d}_{\pi(k)}\}_{k=1}^{K}$, Eq. (7) can be written as:

$$\begin{aligned}\hat{\mathbf{S}} &= \tilde{\mathbf{D}}\tilde{\mathbf{A}} \\ &= \left[\mathbf{d}_{\pi(1)} \cdots \mathbf{d}_{\pi(k)} \cdots \mathbf{d}_{\pi(K)}\right] \cdot \left[\boldsymbol{\beta}_{\pi(1)}^T \cdots \boldsymbol{\beta}_{\pi(k)}^T \cdots \boldsymbol{\beta}_{\pi(K)}^T\right]^T\end{aligned} \tag{11}$$

Then, the span of the first P atoms can be taken as a principal subspace $\mathbf{D}_P^{(\mathbf{S})}$, and the remaining atoms span a noise subspace $\mathbf{D}_{K-P}^{(N)}$ as:

$$\begin{aligned}\mathbf{D}_P^{(\mathbf{S})} &= \mathrm{span}\{\mathbf{d}_{\pi(1)}, \mathbf{d}_{\pi(2)}, \cdots, \mathbf{d}_{\pi(P)}\} \\ \mathbf{D}_{K-P}^{(N)} &= \mathrm{span}\{\mathbf{d}_{\pi(P+1)}, \mathbf{d}_{\pi(P+2)}, \cdots, \mathbf{d}_{\pi(K)}\}\end{aligned} \tag{12}$$

An estimate $\hat{\mathbf{S}}_P$ of the underlying signal $\mathbf{S}$ embedded in the observed data set $\mathbf{X}$ can be obtained on the principal subspace $\mathbf{D}_P^{(\mathbf{S})}$ simply by linear combination:

$$\begin{aligned}\hat{\mathbf{S}}_P &= \mathbf{D}_P^{(\mathbf{S})}.\mathbf{A}_P^{(\mathbf{S})} \\ &= \left[\mathbf{d}_{\pi(1)} \cdots \mathbf{d}_{\pi(k)} \cdots \mathbf{d}_{\pi(P)}\right] \cdot \left[\boldsymbol{\beta}_{\pi(1)}^T \cdots \boldsymbol{\beta}_{\pi(k)}^T \cdots \boldsymbol{\beta}_{\pi(P)}^T\right]^T\end{aligned} \tag{13}$$

Threshold of Atom's Frequency

Determining the number P of atoms spanning the signal subspace $\mathbf{D}_P^{(\mathbf{S})}$ is always a hard topic especially for wide-band signals. Here, P is the threshold of atom's frequency f_k to distinguish a signal subspace and a noise subspace. One of the advantages of 3SD is that this threshold P can be easily chosen without any prior parameter.

For a noiseless signal even with some weak details, such as the image example in Fig. 1a, the atoms' frequencies $f_{\pi(k)}^{\text{image}}$s shown in Fig. 1d (in black line) are almost always high except the zero value. For a signal with strong noise, such as the example in Fig. 1b, the atoms' frequencies $f_{\pi(k)}^{noise}$s shown in Fig. 1d (in red line) are almost always equal to 1 without zero and very few with a value 2 or 3. It is easy to set a threshold P of f_k (dotted line in the Fig. 1d) to separate the signal's atoms from the noise's atoms. By contrast, using the values of atom's energies $\|\boldsymbol{\beta}_k\|_2$s for the two images shown in Fig. 1c, it is rather a puzzle to identify principal bases.

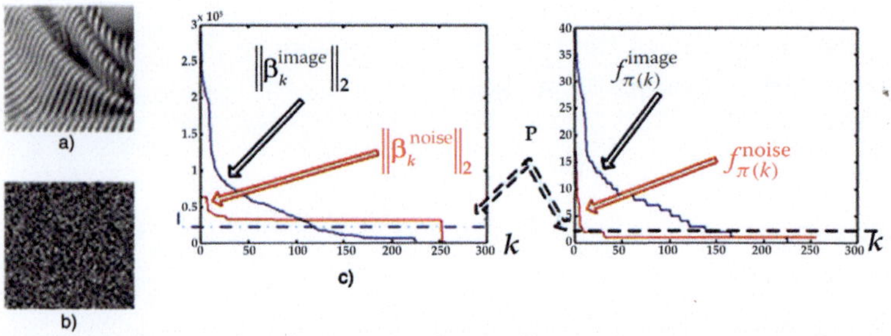

Figure 1: Sparse signal subspaces with criterion of atom's frequency. **a** Image with details. **b**White noise. **c** ℓ^2-norm of $\boldsymbol{\beta}_k$. **d** ℓ^0-norm of $\boldsymbol{\beta}_k$.

For a noisy signal, such as an image example in Fig. 2a, its adaptive over-complete dictionary (Fig. 2b) consists of atoms of principal signal patterns, strong noise patterns, and noisy signal patterns. Principal signal atoms should have higher frequencies, strong noise atoms lower frequencies and noisy signal atoms moderate frequencies. Intuitively, the red line (Fig. 2c) should be a suitable threshold P of the frequencies f_k s. In practical implementation, the value of P could be simply decided relying on the histogram of f_k. As shown in Fig. 2d, one can set the value of f_k associated with the maximum point of its histogram to P as follows:

$$P = \arg\max_k \text{Hist}\,(\|\boldsymbol{\beta}_k\|_0) \quad (14)$$

In fact, the performances in signal analyses by 3SD method are not sensitive to the threshold P, owed to the dependence of the atoms. To illustrate this point, we take three images, Barbara, Lena, and Boat. Their histograms of f_k are shown in Fig. 3a with the maximum points in dotted lines, 121, 97, and 92, respectively. Figure 3b reports the peak signal-to-noise ratio (PSNR) of the retrieved images $\hat{\mathbf{S}}_P$ on the signal principal subspace $\mathbf{D}_P^{(S)}$ with respect to P. We can see that the PSNRs of the results remain the same in a large range around the maximum points (in dotted lines). Consequently, taking the value of f_k associated to the maximum point of its histogram as the threshold P is a reasonable solution.

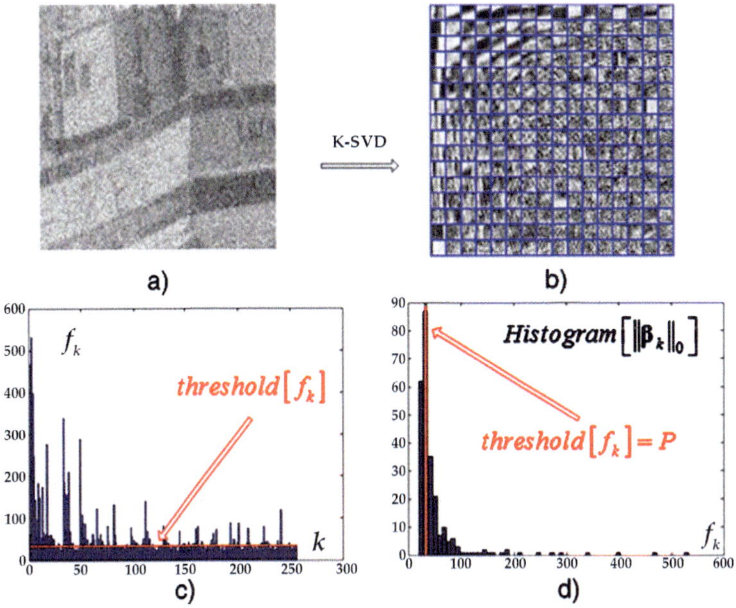

Figure 2: The threshold P of the frequencies f_k s. **a** Noisy image. **b** Over-complete dictionary **D**. **c** Frequency of $\mathbf{d}_k$. **d** Histogram of $\mathbf{d}_k$'s frequency.

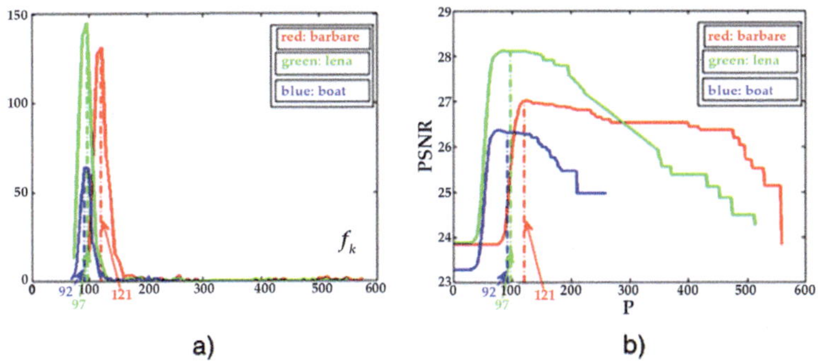

Figure 3: The insensitivity of the threshold P. **a** Histograms of f_k. **b** PSNR of $\hat{\mathbf{S}}_P$ with respect to P.

RESULTS AND DISCUSSION

Signal Decomposition Methods

Taking a part of the noisy Barbara image (Fig. 4a), we show an example of the sparse signal subspace decomposition (3SD) and the corresponding retrieved image (Fig. 4b). For comparison, the traditional sparse decomposition and the PCA-based subspace decomposition are shown in Fig. 4c, d.

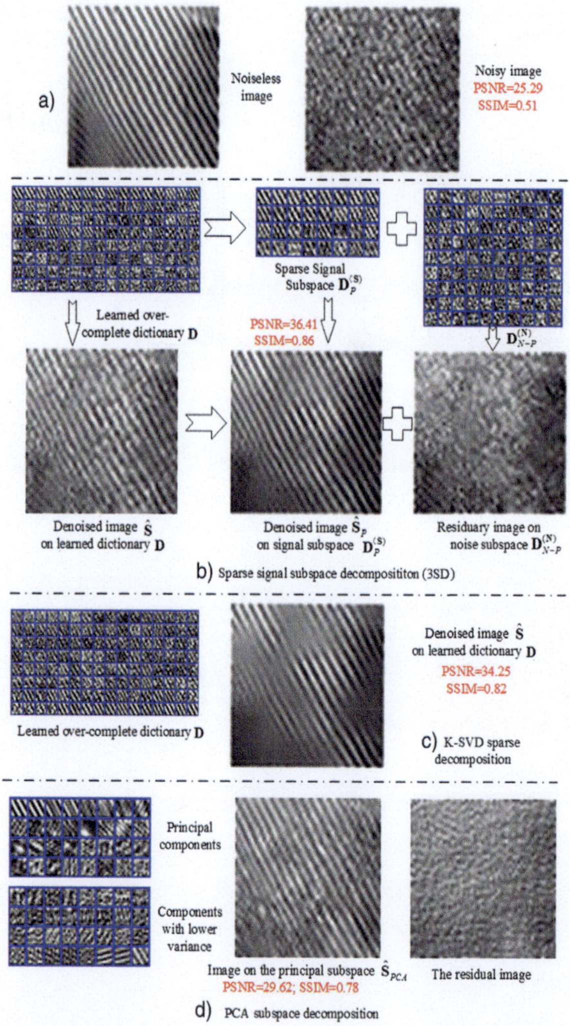

Figure 4: Signal decompositions. a Image sample. b Sparse Subspace decomposition. c Sparse decomposition. d Subspace decomposition.

We use the PSNR to assess the noise removal performance:

$$\mathrm{PSNR} = 20 \cdot \log_{10}\left[\mathrm{MAX}\{\mathbf{S}(i,j)\}\right] - 10 \cdot \log_{10}\left[\mathrm{MSE}\right]$$

$$\mathrm{MSE} = \frac{1}{IJ}\sum_{i=0}^{I-1}\sum_{j=0}^{J-1}\left[\mathbf{S}(i,j) - \hat{\mathbf{S}}(i,j)\right]^2 \qquad (15)$$

and the structural similarity index metric (SSIM) between the denoised image and the pure one to evaluate the preserving detail performance:

$$\mathrm{SSIM}(\mathbf{S},\hat{\mathbf{S}}) = \frac{(2u_{\mathbf{S}}u_{\hat{\mathbf{S}}} + c_1)(2\sigma_{\mathbf{S}\hat{\mathbf{S}}} + c_2)}{\left(u_{\mathbf{S}}^2 + u_{\hat{\mathbf{S}}}^2 + c_1\right)\left(\sigma_{\mathbf{S}}^2 + \sigma_{\hat{\mathbf{S}}}^2 + c_2\right)} \qquad (16)$$

where u_x is the average of x, σ_x^2 is the variance of x, σ_{xy} is the covariance of x and y, and c_1 and c_2 are small variables to stabilize the division with a weak denominator.

Let us look at the proposed sparse signal subspace decomposition on the top of Fig. 4b. The 128 atoms $\mathbf{d}_k$ s of the learned over-complete dictionary $\mathbf{D}$ are shown in descending order of their energies measured by $\|\boldsymbol{\beta}_k\|_2$. The 32 principal signal atoms are chosen from the dictionary $\mathbf{D}$ under the frequency criterion. They are shown in descending order of their frequencies measured by $\|\boldsymbol{\beta}_k\|_0$ composing a signal subspace $\mathbf{D}_{32}^{(\mathbf{S})}$. We can see that some of the principal atoms are not among the first 32 atoms with the largest energy in the over-complete dictionary $\mathbf{D}$. The retrieved images are shown at the bottom of Fig. 4b. The image $\mathbf{S}$ on $\mathbf{D}$ is apparently denoised. The image $\hat{\mathbf{S}}$ on the signal subspace $\mathbf{D}_{32}^{(\mathbf{S})}$ improves obviously by preserving fine details with a high SSIM=0.86 and at suppressing strong noise with a high PSNR=36.41. On the other hand, the residual image on noise subspace $\mathbf{D}_{96}^{(N)}$ contains some very noisy information. This is because the atoms of the over-complete dictionary are not independent.

For the same example, the classical sparse decomposition is shown in Fig. 4c, using the K-SVD algorithm [6] in which the allowed error tolerance ε (in Eq. (4)) is set to a larger value to filter out noise. The retrieved image **S** has a high PSNR=29.62, but it has obviously lost the weak information with SSIM=0.82. This is because signal distortion and residual noise cannot be minimized simultaneously at dictionary learning by Eq. (4).

In another comparison, the PCA-based subspace decomposition is shown in Fig. 4d. The 64 components are orthonormal and the 32 principal components are of the largest variance. The retrieved image by projecting on the signal subspace is rather noisy with PSNR=29.62. This is because it cannot suppress strong noise and preserve weak details of information only using the variance criterion.

Application to Image Denoising

The application of 3SD to image denoising is presented here. A major difficulty of denoising is to separate the underlying signal from the noise. The proposed 3SD method could win this challenge. In the 3SD method, the important components are selected from the over-complete dictionary relying on their occurrence number over the noisy image set. Evidently, the occurrence numbers would be large for the signal, even for weak details, such as edges or textures. On the other hand, the occurrence numbers would be low for different kinds of white Gaussian or non-Gaussian noises, even strong at intensity.

The 3SD algorithm for image denoising is presented as follows:

Input:	Noisy image $\mathbf{X}$
Output:	Denoised image $\hat{\mathbf{S}}$
-	Sparse representation $\{\mathbf{D},\mathbf{A}\}$: using K-SVD algorithm [6] by (4)
-	Identify principal atoms from $\mathbf{D}$ based on $\mathbf{A}$:
	■ Compute the frequencies of atoms $\{\|\boldsymbol{\beta}_k\|_0\}_{k=1}^{K}$ according to (6) and (8)
	■ Get the permutation π sorting the index k of $\{\|\boldsymbol{\beta}_k\|_0\}_{k=1}^{K}$ by (9)
	■ Compute the threshold P by (14)
-	Obtain the signal principal atoms $\{\mathbf{d}_{\pi(k)}\}_{k=1}^{P}$ by (12)
-	Reconstruct image SP^SP^ by (13)

In this application, we intend to preserve faint signal details under a situation of strong noise.

In the experiments, dictionaries used Ds of size 64×256 (K=256 atoms), designed to handle image patches $\mathbf{x}_m$ of size N=64=8×8 pixels.

Image Denoising

A noisy Lena image $\mathbf{X}=\mathbf{S}+\mathbf{V}$ with an additive zero-mean white Gaussian noise $\mathbf{V}$ is used. The standard deviation of noise is $\sigma=35$. A comparison is made between the 3SD method and the K-SVD method [6] which is one of the best denoising methods reported in the recent literatures. From the results shown in Fig. 5, the 3SD method outperforms the K-SVD method by about 1 dB in PSNR and by about 1% in SSIM (depending on how much details in the images and how faint the details). In terms of subjective visual quality, we can see that the corner of the mouth and the nasolabial fold with weak intensities are much better recovered by the 3SD method.

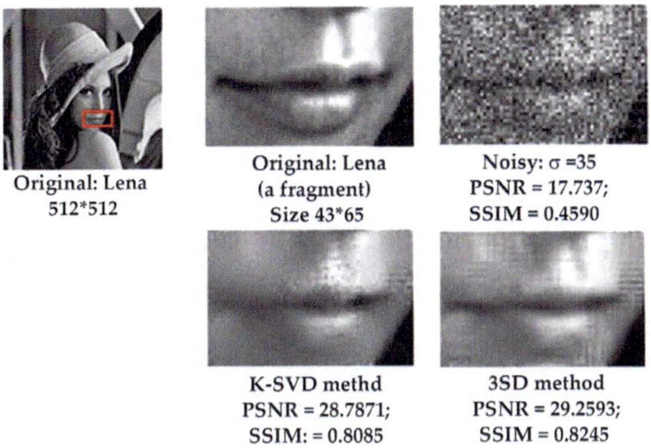

Figure 5: Image denoising comparing the proposed 3SD method with the K-SVD method.

SAR Image Despeckling

In the second experiment, a simulated SAR image with speckle noise is used. Speckle is often modeled as multiplicative noise as $x(i,j)=s(i,j)v(i,j)$ where x, s, and $v$correspond to the contaminated intensity, the original intensity, and the noise level, respectively.

Figure 6 shows the despeckling results of a simulated one-look SAR scenario with a fragment of the Barbara image. A comparison is made with 3SD method and a probabilistic patch-based (PPB) filter based on nonlocal means approach [7] which can cope with non-Gaussian noise. We can see that PPB can well remove speckle noise. However, it also removes the low-intensity details. The 3SD method shows advantages at preserving fine details and at suppressing strong noise.

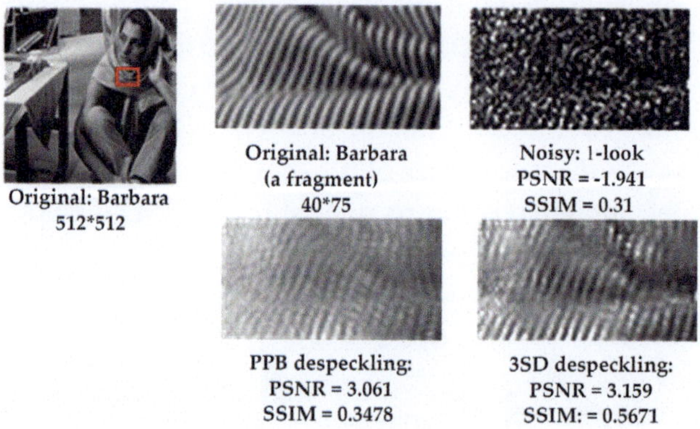

Figure 6: SAR image despeckling comparing the proposed 3SD method with the PPB method.

Comparison with BM3D Method

With a spatial complicated image scene, we make a comparison of the 3SD-based denoising method with the BM3D algorithm [8], one of the best methods especially for image denoising reported in many recent literatures.

The effectiveness of any signal analysis method depends on the different conditions in different applications. For the image denoising application, the signals involved should be homogeneous. Therefore, a procedure of grouping is generally adopted to select homogeneous pixels. In the BM3D method, a block-matching grouping is taken before filtering. We adopt the same grouping technique and then filter each homogeneous group by the proposed 3SD method.

Firstly, we take a 256×256 Barbara image (Fig. 7a) with a strong additive zero-mean white Gaussian noise where σ=70 (Fig. 7b). The denoising result by the BM3D algorithm is shown in Fig. 7c. It displays a quite high performance. The denoising result by the 3SD-based method is shown in Fig. 7d. It demonstrates a higher PSNR and a higher SSIM and a better subjective visual quality over the BM3D algorithm.

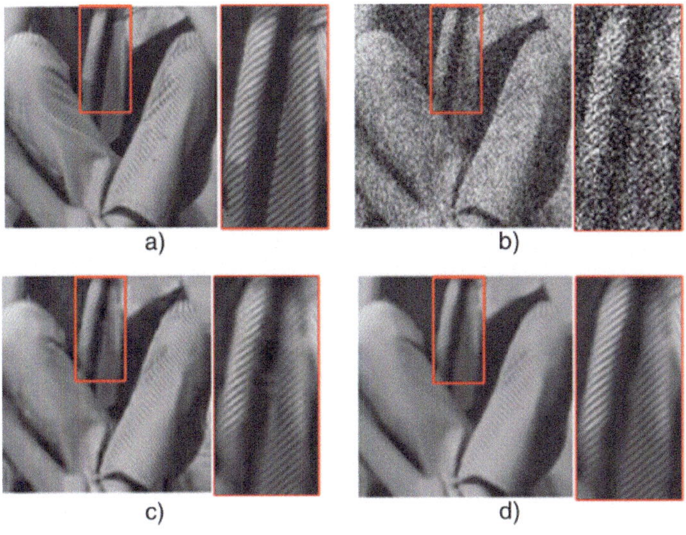

Figure 7: Denoising for spatial complicated image scene comparing BM3D method with 3SD-based method. a Original: Barbara (a fragment) 256∗256. b Noisy: σ=70; PSNR=11.22; SSIM=0.142. c BM3D denoising: PSNR=24.08; SSIM=0.7026. d3SD denoising: PSNR=24.21; SSIM=0.765.

Secondly, we take a simulated one-look SAR image with this 256×256 Barbara image (Fig. 8a) where PSNR=−0.1042. Figure 8b shows the despeckling result by the PPB method [7], in which grouping is realized based on nonlocal similarity and filtering is implemented by averaging each homogeneous group. The despeckled image is too smooth due to average filtering. Figure 8c shows the despeckling result by the BM3D-based method [9], in which grouping is realized based on similar 2-D fragments and filtering is implemented by Wiener shrinkage coefficients from the energy of the 3-D transform coefficients. The despeckled image is much better in PSNR and in SSIM, but it seems a little noisy still due to the used energy criterion which is not effective enough to separate noise elements from the principal elements. Figure 8d shows the despeckling result by the proposed 3SD-based method, in which grouping is realized based on nonlocal similarity [7] and filtering is implemented by the proposed sparse subspace decomposition. The despeckled image demonstrates some advantages of the 3SD method at preserving fine details and at suppressing speckle noise, attributed to the principal subspace decomposition.

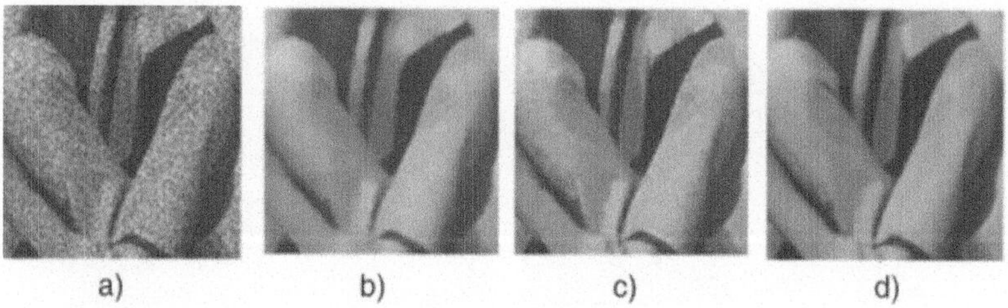

Figure 8: Despeckling for spatial complicated SAR image scene. a Noisy: 1-look; PSNR=−0.1042; SSIM=0.21. b PPB despeckling: PSNR=7.8943; SSIM=0.63. cSAR-BM3D despeckling: PSNR=9.9312; SSIM=0.73. d 3SD-based despeckling: PSNR=10.3316; SSIM=0.74.

CONCLUSIONS

We proposed a method of sparse signal subspace decomposition (3SD). The central idea of the proposed 3SD is to identify principal atoms from an adaptive over-complete dictionary relying on the occurrence frequency of atoms over the data set (Eq. (8)). The atom frequency is measured by zero pseudo-norms of weight vectors of atoms (Eqs. (6) and (8)). The principal subspace is spanned by the maximum frequency atoms (Eq. (12)).

The 3SD method combines the variance criterion, the sparsity criterion, and the component's frequency criterion into a uniform framework. As a result, it can identify more effectively the principal atoms with the three important signal features. On the contrary, PCA uses only variance criterion and sparse coding method uses the variance and the sparsity criterions. In those ways, it is more difficult to distinguish weak information from strong noise.

Another interesting asset of the 3SD method is that it takes benefits from using an over-complete dictionary which reserves details of information and from subspace decomposition which rejects strong noise. On the contrary, some undercomplete dictionary methods [10] and some sparse shrinkage methods [11, 12] might lose weak information when suppressing noise.

Moreover, the 3SD method is very simple with a linear retrieval operation (Eq. (13)). It does not require any prior knowledge on distribution or parameter to determine a threshold (Eq. (14)). On the contrary, some sparse shrinkage methods, such as [11], necessitate non-linear processing with some prior distributions of signals.

The proposed 3SD could be interpreted as a PCA in sparse decomposition, so it admits straightforward extension to applications of feature extraction, inverse problems, or machine learning.

ACKNOWLEDGEMENTS

The idea of the sparse signal subspace decomposition here arises through a lot of deep discussions with Professor Henri Maitre at Telecom ParisTech in France; he also gave suggestion on the structure of the manuscript.

AUTHORS' CONTRIBUTIONS

HS proposed the idea, designed the algorithms and experiments, and drafted the manuscript. CWS gave suggestion on the design of the algorithm, realized the algorithms, and carried out the experiments. DLR gave suggestions on the mathematical expressions of the manuscript and experiment analysis as well as explanation and helped draft the manuscript. All authors read and approved the final manuscript.

REFERENCES

1. K Hermus, P Wambacq, HV Hamme, A review of signal subspace speech enhancement and its application to noise robust speech recognition. EURASIP J. Adv. Signal Process. 888–896 (2007). doi:10.1155/2007/45821.
2. DL Donoho, IM Johnstone, G Kerkyacharian, D Picard, Wavelet shrinkage: asymptopia?J. R. Stat. Soc. Ser. B. 57:, 301–369 (1995). http://citeseer.ist.psu.edu/viewdoc/summary?doi=10.1.1.162.1643.
3. DW Tufts, R Kumaresan, I Kirsteins, Data adaptive signal estimation by singular value decomposition of a data matrix. IEEE Proc. 70(6), 684–685 (1982). doi:10.1109/PROC.1982.12367.
4. M Elad, M Aharon, Image denoising via sparse and redundant representations over learned dictionaries. IEEE Trans. Image Process. 15(12), 3736–3745 (2006). doi:10.1109/TIP.2006.881969.
5. G Tartavel, Y Gousseau, G Peyré, Variational texture synthesis with sparsity and spectrum constraints. J. Math. Imaging Vis. 52(1), 124–144 (2015). doi:10.1007/s10851-014-0547-7.
6. M Aharon, M Elad, A Bruckstein, K-SVD: an algorithm for designing overcomplete dictionaries for sparse representation. IEEE Trans. Signal Process. 54(11), 4311–4322 (2006). doi:10.1109/TSP.2006.881199.
7. CA Deledalle, L Denis, F Tupin, Iterative weighted maximum likelihood denoising with probabilistic patch-based weights. IEEE Trans. Image Process. 18(12), 2661–2672 (2009). doi:10.1109/TIP.2009.2029593.
8. K Dabov, A Foi, V Katkovnik, K Egiazarian, Image denoising by sparse 3-D transform-domain collaborative filtering. IEEE Trans. Image Process. 16:, 2080–2095 (2007). doi:10.1109/TIP.2007.901238.
9. S Parrilli, M Poderico, CV Angelino, L Verdoliva, A nonlocal SAR image denoising algorithm based on LLMMSE wavelet shrinkage. IEEE Trans. Geosci. Remote Sens. 50:, 606–616 (2012). doi:10.1109/TGRS.2011.2161586.
10. F Porikli, R Sundaresan, K Suwa, SAR depeckling by sparse reconstruction on affinity nets. EUSAR2012. 18:, 796–799 (2012). https://www.researchgate.net/publication/232905473_SAR_Despeckling_by_Sparse_Reconstruction_on_Affinity_Nets_SRAN.
11. A Hyvarinen, P Hoyer, E Oja, Sparse code shrinkage for image denoising. IEEE World Congr. Comput. Intell. 2:, 59–864 (1998).

doi:10.1109/IJCNN.1998.685880.

12. R Malutan, R Terebes, C Germain, Speckle noise removal in ultrasound images using sparse code shrinkage. 5th IEEE Conf. E-Health Bioeng. Conf. 2:, 1–4 (2015). doi:10.1109/EHB.2015.7391394.

Chapter 8

Lower Bounds for the Low-rank Matrix Approximation

Jicheng Li[1] , Zisheng Liu[1,2] and Guo Li[3]

[1]School of Mathematics and Statistics, Xi'an Jiaotong University, No. 28, Xianning West Road, Xi'an, 710049, China

[2]School of Statistics, Henan University of Economics and Law, No. 180, Jinshui East Road, Zhengzhou, 450046, China

[3]College of Mathematics and Statistics, Shenzhen University, No. 3688, Nanhai Ave, Shenzhen, 518060, China

ABSTRACT

Low-rank matrix recovery is an active topic drawing the attention of many researchers. It addresses the problem of approximating the observed data matrix by an unknown low-rank matrix. Suppose that A is a low-rank matrix approximation of D, where D and A are $m\times n$ matrices. Based on a useful decomposition of $D^{\dagger}-A^{\dagger}$, for the unitarily invariant norm $\|\cdot\|$, when

Citation (APA): Li, J., Liu, Z., & Li, G. (2017). Lower bounds for the low-rank matrix approximation. Journal of inequalities and applications, 2017(1), 288. (14 pages), DOI: https://doi.org/10.1186/s13660-017-1564-z

$\|D\| \geq \|A\|$ and $\|D\| \leq \|A\|$, two sharp lower bounds of $D-A$ are derived respectively. The presented simulations and applications demonstrate our results when the approximation matrix A is low-rank and the perturbation matrix is sparse.

Keywords: low-rank matrix, approximation, error estimation, pseudo-inverse, matrix norms

INTRODUCTION

In mathematics, low-rank approximation is a minimization problem, in which the cost function measures the fit between a given matrix (the data) and an approximating matrix (the optimization variable), subject to a constraint that the approximating matrix has reduced rank. The problem is used for mathematical modeling and data compression. The rank constraint is related to a constraint on the complexity of a model that fits the data.

Low-rank approximation of a linear operator is ubiquitous in applied mathematics, scientific computing, numerical analysis, and a number of other areas. For example, a low-rank matrix could correspond to a low-degree statistical model for a random process (e.g., factor analysis), a low-order realization of a linear system [1], or a low-dimensional embedding of data in the Euclidean space [2], the image and computer vision [3, 4, 5], bioinformatics, background modeling and face recognition [6], latent semantic indexing [7, 8], machine learning [9, 10, 11, 12] and control [13] etc. These data may have thousands or even billions of dimensions, and a large number of samples may have the same or similar structure. As we know, the important information lies in some low-dimensional subspace or low-dimensional manifold, but interfered with some perturbative components (sometimes interfered by the sparse component).

Let $D \in \mathbb{R}^{m\times n}$ be an observed data matrix which is combined as

$$D = A + E, \qquad (1)$$

where $A \in \mathbb{R}^{m\times n}$ is the low-rank component and $E \in \mathbb{R}^{m\times n}$ is the perturbation component of D. The singular value decomposition (SVD [14]) is a method for dealing with such high-dimensional data. If the matrix E is small, the classical principal components analysis (PCA [15, 16, 17]) can seek the best rank-r estimation of A by solving the following constrained optimization via SVD of D and then projecting the columns of D onto the subspace spanned by the r principal left singular vectors of D:

$$\min_{E} \|E\|_F$$

$$\text{s.t.} \quad \text{Rank}(A) \leq r,$$

$$\|D - A\|_F \leq \epsilon, \qquad (2)$$

where $r \ll \min(m,n)$ is the target dimension of the subspace, ϵ is an upper bound on the perturbative component $\|E\|_F$ and $\|\cdot\|_F$ is the Frobenius norm. Despite its many advantages, the traditional PCA suffers from the fact that the estimation $\hat{A}$ obtained by classical PCA can be arbitrarily far from the true A, when E is sufficiently sparse (relative to the rank of A). The reason for this poor performance is precisely that the traditional PCA makes sense for Gaussian noise and not for sparse noise. Recently, robust PCA (RPCA [6]) is a family of methods that aims to make PCA robust to large errors and outliers. That is, RPCA is an upgrade of PCA.

There are some reasons for the study of lower bound of a low-rank matrix approximation problem. Firstly, as far as we know, there is no literature to consider the lower bound of the low-rank matrix approximation problem. In our paper, we first put forward the lower bound. Secondly, for the low-rank approximation, when a perturbation E exists, there is an approximation error which cannot be avoided, that is, the approximation error cannot equal 0, but tends to 0. Thirdly, from our main results, we can clearly find the influence of the spectral norm ($\|\cdot\|_2$) on the low-rank matrix approximation. For example, for our main result of Case II, when the maximum eigenvalue of the matrix D is larger, the approximation error of $(D-A)$is smaller. In addition, the lower bound can verify whether the solution obtained by algorithms is optimal. For details, please refer to the experiments Section 4 of our paper. Therefore, it is necessary and significant to study the lower bound of the low-rank matrix approximation problem.

Remark 1.1

PCA and RPCA are methods for the low-rank approximation problem when perturbation item exists. Our aim is to prove that no matter what method is used, the lower bound of error always exists and it cannot be avoided with the perturbation item E. Considering the existence of error, this paper focuses on the specific situation of this lower bound.

Notations

For a matrix $A \in \mathbb{R}^{m\times n}$, let $\|A\|_2$ and $\|A\|_*$ denote the spectral norm and the nuclear norm (i.e., the sum of its singular values), respectively. Let $\|\cdot\|$ be a unitarily invariant norm. The pseudo-inverse and the conjugate transpose of A are denoted by $A^\dagger$ and A^H, respectively. We consider the singular value decomposition (SVD) of a matrix A of rank r

$$A = USV^H, \quad S = \mathrm{diag}(\{\sigma_i\}), 1 \le i \le r,$$

where U and V are m×r and n×r matrices with orthonormal columns, respectively, and σ_i is the positive singular values. We always assume that the SVD of a matrix is given in the reduced form above. Furthermore, $\langle A,B\rangle=\mathrm{trace}(A^H B)$denotes the standard inner product, then the Frobenius norm is

$$\|A\|_F = \sqrt{\langle A,A\rangle} = \sqrt{\mathrm{tr}(A^H A)} = \left(\sum_{i=1}^{m}\sum_{j=1}^{n} A_{ij}^2\right)^{\frac{1}{2}} = \left(\sum_{i=1}^{r} \sigma_i^2\right)^{\frac{1}{2}}.$$

Organization

In this paper, we study a perturbation theory for low-rank matrix approximation. When $\|D\| \ge \|A\|$ or $\|D\| \le \|A\|$, two sharp lower bounds of D−A are derived for a unitarily invariant norm respectively. This work is organized as follows. In Section 2, we provide a review of relevant linear algebra and some preliminary results. In Section 3, under different norms, two sharp lower bounds of D−A are given for the low-rank approximation problem and some proofs of Theorem 3.5 are presented. In Section 4, example and applications are given to verify the provided lower bounds. Finally, we conclude the paper with a short discussion.

PRELIMINARIES

In order to prove our main results, we mention the following results for our further discussions.

Unitarily Invariant Norm

An important property of a Euclidean space is that shapes and distance do not change under rotation. In particular, for any vector x and for any unitary matrices U, we have

$$\|Ux\|_2 = \|x\|_2.$$

An analogous property is shared by the spectral and Frobenius norms: namely, for any unitary matrices U and V, the product UAVHUAVH is defined by

$$\left\|UAV^H\right\|_p = \|A\|_p, \quad p = 2, F.$$

These examples suggest the following definition.

Definition 2.1

([18])

A norm $\|\cdot\|$ on $\mathbb{C}^{m\times n}$ is unitarily invariant if it satisfies

$$\left\|UAV^H\right\| = \|A\|$$

for any unitary matrices U and V. It is normalized if

$$\|A\| = \|A\|_2$$

whenever A is of rank one.

Remark 2.2

Let $\Sigma = UAV^H$ be the singular value decomposition of the matrix A with order n. Let $\|\cdot\|$ be a unitarily invariant norm. Since U and V are unitary,

$$\|A\| = \|\Sigma\|.$$

Thus $\|A\|$ is a function of the singular values of A.

The 2-norm plays a special role in the theory of unitarily invariant norms as the following theorem shows.

Theorem 2.3

([18])

Let $\|\cdot\|$ be a family of unitarily invariant norm. Then

$$\|AB\| \leq \|A\|\,\|B\|_2 \qquad (3)$$

and

$$\|AB\| \leq \|A\|_2\,\|B\|. \qquad (4)$$

Moreover, if Rank(A)=1, then

$$\|A\| = \|A\|_2.$$

We have observed that the spectral and Frobenius norms are unitarily invariant. However, not all norms are unitarily invariant as the following example shows.

Example 2.4

Let

$$A = \begin{pmatrix} 1 & 1 \\ 1 & 1 \end{pmatrix},$$

obviously, $\|A\|_{\infty=2}$, but for a unitary matrix

$$U = \begin{pmatrix} \frac{1}{\sqrt{2}} & \frac{1}{\sqrt{2}} \\ -\frac{1}{\sqrt{2}} & \frac{1}{\sqrt{2}} \end{pmatrix},$$

we have

$$\|UA\|_\infty = \left\| \begin{pmatrix} \frac{2}{\sqrt{2}} & \frac{2}{\sqrt{2}} \\ 0 & 0 \end{pmatrix} \right\|_\infty = \frac{4}{\sqrt{2}}.$$

Remark 2.5

It is easy to verify that the nuclear norm $\|\cdot\|_*$ is a unitarily invariant norm.

Projection

Let $\mathbb{C}^m$ and $\mathbb{C}^n$ be m and n-dimensional inner product spaces over the complex field, respectively, and $A \in \mathbb{C}^{m\times n}$ be a linear transformation from $\mathbb{C}^n$ into $\mathbb{C}^m$.

Definition 2.6

([18])

The column space (range) of A is denoted by

$$\mathcal{R}(A) = \{x \in \mathbb{C}^m | x = Ay, y \in \mathbb{C}^n\} \qquad (5)$$

and the null space of A by

$$\mathcal{N}(A) = \{y \in \mathbb{C}^n | Ay = 0\}. \qquad (6)$$

Further, we let $\perp$ denote the orthogonal complement and get $\mathcal{R} = \mathcal{N}(A^H)^\perp$ and $\mathcal{N}(A) = \mathcal{R}(A^H)^\perp$.

The following properties [18] of the pseudo-inverse are easily established.

Theorem 2.7

([18])

For any matrix A, the following hold.

1. If $A \in \mathbb{C}^{m\times n}$ has rank n, then $A^{\dagger} = (A^H A)^{-1} A^H$ *and* $A^{\dagger}A = I^{(n)}$.
2. If $A \in \mathbb{C}^{m\times n}$ has rank m, then $A^{\dagger} = A^H (AA^H)^{-1}$ *and* $AA^{\dagger} = I^{(m)}$.

Here $I^{(n)} \in \mathbb{R}^{n\times n}$ is the identity matrix.

Theorem 2.8

([18])

For any matrix A, $P_A = AA^{\dagger}$ is the orthogonal projector onto $\mathcal{R}(A)$, $P_{A^H} = A^{\dagger}A$ is the orthogonal projector onto $\mathcal{R}(A^H)$, $I - P_{A^H}$ is the orthogonal projector onto $\mathcal{N}(A)$.

The Decomposition of $D^{\dagger}-A^{\dagger}$

In this section, we focus on the decomposition of $D^{\dagger}-A^{\dagger}$ and a general bound of the perturbation theory for pseudo-inverses. Firstly, according to the orthogonal projection, we can deduce the following lemma.

Lemma 2.9

For any matrix A, $P_A = AA^{\dagger}$ *and* $P_{A^H} = A^{\dagger}A$, then we have

$$P_A^{\perp}A = 0, \qquad AP_{A^H}^{\perp} = 0, \qquad P_{A^H}^{\perp}A^H = 0, \qquad A^{\dagger}P_A^{\perp} = 0. \tag{7}$$

Proof

Since $P_A^{\perp} = I - P_A$ and $P_{A^H}^{\perp} = I - P_{A^H}$, then we have that

$$P_A^{\perp}A = (I - P_A)A = A - AA^{\dagger}A = A - AA^H\left(AA^H\right)^{-1}A = 0,$$

$$AP_{A^H}^{\perp} = A(I - P_{A^H}) = A - AA^{\dagger}A = A - A\left(A^H A\right)^{-1}A^H A = 0,$$

$$P_{A^H}^{\perp}A^H = A^H - P_{A^H}A^H = A^H - A^{\dagger}AA^H = A^H - \left(A^H A\right)^{-1}A^H AA^H = 0,$$

$$A^{\dagger}P_A^{\perp} = A^{\dagger}(I - P_A) = A^{\dagger} - A^{\dagger}AA^{\dagger} = A^{\dagger} - \left(A^H A\right)^{-1}A^H AA^{\dagger} = 0.$$

The proof is completed.

Using Lemma 2.9, the decompositions of $D^{\dagger}-A^{\dagger}$ are developed by Wedin [19].

Theorem 2.10

([19])

Let D=A+E, then the difference $D^{\dagger}-A^{\dagger}$ is given by the expressions

$$D^{\dagger} - A^{\dagger} = -A^{\dagger}ED^{\dagger} - A^{\dagger}P_D^{\perp} + P_{A^H}^{\perp}D^{\dagger}, \tag{8}$$

$$D^{\dagger} - A^{\dagger} = -A^{\dagger}P_A E P_{D^H}D^{\dagger} - A^{\dagger}P_A P_D^{\perp} + P_{A^H}^{\perp}P_{D^H}D^{\dagger}, \tag{9}$$

$$D^{\dagger} - A^{\dagger} = -D^{\dagger}P_D E P_{A^H} A^{\dagger} + (D^H D)^{\dagger} P_{D^H} E^H P_A^{\perp}$$
$$- P_{D^H}^{\perp} E P_A (AA^H)^{\dagger}. \quad (10)$$

By Lemma 2.9, using $P_A = AA^{\dagger}, P_{A^H} = A^{\dagger}A, P_A^{\perp} = I - P_A, P_{A^H}^{\perp} = I - P_{A^H}$, these expressions can be verified.

In previous work [19], Wedin developed a general bound of the perturbation theory for pseudo-inverses. Theorem 2.11 is based on a useful decomposition of $D^{\dagger}-A^{\dagger}$, where D and A are m×n matrices. Sharp estimates of $\|D^{\dagger}-A^{\dagger}\|$ are derived for a unitarily invariant norm. In [20], Chen et al. presented some new perturbation bounds for the orthogonal projections $\|P_D-P_A\|$.

Theorem 2.11

([19])

Suppose D=A+E, then the error of $D^{\dagger}-A^{\dagger}$ has the following bound:

$$\|D^{\dagger} - A^{\dagger}\| \le \gamma \max\{\|A^{\dagger}\|_2^2, \|D^{\dagger}\|_2^2\}\|E\|, \quad (11)$$

where γ is given in Table 1.

Table 1: Value options for γ

$\|\cdot\|$	**Arbitrary**	**Spectral**	**Frobenius**
γ	3	1+5√21+52	2–√2

Remark 2.12

For the spectral norm, by formula (11) we can achieve $\gamma = \frac{1+\sqrt{5}}{2}$. When $\|\cdot\|$ is the Frobenius norm, by formula (12), we have $\gamma = \sqrt{2}$. Similarly, for an arbitrary unitarily invariant norm, according to formula (13), we can deduce γ=3.

Remark 2.13

From Theorem 2.11, since E=D−A, in fact, if Rank(A)≤Rank(D), then (11) gives the lower bound of the low-rank matrix approximation:

$$\|D - A\| \ge \frac{\|D^{\dagger} - A^{\dagger}\|}{\gamma \max\{\|A^{\dagger}\|_2^2, \|D^{\dagger}\|_2^2\}}. \quad (12)$$

In the following section, based on Theorem 2.11, we provide two lower error bounds of D−A for a unitarily invariant norm.

OUR MAIN RESULTS

In this section, we consider the lower bound theory for the low-rank matrix approximation based on a useful decomposition of $D^{\dagger}-A^{\dagger}$. When Rank(A)≤Rank(D), some sharp lower bounds of D−A are derived in terms of a unitarily invariant norm. In order to prove our result, some lemmas are listed below.

Lemma 3.1

([18])

Let D=A+E, the projections P_D and P_A satisfy

$$P_D P_A^{\perp} = \left(D^{\dagger}\right)^H P_{D^H} E^H P_A^{\perp} = \left(P_A^{\perp} P_D\right)^H, \tag{13}$$

therefore

$$\left\|P_D P_A^{\perp}\right\| \leq \left\|D^{\dagger}\right\|_2 \|E\|. \tag{14}$$

If Rank(A)≤Rank(D), then

$$\left\|P_D^{\perp} P_A\right\| \leq \left\|P_D P_A^{\perp}\right\|. \tag{15}$$

Lemma 3.2

([21])

Let $A, D \in \mathbb{C}^{m\times n}$, $\mathrm{Rank}(A) = r$, $\mathrm{Rank}(D) = s$, $r \leq s$, then there exists a unitary matrix $Q \in \mathbb{C}^{m\times m}$ such that

$$QP_AQ^H = \begin{pmatrix} I^{(r)} & 0 & 0 \\ 0 & 0 & 0 \\ 0 & 0 & 0 \end{pmatrix} \quad and \quad QP_DQ^H = \begin{pmatrix} \Gamma_r^2 & 0 & \Gamma_r\Sigma_r \\ 0 & I^{(s-r)} & 0 \\ \Sigma_r\Gamma_r & 0 & \Sigma_r^2 \end{pmatrix}, \tag{16}$$

where

$$\Gamma_r = \begin{pmatrix} \Gamma_1 & 0 \\ 0 & I \end{pmatrix} \quad and \quad \Sigma_r = \begin{pmatrix} \Sigma_1 & 0 \\ 0 & 0 \end{pmatrix},$$

$\Gamma_1 = \mathrm{diag}(\gamma_1, \ldots, \gamma_{r_1}), 0 \leq \gamma_1 \leq \cdots \leq \gamma_{r_1}$ *and* $\Sigma_1 = \mathrm{diag}(\sigma_1, \ldots, \sigma_{r_1}), 0 \leq \sigma_1 \leq \cdots \leq \sigma_{r_1}$. More*over,* γ_i *and* σ_i *satisfy* $\gamma_i^2 + \sigma_i^2 = 1, i = 1, \ldots, r_1$

According to Lemma 3.2, we can easily get the following result.

Lemma 3.3

Let $A, D \in \mathbb{C}^{m\times n}$, $\mathrm{Rank}(A) = r$, $\mathrm{Rank}(D) = s$, $r \leq s$, then we have

$$\left\|P_A^{\perp} P_D\right\| = \left\|P_D P_A^{\perp}\right\|. \tag{17}$$

Proof

Since

$$P_A^{\perp} = I - P_A = Q^H \begin{pmatrix} 0 & 0 & 0 \\ 0 & I^{(s-r)} & 0 \\ 0 & 0 & I^{(m-s)} \end{pmatrix} Q, \tag{18}$$

and

$$P_D = Q^H \begin{pmatrix} \Gamma_r^2 & 0 & \Gamma_r \Sigma_r \\ 0 & I^{(s-r)} & 0 \\ \Sigma_r \Gamma_r & 0 & \Sigma_r^2 \end{pmatrix} Q, \tag{19}$$

then

$$P_A^{\perp} P_D = Q^H \begin{pmatrix} 0 & 0 & 0 \\ 0 & I^{(s-r)} & 0 \\ \Sigma_r \Gamma_r & 0 & \Sigma_r^2 \end{pmatrix} Q \tag{20}$$

and

$$P_D P_A^{\perp} = Q^H \begin{pmatrix} 0 & 0 & \Gamma_r \Sigma_r \\ 0 & I^{(s-r)} & 0 \\ 0 & 0 & \Sigma_r^2 \end{pmatrix} Q. \tag{21}$$

Therefore, they have the same singular values which yield that $\|P_A^{\perp} P_D\| = \|P_D P_A^{\perp}\|$.

This is a useful lemma that we will use in the proof of the main result. In order to prove our main theorem, two lower bounds of $D-A$ are required by the following lemma.

Lemma 3.4

For the unitarily invariant norm, if $\text{Rank}(A) \leq \text{Rank}(D)$, then the lower bound of $D-A$ satisfies:

Case I: For $\|D\| \geq \|A\|$, we have

$$\|D - A\| \geq \|D\| - \|A\| - \left\|D^{\dagger} - A^{\dagger}\right\| \|D\|_2 \|A\|_2. \tag{22}$$

Case II: For $\|D\| \leq \|A\|$, we have

$$\|D - A\| \geq \|A\| - \|D\| - \left\|D^{\dagger} - A^{\dagger}\right\| \|D\|_2^2. \tag{23}$$

Proof

Case I: Since $\|D\| \geq \|A\|$, we have $\|D-A\| \geq \|D\| - \|A\|$. Using Theorem 2.3and Lemma 3.1, we have $\|AB\| \leq \|A\|_2 \|B\|$ and $\|P_D^{\perp} P_A\| \leq \|P_D P_A^{\perp}\|$, respectively. By Lemma 2.9, we have $P_D^\perp D = 0, AP_{A^H}^\perp = 0$ and $A^\dagger P_A^\perp = 0$, this also yields

$$\begin{aligned}\|D-A\| &\geq \|D\| - \|A\| = \|D\| - \left\|(P_D + P_D^\perp)(P_A + P_A^\perp)A\right\| \\ &= \|D\| - \left\|P_D P_A A + P_D^\perp P_A A\right\| \quad \text{(by Lemma 2.9)} \\ &= \|D\| - \|P_D P_A A\| - \left\|P_D^\perp P_A A\right\| \quad (\text{by } P_D \perp P_D^\perp) \\ &\geq \|D\| - \|A\| - \left\|P_D^\perp P_A\right\| \|A\|_2 \quad \text{(by Lemma 3.1)} \\ &\geq \|D\| - \|A\| - \left\|P_D P_A^\perp\right\| \|A\|_2 \\ &= \|D\| - \|A\| - \left\|D(D^\dagger - A^\dagger)P_A^\perp\right\| \|A\|_2 \\ &\geq \|D\| - \|A\| - \left\|D^\dagger - A^\dagger\right\| \|D\|_2 \|A\|_2 \quad \text{(by Theorem 2.3).}\end{aligned}$$

Case II: Since $\|D\| \leq \|A\|$, we have $\|D-A\| \geq \|A\| - \|D\|$. Similarly, by Lemma 3.3, using $\|P_A^\perp P_D\| = \|P_D P_A^\perp\|$, we have

$$\begin{aligned}\|D-A\| &\geq \|A\| - \|D\| = \|A\| - \left\|(P_A + P_A^\perp)(P_D + P_D^\perp)D\right\| \\ &= \|A\| - \left\|P_A P_D D + P_A^\perp P_D D\right\| \quad \text{(by Lemma 2.9)} \\ &= \|A\| - \|P_A P_D D\| - \left\|P_A^\perp P_D D\right\| \quad (\text{by } P_A \perp P_A^\perp) \\ &\geq \|A\| - \|D\| - \left\|P_A^\perp P_D\right\| \|D\|_2 \\ &\geq \|A\| - \|D\| - \left\|P_D P_A^\perp\right\| \|D\|_2 \quad \text{(by Lemma 3.3)} \\ &= \|A\| - \|D\| - \left\|D(D^\dagger - A^\dagger)P_A^\perp\right\| \|D\|_2 \\ &\geq \|A\| - \|D\| - \left\|D^\dagger - A^\dagger\right\| \|D\|_2^2 \quad \text{(by Theorem 2.3).}\end{aligned}$$

We complete the proof of Lemma 3.4.

Our main results can be described as the following theorem.

Theorem 3.5

Suppose that $D = A + E$, $\text{Rank}(A) \leq \text{Rank}(D)$, for the unitarily invariant norm$\|\cdot\|$, the error of D−A has the following bounds.

Case I: For $\|D\| \geq \|A\|$, we have

$$\|D-A\| \geq \frac{\|D\| - \|A\|}{1 + \gamma \max\{\|A^\dagger\|_2^2, \|D^\dagger\|_2^2\}\|D\|_2 \|A\|_2}. \qquad (24)$$

Case II: For $\|D\|\le\|A\|$, we have

$$\|D-A\| \ge \frac{\|A\| - \|D\|}{1 + \gamma \max\{\|A^{\dagger}\|_2^2, \|D^{\dagger}\|_2^2\}\|D\|_2^2}, \tag{25}$$

where the value options for γ are the same as in Table 1.

Proof

Case I: For $\|D\|\ge\|A\|$, by Theorem 2.11 and Lemma 3.4 (22), we can deduce

$$\begin{aligned}\|D\| - \|A\| &\le \|D-A\| + \left\|D^{\dagger} - A^{\dagger}\right\| \|D\|_2 \|A\|_2 \\ &\le \|D-A\| + \gamma \max\left\{\left\|A^{\dagger}\right\|_2^2, \left\|D^{\dagger}\right\|_2^2\right\} \|D-A\| \|D\|_2 \|A\|_2,\end{aligned}$$

this yields

$$\|D-A\| \ge \frac{\|D\| - \|A\|}{1 + \gamma \max\{\|A^{\dagger}\|_2^2, \|D^{\dagger}\|_2^2\}\|D\|_2 \|A\|_2}. \tag{26}$$

Case II: Similarly, for $\|D\|\le\|A\|\|D\|\le\|A\|$, by Theorem 2.11 and Lemma 3.4 (23), we can deduce

$$\|D-A\| \ge \frac{\|D\| - \|A\|}{1 + \gamma \max\{\|A^{\dagger}\|_2^2, \|D^{\dagger}\|_2^2\}\|D\|_2 \|A\|_2}.$$

this yields

$$\|D-A\| \ge \frac{\|A\| - \|D\|}{1 + \gamma \max\{\|A^{\dagger}\|_2^2, \|D^{\dagger}\|_2^2\}\|D\|_2^2}, \tag{27}$$

where the value options for γ are the same as in Table 1. In summary, we prove the lower bounds of Theorem 3.5.

Remark 3.6

From the main theorem, we can see that if $\|D\|=\|A\|$, then $\|D-A\|=0$. However, in the problem of low-rank matrix approximation, $\|D\|$ is not necessarily equal to $\|A\|$, so the approximation error is present. Furthermore, when $\|D\|$ is close to $\|A\|$, simulations demonstrate that the error has a very small magnitude (see Section 4).

In this section, we discuss the error bounds under different conditions for the unitarily invariant norm. Based on a useful decomposition of $D^{\dagger}-A^{\dagger}$, for $\|D\|\ge\|A\|$ and $\|D\|\le\|A\|$, we have bounds (26) and (27), respectively. The two error bounds are useful in low-rank matrix approximation. The following experiments illustrate our results when the approximation matrix A is low-rank and the perturbation matrix E is sparse.

EXPERIMENTS

The Singular Value Thresholding Algorithm

Our results are obtained by a singular value thresholding (SVT [22]) algorithm. This algorithm is easy to implement and surprisingly effective both in terms of computational cost and storage requirement when the minimum nuclear norm solution is also the lowest-rank solution. The specific algorithm is described as follows.

For the low-rank matrix approximation problem which is contaminated with perturbation item E, we observe that the data matrix D=A+E. To approximate D, we can solve the convex optimization problem

$$\min \|A\|_*$$

$$\text{s.t.} \quad \|D-A\|_F \leq \varepsilon, \qquad (28)$$

where $\|\cdot\|_*$ denotes the nuclear norm of a matrix (i.e., the sum of its singular values).

For solving (28), we introduce the soft-thresholding operator D_τ [22] which is defined as

$$\mathcal{D}_\tau(A) := U\mathcal{D}_\tau(S)V^*, \quad \mathcal{D}_\tau(S) = \text{diag}(\{(\sigma_i - \tau)_+\}),$$

where $(\sigma_i - \tau)_+ = \max\{0, \sigma_i - \tau\}$. In general, this operator can effectively shrink some singular values toward zero. The following theorem is with respect to the shrinkage operators [22, 23, 24], which will be used at each iteration of the proposed algorithms.

Theorem 4.1

([22])

For each $\tau > 0$ *and* $W \in \mathbb{R}^{m\times n}$, the singular value shrinkage operator $D_\tau(\cdot)$ obeys

$$\mathcal{D}_\tau(W) = \arg\min_A \tau\|A\|_* + \frac{1}{2}\|A - W\|_F^2,$$

where $\mathcal{D}_\tau(W) := U\mathcal{D}_\tau(S)V^*$.

By introducing a Lagrange multiplier Y to remove the inequality constraint, one has the augmented Lagrangian function of (28)

$$L(A, Y) = \|A\|_* - \langle Y, A-D\rangle + \frac{\tau}{2}\|A-D\|_F^2.$$

The iterative scheme of the classical augmented Lagrangian multipliers method is

$$\begin{cases} A^{k+1} \in \arg\min_A L(A, Y^k), \\ Y^{k+1} = Y^k - \tau(D - A^{k+1}). \end{cases} \tag{29}$$

Based on the optimality conditions, (29) is equivalent to

$$\begin{cases} \mathbf{0} \in \frac{1}{\tau}\partial(\|A^{k+1}\|_*) + A^{k+1} - (D + \frac{1}{\tau}Y^k), \\ Y^{k+1} = Y^k - \tau(D - A^{k+1}), \end{cases} \tag{30}$$

where $\partial(\cdot)$ denotes the subgradient operator of a convex function. Then, by Theorem 4.1 above, we have the iterative solution

$$\begin{cases} A^{k+1} = \mathcal{D}_{1/\tau}(D + \frac{1}{\tau}Y^k), \\ Y^{k+1} = Y^k - \tau(D - A^{k+1}). \end{cases} \tag{31}$$

The SVT approach works as described in Algorithm 1.

Algorithm 1

Task: Approximate the solution of (28).

Input: Observation matrix $D = A + E$, weight τ. $Y^0 = \text{zeros}(m, n)$

while the termination criterion is not met, **do**

$A^{k+1} = \mathcal{D}_{1/\tau}(D + \frac{1}{\tau}Y^k)$,

$Y^{k+1} = Y^k - \tau(D - A^{k+1})$,

$k \leftarrow k + 1$.

end while

Output: $A \leftarrow A^{k+1}$.

Simulations

In this section, we use the SVT algorithm for the low-rank matrix approximation problem. Let $D = A + E \in \mathbb{R}^{m \times n}$ be the available data. Simply, we restrict our examples to square matrices (m=n). We draw A according to the independent random matrices and generate the perturbation matrix E to be sparse, which satisfies the i.i.d. Gaussian distribution. Specially, the rank of the matrix A and the sparse entries of the perturbation matrix E are selected to be 5%m and 5%m^2, respectively.

Table 2 reports the results obtained by lower bounds (24), (25) and (12), respectively. Bounds (24) and (25) are our new result, bound (12) is the previous result. Then, comparing the bounds with each other by numerical experiments, we find that lower bounds (24), (25) are smaller than lower bound (12).

Table 2: Lower bound comparison results

	Bound (24)		**Bound (25)**		**Bound (12)**	
$m = n$	$\|\cdot\|_2$	$\|\cdot\|_F$	$\|\cdot\|_2$	$\|\cdot\|_F$	$\|\cdot\|_2$	$\|\cdot\|_F$
100	8.13e-7	1.89e-7	1.54e-7	3.31e-7	1.01e-4	1.27e-4
500	5.11e-8	3.71e-8	4.22e-8	4.62e-8	4.23e-4	5.22e-4
1,000	3.76e-8	2.14e-8	1.01e-8	1.19e-8	5.57e-4	7.48e-4

Applications

In this section, we use the SVT algorithm for the low-rank image approximation. From Figures 1 and 2, comparing with the original image (a), the low-rank image (b) loses some details. We can hardly get any detailed information from incomplete image (c). However, the output image (d) $=A^k$, which is obtained by the SVT algorithm, can recover the details of the low-rank image (b). If we denote image (b) to be a low-rank matrix A, then image (c) is the observed data matrix D which is perturbed by a sparse matrix E, that is,

image (c) = image (b)+E..

(a)

(b)

Figure 1: Cameraman. (a) Original 256×256 image with full rank. (b) Original image truncated to be rank 50. (c) 50% randomly masked of (b). (d) Recovered image from (c).

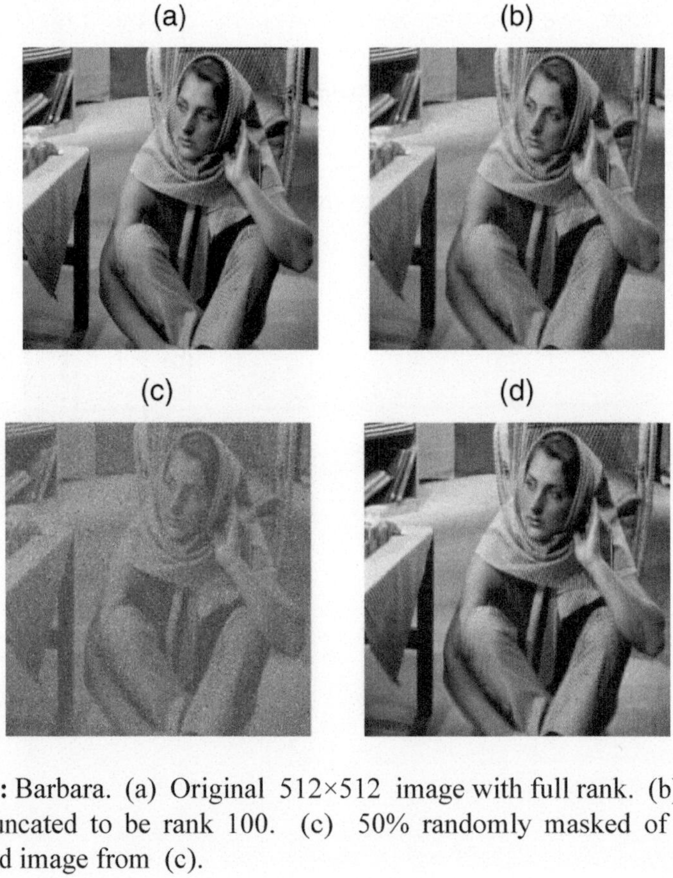

Figure 2: Barbara. (a) Original 512×512 image with full rank. (b) Original image truncated to be rank 100. (c) 50% randomly masked of (b). (d) Recovered image from (c).

Using the SVT algorithm for the low-rank image approximation problem, the lower bound comparison results are shown in Table 3. We calculate $\|E\|_F = \|D-A\|_F$ Fare 8.71e-2 and 7.23e-2 for images Cameraman and Barbara, respectively. But for F-norm of our lower bound (25), we can see that they are 2.59e-5 and 1.09e-5 for images Cameraman and Barbara, respectively. That is to say, our error bounds can verify that the SVT algorithm still can be improved.

Table 3: Lower bound comparison results of low-rank image approximation

	Cameraman	**Barbara**
$\|E\|_F$	8.71e-2	7.23e-2
Bound (25)	2.59e-5	1.09e-5
Iters	200	200

CONCLUSION

Low-rank matrix approximation problem is a field which arises in a number of applications in model selection, system identification, complexity theory, and optics. Based on a useful decomposition of $D^{\dagger}-A^{\dagger}$, this paper reviewed the previous work and provided two sharp lower bounds for the low-rank matrices recovery problem with a unitarily invariant norm.

From our main Theorem 3.5, we can see that if $\|D\|=\|A\|$, then $\|D-A\|=0$. However, in the problem of low-rank matrix approximation, $\|D\|$ is not necessarily equal to $\|A\|$, so the approximation error is present. Furthermore, from the main results, we can clearly find the influence of the spectral norm ($\|\cdot\|_2$) on the low-rank matrix approximation. For example, in Case II, when the maximum eigenvalue of the matrix D is larger, the error of D−A is smaller.

Finally, we use the SVT algorithm for the low-rank matrix approximation problem. Table 2 shows that our lower bounds (24), (25) are smaller than lower bound (12). Simulation results demonstrate that the lower bounds have a very small magnitude. In applications section, we use the SVT algorithm for the low-rank image approximation problem, the lower bounds comparison results are shown in Table 3. From the comparison results, we find that our lower bounds can verify whether the SVT algorithm can be improved.

ACKNOWLEDGEMENTS

This work is partially supported by the National Natural Science Foundation of China under grant No. 11671318, and the Fundamental Research Funds for the Central Universities (Xi'an Jiaotong University, Grant No. xkjc2014008).

AUTHORS' CONTRIBUTIONS

All authors worked in coordination. All authors carried out the proof, read and approved the final version of the manuscript.

REFERENCES

1. Fazel, M, Hindi, H, Boyd, S: A rank minimization heuristic with application to minimum order system approximation. In: Proceedings of the American Control Conference, vol. 6, pp. 4734-4739 (2002)
2. Linial, N, London, E, Rabinovich, Y: The geometry of graphs and some of its algorithmic applications. Combinatorica 15, 215-245 (1995)
3. Tomasi, C, Kanade, T: Shape and motion from image streams under orthography: a factorization method. Int. J. Comput. Vis. 9, 137-154 (1992)
4. Chen, P, Suter, D: Recovering the missing components in a large noisy low-rank matrix: application to SFM. IEEE Trans. Pattern Anal. Mach. Intell. 26(8), 1051-1063 (2004)
5. Liu, ZS, Li, JC, Li, G, Bai, JC, Liu, XN: A new model for sparse and low-rank matrix decomposition. J. Appl. Anal. Comput. 2, 600-617 (2017)
6. Wright, J, Ganesh, A, Shankar, R, Yigang, P, Ma, Y: Robust principal component analysis: exact recovery of corrupted low-rank matrices via convex optimization. In: Twenty-Third Annual Conference on Neural Information Processing Systems (NIPS 2009) (2009)
7. Deerwester, S, Dumains, ST, Landauer, T, Furnas, G, Harshman, R: Indexing by latent semantic analysis. J. Am. Soc. Inf. Sci. Technol. 41(6), 391-407 (1990)
8. Papadimitriou, C, Raghavan, P, Tamaki, H, Vempala, S: Latent semantic indexing, a probabilistic analysis. J. Comput. Syst. Sci. 61(2), 217-235 (2000)
9. Argyriou, A, Evgeniou, T, Pontil, M: Multi-task feature learning. Adv. Neural Inf. Process. Syst. 19, 41-48 (2007)
10. Abernethy, J, Bach, F, Evgeniou, T, Vert, JP: Low-rank matrix factorization with attributes. arXiv:cs/0611124 (2006)
11. Amit, Y, Fink, M, Srebro, N, Ullman, S: Uncovering shared structures in multiclass classification. In: Proceedings of the 24th International Conference on Machine Learning, pp. 17-24. ACM, New York (2007)
12. Zhang, HY, Lin, ZC, Zhang, C, Gao, J: Robust latent low rank representation for subspace clustering. Neurocomputing 145, 369-373 (2014)
13. Mesbahi, M, Papavassilopoulos, GP: On the rank minimization

problem over a positive semidefinite linear matrix inequality. IEEE Trans. Autom. Control 42, 239-243 (1997)

14. Golub, GH, Van Loan, CF: Matrix Computations, 4th edn. Johns Hopkins University Press, Baltimore (2013)
15. Eckart, C, Young, G: The approximation of one matrix by another of lower rank. Psychometrika 1(3), 211-218 (1936)
16. Hotelling, H: Analysis of a complex of statistical variables into principal components. J. Educ. Psychol. 24(6), 417-520 (1932)
17. Jolliffe, I: Principal Component Analysis. Springer, Berlin (1986)
18. Stewart, GW, Sun, JG: Matrix Perturbation Theory. Academic Press, New York (1990)
19. Wedin, P-Å: Perturbation theory for pseudo-inverses. BIT Numer. . 13(2), 217-232 (1973)
20. Chen, YM, Chen, XS, Li, W: On perturbation bounds for orthogonal projections. Numer. Algorithms 73, 433-444 (2016)
21. Sun, JG: Matrix Perturbation Analysis, 2nd edn. Science Press, Beijing (2001)
22. Cai, J-F, Candès, EJ, Shen, ZW: A singular value thresholding algorithm for matrix completion. SIAM J. Optim. 20(4), 1956-1982 (2010)
23. Candès, EJ, Li, X, Ma, Y, Wright, J: Robust principal component analysis? J. ACM 58(3), 1-37 (2011)
24. Tao, M, Yuan, XM: Recovering low-rank and sparse components of matrices from incomplete and noisy observations. SIAM J. Optim. 20(1), 57-81 (2011)

Chapter 9

A Reduced-rank Approach for Implementing Higher-order Volterra Filters

Eduardo L. O. Batista and Rui Seara

LINSE—Circuits and Signal Processing Laboratory, Department of Electrical and Electronics Engineering, Federal University of Santa Catarina, Campus Universitário Trindade, Florianópolis, Brazil

ABSTRACT

The use of Volterra filters in practical applications is often limited by their high computational burden. To cope with this problem, many strategies for implementing Volterra filters with reduced complexity have been proposed in the open literature. Some of these strategies are based on reduced-rank approaches obtained by defining a matrix of filter coefficients and applying the singular value decomposition to such a matrix. Then, discarding the smaller singular values, effective reduced-complexity Volterra implementations can be obtained. The application of this type of approach to

Citation (APA): Batista, E. L., & Seara, R. (2016). A reduced-rank approach for implementing higher-order Volterra filters. EURASIP Journal on Advances in Signal Processing, 2016(1), 118. (8 pages), DOI: https://doi.org/10.1186/s13634-016-0420-5

higher-order Volterra filters (considering orders greater than 2) is however not straightforward, which is especially due to some difficulties encountered in the definition of higher-order coefficient matrices. In this context, the present paper is devoted to the development of a novel reduced-rank approach for implementing higher-order Volterra filters. Such an approach is based on a new form of Volterra kernel implementation that allows decomposing higher-order kernels into structures composed only of second-order kernels. Then, applying the singular value decomposition to the coefficient matrices of these second-order kernels, effective implementations for higher-order Volterra filters can be obtained. Simulation results are presented aiming to assess the effectiveness of the proposed approach.

Keywords: Nonlinear filtering, Reduced-rank implementation, Volterra filter

INTRODUCTION

The first challenge in filtering applications involving nonlinear systems is to choose an adequate model of the nonlinear filter [1]. To meet this challenge, one important filter characteristic that needs to be considered is the trade-off between implementation complexity and approximation capability. The well-known Volterra filter [1] represents one extreme of this trade-off, since its universal approximation capability [2, 3, 4] comes at the cost of a high computational complexity (which is due to the large number of coefficients required for the implementation) [1, 5, 6, 7, 8, 9]. In this context, one topic that has drawn attention from researchers in the last decades is the development of Volterra implementations having an enhanced trade-off between computational complexity and approximation capability.

Several different approaches have been proposed in the open literature aiming to obtain reduced-complexity implementations of Volterra filters. Some of these approaches are based on sparse Volterra implementations that are obtained by zeroing the less significant filter coefficients [9]. Examples of these implementations are the Volterra delay filter [10], the power filter [11], and general diagonally-pruned implementations [12, 13, 14]. Other approaches combine interpolation and sparse implementations for the sake of performance [15, 16, 17]. Frequency-domain approaches also have been used for obtaining effective reduced-complexity Volterra implementations [18, 19]. All aforementioned approaches are, in some sense, based on the use of predefined forms of basis vectors for identifying and discarding the less significant coefficients of a Volterra filter. In contrast, the approaches from [20, 21, 22, 23, 24, 25] involve the identification of particular basis vectors that can then be exploited aiming to reduce the complexity of a Volterra filter. These approaches are typically based on the definition of coefficient matrices, which are decomposed aiming to obtain the basis vectors. The singular value decomposition is often used for carrying out such a matrix decomposition and, as a result, the obtained basis vectors are singular vectors of the coefficient matrix considered. Thus, discarding the singular vectors related to the smaller singular values, effective reduced-complexity reduced-rank Volterra implementations are obtained.

The first reduced-rank approaches used for implementing Volterra filters are focused on second-order Volterra kernels [20, 21, 22]. This is due to the fact that the second-order Volterra coefficients have two indices, which makes the definition of a second-order coefficient matrix a straightforward task. For higher-order filters, matrix-based reduced-rank approaches are usually obtained by considering non-trivial definitions of (often rectangular) coefficient matrices [24], which occasionally lead to ineffective reduced-rank implementations. In this context, the present paper is focused on the development of a novel reduced-rank approach for implementing higher-order Volterra filters. This approach is based on a new form of Volterra kernel implementation that allows converting a higher-order Volterra kernel into a structure composed of second-order kernels. Then, applying second-order reduced-rank implementation strategies to such a structure, effective reduced-rank implementations for higher-order Volterra filters can be achieved.

The remainder of this paper is organized as described in the following. Section 2revisits the Volterra filters, discussing the redundancy-removed and matrix-form representations of kernel input-output relationships. Also, in Section 2, the reduced-rank implementations of Volterra filters are briefly described. Section 3 is dedicated to the contributions of this paper, comprising a new form of kernel implementation and the description of the proposed approach for implementing reduced-rank Volterra filters. Finally, Sections 4 and 5 are dedicated to present the experimental results and the concluding remarks, respectively.

The mathematical notation considered in this paper is based on the standard practice of using lowercase boldface letters for vectors, uppercase boldface letters for matrices, and both italic Roman and Greek letters for scalar quantities. Moreover, superscript T stands for transpose, ⊗ represents the Kronecker product, and $\|\cdot\|_2$ denotes a quadratic norm. Additionally, underbars specify variables related to the redundancy-removed Volterra representation and overbars indicate variables related to the proposed approach.

VOLTERRA FILTERS AND REDUCED-RANK IMPLEMENTATIONS

A truncated Pth-order Volterra filter is composed of P kernels, each corresponding to a certain order of polynomial nonlinearity [1]. The output y(n) of such a filter is obtained from

$$y(n) = \sum_{p=1}^{P} y_p(n) \tag{1}$$

with $y_p(n)$ representing the output of the pth-order kernel. In its standard form, the input-output relationship of the pth-order kernel is given by

$$y_p(n) = \sum_{m_1=0}^{N-1} \cdots \sum_{m_p=0}^{N-1} h_{p(m_1,m_2,\ldots,m_p)} \prod_{k=1}^{p} x(n-m_k) \tag{2}$$

with x(n) denoting the input signal and $h_{p(m1,m2,\ldots,mp)}$, the pth-order coefficients. In this work, the standard kernels are assumed to be symmetric, which means that $h_{p(m_1,m_2,\ldots,m_p)} = h_{p(m_2,m_1,\ldots,m_p)} = \cdots = h_{p(m_p,m_{p-1},\ldots,m_1)}$ (i.e., all pth-order coefficients with permutated indices have identical values). This assumption is used without loss of generality, since any standard Volterra kernel can be represented in symmetric form [1, 25].

The first-order kernel of the Volterra filter is a linear kernel whose input-output relationship

$$y_1(n) = \sum_{m_1=0}^{N-1} h_{p(m_1)} x(n-m_1) \tag{3}$$

is that of a standard finite impulse response (FIR) filter with N coefficients. Thus, (3) can be rewritten in a vector form as

$$y_1(n) = \mathbf{h}_1^{\mathrm{T}} \mathbf{x}_1(n) \tag{4}$$

with

$$\mathbf{h}_1 = [h_{1(0)} \quad h_{1(1)} \quad \cdots \quad h_{1(N-1)}]^{\mathrm{T}} \tag{5}$$

and

$$\mathbf{x}_1(n) = [x(n)\ x(n-1)\ \cdots\ x(n-N+1)]^{\mathrm{T}}. \tag{6}$$

The other kernels (with p≥2) are nonlinear kernels whose outputs depend on cross products of samples of the input signal. As described in [1], the input-output relationships of the nonlinear kernels can also be expressed in vector form. Thus, for the pth-order kernel, one has

$$y_p(n) = \mathbf{h}_p^{\mathrm{T}} \mathbf{x}_p(n), \tag{7}$$

where $\mathbf{h}_p$ is the pth-order coefficient vector (composed of coefficients $h_{p(m1,m2,\ldots,mp)}$ with $m_1,\ldots,m_p$ ranging from 0 to N−1) and $\mathbf{x}_p(n)=\mathbf{x}_1(n)\otimes\mathbf{x}_{p-1}(n)$ is the pth-order input vector.

Note, from (2) and (7), that the standard pth-order Volterra kernel has one coefficient for each pth-order cross product of input samples, resulting in a number of coefficients given by

$$N_p = N^p. \tag{8}$$

This number increases exponentially with both the memory size and the order of the kernel. As a consequence, the computational cost for implementing a Volterra filter may become prohibitive even in applications involving kernels with relatively small memory size.

Redundancy-removed Implementation

The large number of coefficients required to implement Volterra filters can be reduced by exploiting the redundancy of part of the coefficients of the standard nonlinear kernels [1, 15]. Such redundancy arises from the fact that coefficients with permutated indices (e.g., $h_{2(0,1)}$ and $h_{2(1,0)}$) are multiplied by the same cross product of the input signal (e.g., x(n)x(n−1) in the case of $h_{2(0,1)}$ and $h_{2(1,0)}$) when the kernel output is evaluated. Thus, merging redundant coefficients into a single coefficient, the input-output relationship of the pth-order kernel, given by (2), can be rewritten as

$$y_p(n) = \sum_{m_1=0}^{N-1} \cdots \sum_{m_p=m_{p-1}}^{N-1} \underline{h}_{p(m_1,m_2,\ldots,m_p)} \prod_{k=1}^{p} x(n-m_k) \tag{9}$$

with $\underline{h}_{p(m_1,m_2,\ldots,m_p)}$ denoting the pth-order coefficients of the redundancy-removed kernel. Such representation of the kernel input-output relationship, known as triangular [1] or even redundancy-removed [15] representation, results in a number of coefficients given by

$$\underline{N}_p = \frac{(N+p-1)!}{(N-1)!p!}. \tag{10}$$

Moreover, it is important to highlight that the reduction in the number of coefficients from (8) to (10) obtained by using the redundancy-removed implementation comes without loss of generality (i.e., a given kernel can be equivalently implemented by using either the standard or the redundancy-removed implementation).

As in the case of the standard Volterra kernels, the input-output relationship of the redundancy-removed ones can also be represented in vector form. Thereby, we have [9]

$$y_p(n) = \underline{\mathbf{h}}_p^{\mathrm{T}} \underline{\mathbf{x}}_p(n), \tag{11}$$

where $\underline{\mathbf{h}}_p$ is the pth-order redundancy-removed coefficient vector, which is composed of coefficients $\underline{h}_{p(m_1,m_2,\ldots,m_p)}$, and $\underline{\mathbf{x}}_p(n) = \mathbf{L}_p[\mathbf{x}_{p-1}(n) \otimes \mathbf{x}_1(n)]$ is the pth-order redundancy-removed input vector with $\mathbf{L}_p$ denoting the pth-order elimination matrix [9]. For instance, considering the second-order kernel, (9) results in

$$y_2(n) = \sum_{m_1=0}^{N-1} \sum_{m_2=m_1}^{N-1} \underline{h}_{2(m_1,m_2)} x(n-m_1)x(n-m_2), \tag{12}$$

whereas (11) results in

$$y_2(n) = \underline{\mathbf{h}}_2^{\mathrm{T}} \underline{\mathbf{x}}_2(n) \tag{13}$$

with

$$\begin{aligned}\underline{\mathbf{h}}_2 =&[\underline{h}_{2(0,0)} \ \underline{h}_{2(0,1)} \ \cdots \ \underline{h}_{2(0,N-1)} \\ &\underline{h}_{2(1,1)} \ \cdots \ \underline{h}_{2(N-1,N-1)}]^{\mathrm{T}}\end{aligned} \tag{14}$$

and

$$\begin{aligned}\underline{\mathbf{x}}_2(n) =&[x^2(n) \ x(n)x(n-1) \ \cdots \ x(n)x(n-N+1) \\ &x^2(n-1) \ \cdots \ x^2(n-N+1)].\end{aligned} \tag{15}$$

Matrix-form Kernel Representation

The input-output relationship of nonlinear Volterra kernels can also be formulated as a function of coefficient matrices instead of coefficient vectors. This type of representation is especially suited for the development of reduced-rank implementations, as will be shown in the next section. For the standard second-order kernel, a matrix-form representation can be obtained by considering m_1 and m_2 as coordinates of a Cartesian space to define the following second-order coefficient matrix:

$$\mathbf{H}_2 = \begin{bmatrix} h_{2(0,0)} & h_{2(0,1)} & \cdots & h_{2(0,N-1)} \\ h_{2(1,0)} & h_{2(1,1)} & \cdots & h_{2(1,N-1)} \\ \vdots & \vdots & \ddots & \vdots \\ h_{2(N-1,0)} & h_{2(N-1,1)} & \cdots & h_{2(N-1,N-1)} \end{bmatrix}. \quad (16)$$

Then, from (16), the input-output relationship of the second-order kernel can be written as

$$y_2(n) = \mathbf{x}_1^{\mathrm{T}}(n)\mathbf{H}_2\mathbf{x}_1(n). \quad (17)$$

In the case of the redundancy-removed second-order kernel, the following coefficient matrix can be defined:

$$\underline{\mathbf{H}}_2 = \begin{bmatrix} \underline{h}_{2(0,0)} & \underline{h}_{2(0,1)} & \cdots & \underline{h}_{2(0,N-1)} \\ 0 & \underline{h}_{2(1,1)} & \cdots & \underline{h}_{2(1,N-1)} \\ \vdots & \vdots & \ddots & \vdots \\ 0 & 0 & \cdots & \underline{h}_{2(N-1,N-1)} \end{bmatrix}. \quad (18)$$

Now, considering (18), (13) can be rewritten as

$$y_2(n) = \mathbf{x}_1^{\mathrm{T}}(n)\underline{\mathbf{H}}_2\mathbf{x}_1(n). \quad (19)$$

The approach presented in [24] generalizes the matrix-form representations from (16) to (19) for higher-order kernels. Such an approach is based on defining an $\underline{N}_{p_1} \times \underline{N}_{p_2}$ pth-order coefficient matrix $\hat{\mathbf{H}}_{p,p_1,p_2}$ with $p=p_1+p_2$. This matrix contains the $\underline{N}_p$ coefficients of the pth-order redundancy-removed kernel arranged in such a way that the kernel output can be written as

$$y_p(n) = \underline{\mathbf{x}}_{p_1}^{\mathrm{T}}(n)\hat{\mathbf{H}}_{p,p_1,p_2}\underline{\mathbf{x}}_{p_2}(n), \quad (20)$$

where $\underline{\mathbf{x}}_{p_1}(n)$ and $\underline{\mathbf{x}}_{p_2}(n)$ are the redundancy-removed input vectors with orders p_1 and p_2, respectively. It is important to mention that the number of coefficients $\underline{N}_p$ of the pth-order kernel is smaller than the number of entries $(\underline{N}_{p_1} \times \underline{N}_{p_2})$ of $\hat{\mathbf{H}}_{p,p_1,p_2}$. As a result, several elements of $\hat{\mathbf{H}}_{p,p_1,p_2}$ are in fact equal to zero [24].

Reduced-rank Implementations

As described in [20, 21, 22, 24], approaches based on low-rank approximations can be used for obtaining effective reduced-complexity Volterra implementations. Most of these approaches are based on using the singular value decomposition along with matrix-form representations of Volterra kernels. For instance, the approach used for implementing second-order kernels described in [22] is based on the application of the singular value decomposition to the standard second-order coefficient matrix $\mathbf{H}_2$ given by (16). Since this matrix is symmetric, its left singular vectors are equal to its right singular vectors and, as a result, one obtains

$$\mathbf{H}_2 = \sum_{k=0}^{N-1} \lambda_k \tilde{\mathbf{h}}_{2,k} \tilde{\mathbf{h}}_{2,k}^{\mathrm{T}}, \tag{21}$$

where λ_k and $\tilde{\mathbf{h}}_{2,k}$ are, respectively, the kth singular value and the kth singular vector of $\mathbf{H}_2$. Now, substituting (21) into (17), one gets

$$y_2(n) = \sum_{k=0}^{N-1} \lambda_k [\mathbf{x}_1^{\mathrm{T}}(n) \tilde{\mathbf{h}}_{2,k}]^2 . \tag{22}$$

Note that $\mathbf{x}_1^{\mathrm{T}}(n)\tilde{\mathbf{h}}_{2,k}$ corresponds to the input-output relationship of an FIR filter with coefficient vector given by $\tilde{\mathbf{h}}_{2,k}$. Thus, (22) is in fact the input-output relationship of the structure shown in Fig. 1, which is a parallel structure of N FIR filters with their squared outputs multiplied by the singular values of $\mathbf{H}_2$. Moreover, since the singular vectors $\tilde{\mathbf{h}}_{2,k}$ (with k=0,…,N−1) are unit vectors, the branches of the structure from Fig. 1 involving small values of λ_k can be disregarded, resulting in a reduction of computational complexity with low impact on the implementation precision.

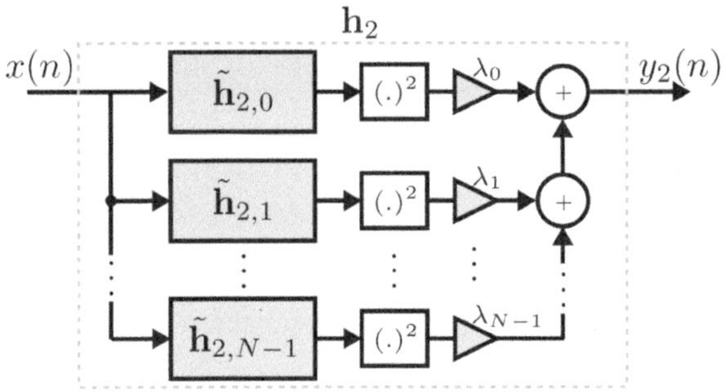

Figure 1: Block diagram of the implementation of a second-order kernel based on the singular value decomposition.

The implementation depicted in Fig. 1 is based on the standard matrix-form representation of the input-output relationship of the second-order kernel. The redundancy-removed matrix-form representation given by (19) can also be used for obtaining a reduced-rank implementation. In this case, since the left and right singular vectors are not the same (due to the fact that $\underline{\mathbf{H}}_2$ is not symmetric), the resulting input-output relationship is

$$y_2(n) = \sum_{k=0}^{N-1} \lambda_k [\mathbf{x}_1^{\mathrm{T}}(n)\tilde{\mathbf{h}}_{2\mathrm{l},k}]\,[\mathbf{x}_1^{\mathrm{T}}(n)\tilde{\mathbf{h}}_{2\mathrm{r},k}] \tag{23}$$

with $\tilde{\mathbf{h}}_{2\mathrm{l},k}$ representing the kth left singular vector of $\underline{\mathbf{H}}_2$ and $\tilde{\mathbf{h}}_{2\mathrm{r},k}$ denoting the kth right singular vector. By comparing (22) and (23), one can notice that the latter (which is based on the redundancy-removed implementation) results in a structure with higher computational cost than the structure resulting from the former (based on the standard representation). Thus, one verifies that the standard representation is in fact more advantageous than the less costly redundancy-removed one for obtaining this type of reduced-rank implementations of second-order kernels.

In the case of higher-order kernels (with $p \geq 3$), one appealing approach for obtaining reduced-rank implementations is the one used in [24] to obtain the so-called parallel-cascade Volterra structures. Such an approach is based on the application of the singular value decomposition to the coefficient matrix of the general matrix-form kernel representation given by (20). For instance, considering the case of a third-order kernel, (20) becomes

$$y_3(n) = \underline{\mathbf{x}}_1^{\mathrm{T}}(n)\hat{\mathbf{H}}_{3,1,2}\underline{\mathbf{x}}_2(n) \tag{24}$$

with $\underline{\mathbf{x}}_1(n) = \mathbf{x}_1(n)$. By applying the singular value decomposition to $\hat{\mathbf{H}}_{3,1,2}$

, one obtains

$$\hat{\mathbf{H}}_{3,1,2} = \sum_{k=0}^{N-1} \sigma_k \hat{\mathbf{h}}_{3,1,k} \hat{\mathbf{h}}_{3,2,k}^{\mathrm{T}}, \tag{25}$$

where σ_k is the kth singular value of $\hat{\mathbf{H}}_{3,1,2}$, $\hat{\mathbf{h}}_{3,1,k}$ is the kth left singular vector, and $\hat{\mathbf{h}}_{3,2,k}$ is the kth right singular vector. Then, substituting (25) into (24) and manipulating the resulting expression, one gets

$$y_3(n) = \sum_{k=0}^{N-1} \sigma_k [\, \hat{\mathbf{h}}_{3,1,k}^{\mathrm{T}} \underline{\mathbf{x}}_1(n)] \, [\, \hat{\mathbf{h}}_{3,2,k}^{\mathrm{T}} \underline{\mathbf{x}}_2(n)] \, . \tag{26}$$

From (4), one can notice that $\hat{\mathbf{h}}_{3,1,k}^{\mathrm{T}} \underline{\mathbf{x}}_1(n)$ corresponds to the filtering of the input signal by a first-order kernel (FIR filter) with coefficient vector $\hat{\mathbf{h}}_{3,1,k}$. Similarly, considering (13), one notices that $\hat{\mathbf{h}}_{3,2,k}^{\mathrm{T}} \underline{\mathbf{x}}_2(n)$ corresponds to the filtering of the input signal by a second-order kernel with coefficient vector $\hat{\mathbf{h}}_{3,2,k}$. Thus, (26) in fact corresponds to the structure depicted in Fig. 2, which is composed of a set of N branches, with the output of the kth branch given by the product of the outputs of two kernels (a first-order kernel and a second-order one), weighted by the kth singular value σ_k of $\hat{\mathbf{H}}_{3,1,2}$. As in the case of the structure of Fig. 1, reduced-complexity reduced-rank Volterra implementations can be obtained from the parallel-cascade structure of Fig. 2, removing the branches related to the smallest singular values of $\hat{\mathbf{H}}_{3,1,2}$. In addition, as mentioned in [24], the second-order blocks of the structure of Fig. 2 can be further decomposed by using reduced-rank approaches, which allows a more detailed singular-value-dependent kernel pruning. However, this pruning is not a straightforward task due to the hierarchical nature of the resulting structure (i.e., it involves reduced-rank decompositions with different levels of importance).

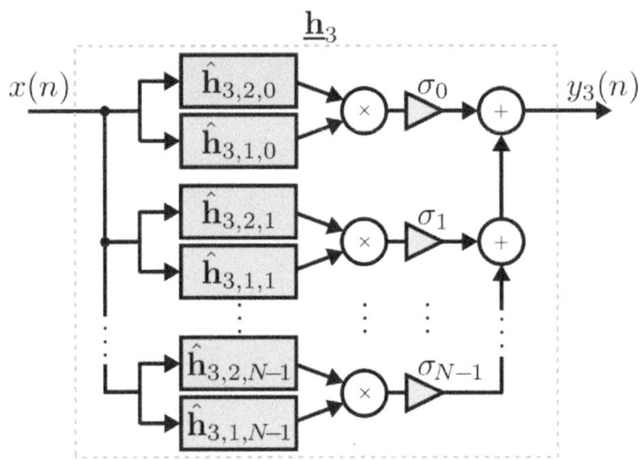

Figure 2: Block diagram of the parallel-cascade implementation of a third-order kernel.

NOVEL REDUCED-RANK APPROACH FOR IMPLEMENTING VOLTERRA FILTERS

This section is devoted to the development of a novel reduced-rank approach for implementing Volterra filters. To this end, a new strategy for implementing higher-order kernels using a parallel structure composed of lower-order kernels is first discussed. Then, such a strategy is exploited along with the singular value decomposition to obtain the proposed reduced-rank implementation approach.

Kernel Implementation Redesigned

In this section, the aim is to develop a new form of Volterra kernel implementation that allows factorizing higher-order kernels in terms of lower-order ones. For the second-order kernel, such a type of implementation is obtained by rewriting the second-order redundancy-removed input-output relationship as

$$y_2(n) = \sum_{m_1=0}^{N-1} x(n-m_1) \sum_{m_2=m_1}^{N-1} \underline{h}_{2(m_1,m_2)} x(n-m_2). \tag{27}$$

The rightmost summation term in (27) corresponds to the input-output relationship of a first-order kernel (an FIR filter) with memory size N−m $_1$, coefficient vector

$$\bar{\mathbf{h}}_{2,m_1} = [\underline{h}_{2(m_1,m_1)} \;\; \underline{h}_{2(m_1,m_1+1)} \;\; \cdots \;\; \underline{h}_{2(m_1,N-1)}]^{\mathrm{T}} \tag{28}$$

and input vector

$$\bar{\mathbf{x}}_{2,m_1}(n) = [x(n-m_1) \;\; x(n-m_1-1) \;\; \cdots \;\; x(n-N+1)]^{\mathrm{T}}. \tag{29}$$

Thus, (27) can be rewritten as

$$y_2(n) = \sum_{m_1=0}^{N-1} x(n-m_1)\bar{\mathbf{h}}_{2,m_1}^{\mathrm{T}}\bar{\mathbf{x}}_{2,m_1}(n). \tag{30}$$

Note that, in (30), the output of the second-order kernel is evaluated by summing the outputs of N first-order kernels multiplied by delayed samples of the input signal. Therefore, (30) can be seen as a decomposition of the second-order kernel into a parallel structure composed of first-order kernels.

In the case of the third-order kernel, the redundancy-removed input-output relationship can be written as

$$y_3(n) = \sum_{m_1=0}^{N-1} x(n-m_1) \times \sum_{m_2=m_1}^{N-1} \sum_{m_3=m_2}^{N-1} \underline{h}_{3(m_1,m_2,m_3)} x(n-m_2)x(n-m_3). \tag{31}$$

Now, considering (12), one can notice that the double summation in (31) corresponds to the output of a second-order kernel with coefficient vector

$$\bar{\mathbf{h}}_{3,m_1} = [\underline{h}_{3(m_1,m_1,m_1)} \;\; \underline{h}_{3(m_1,m_1,m_1+1)} \;\; \cdots \;\; \underline{h}_{3(m_1,m_1,N-1)} \;\; \underline{h}_{3(m_1,m_1+1,m_1+1)} \;\; \cdots \;\; \underline{h}_{3(m_1,N-1,N-1)}]^{\mathrm{T}} \tag{32}$$

and input vector

$$\bar{\mathbf{x}}_{3,m_1}(n) = [x^2(n-m_1) \;\; x(n-m_1)x(n-m_1-1) \;\; \cdots \;\; x(n-m_1)x(n-N+1) \;\; x^2(n-m_1-1) \;\; \cdots \;\; x^2(n-N+1)]^{\mathrm{T}}. \tag{33}$$

Consequently, from (13), (31) can be rewritten as

$$y_3(n) = \sum_{m_1=0}^{N-1} x(n-m_1)\bar{\mathbf{h}}_{3,m_1}^{\mathrm{T}}\bar{\mathbf{x}}_{3,m_1}(n) \tag{34}$$

which corresponds to the decomposition of the third-order kernel into a structure composed of second-order kernels.

Similarly to (30) and (34), for the pth-order kernel, the following input-output relationship can be obtained:

$$y_p(n) = \sum_{m_1=0}^{N-1} x(n - m_1)\bar{\mathbf{h}}_{p,m_1}^{\mathrm{T}} \bar{\mathbf{x}}_{p,m_1}(n), \tag{35}$$

where the product $\bar{\mathbf{h}}_{p,m_1}^{\mathrm{T}} \bar{\mathbf{x}}_{p,m_1}(n)$ corresponds to the input-output relationship of a (p−1)th-order kernel with memory size $N-m_1$. Thus, from (35), we can infer that any pth-order kernel can be decomposed into N kernels with order p−1, as illustrated in Fig. 3.

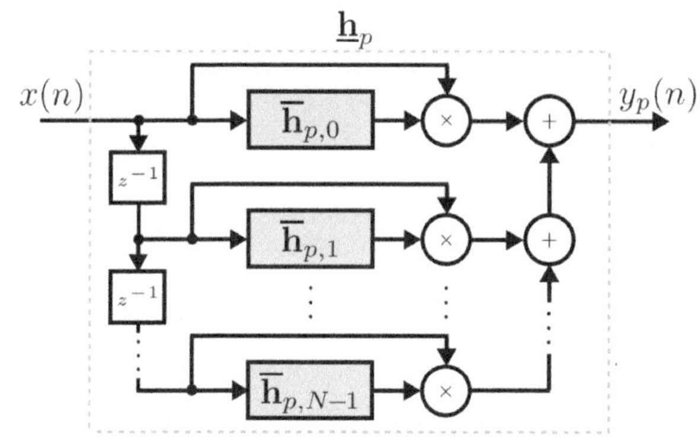

Figure 3: Block diagram of a pth-order kernel implementation using a parallel structure composed of (p−1)th-order kernels.

Proposed Approach

Aiming to develop the proposed reduced-rank approach for implementing Volterra filters, we first consider that the kernel implementation strategy introduced in Section 3.1 can be used to decompose any higher-order kernel (with p≥3) into a parallel structure composed exclusively of second-order kernels. This decomposition is straightforward for the third-order kernel, since, by using p=3 in (35), one obtains a structure composed of N second-order kernels, as described by (31)–(34). In the case of the fourth-order kernel, (35) can be used first for obtaining a structure composed of N third-order kernels and then for decomposing each of these third-order kernels into second-order kernels. As a result, a structure composed of N^2 second-order kernels is obtained for the implementation of the fourth-order kernel. Following this rationale, one can notice that, by using (35) to carry out successive kernel decompositions, an implementation composed of $N^{(p-2)}$ second-order kernels can always be obtained for a pth-order kernel.

Now, the idea behind the proposed reduced-rank approach for implementing Volterra filters is to exploit the fact that any pth-order kernel can be decomposed into a parallel structure of second-order kernels as previously described. Taking into account this fact, a reduced-rank implementation for the pth-order kernel can be obtained by applying, to each second-order kernel resulting from the kernel decomposition, the strategy used for obtaining the reduced-rank implementation of Fig. 1 (see Section 2.3). Thus, one obtains a structure composed of $N^{(p-2)}$ blocks, each having the form of the structure of Fig. 1. Then, disregarding the branches of these blocks related to the smaller singular values, a reduced-complexity reduced-rank kernel implementation is obtained. In this context, the proposed approach can be summarized as follows:

i) Exploit the strategy described in Section 3.1 (see Fig. 3) to obtain an implementation of the pth-order kernel of interest in the form of a parallel structure composed of second-order kernels.

ii) Use the standard matrix-form representation (see Section 2.2) to represent all second-order kernels that compose the kernel of interest.

iii) Obtain reduced-rank implementations of such second-order kernels by using the singular value decomposition as described in Section 2.3.

iv) Remove (prune) the branches of the resulting structure related to the smaller singular values of the involved second-order coefficient matrices.

The proposed reduced-rank approach for implementing Volterra filters consists of the application of these four steps to all kernels, which allows obtaining effective reduced-complexity Volterra filter implementations.

SIMULATION RESULTS

This section aims to assess the effectiveness of the proposed reduced-rank approach for obtaining reduced-complexity implementations of Volterra filters. To this end, the proposed approach is compared with the parallel-cascade (PC) one from [24] in the context of the implementation of third-order and forth-order kernels whose coefficients are known in advance. The effectiveness of these approaches is assessed in terms of normalized misalignment [26], which is defined as

$$\mathcal{M} = 10 \log_{10}\left(\frac{||\underline{\mathbf{h}}_k - \underline{\mathbf{h}}_{rr}||_2^2}{||\underline{\mathbf{h}}_k||_2^2}\right) \tag{36}$$

with $\underline{\mathbf{h}}_k$ denoting the coefficient vector of the kernel to be implemented, and $\underline{\mathbf{h}}_{rr}$, the coefficient vector obtained by using the reduced-rank approach. A hierarchical branch-level pruning is applied to the parallel structures obtained by using the approaches considered, which means that one branch is removed at a time, with the branches related to the smallest singular values removed first. After the removal of each branch, both the normalized misalignment and the number of required arithmetic operations are evaluated, resulting in the curves used here for comparing the different approaches. Markers are presented along these curves, indicating each evaluated pair of normalized misalignment and number of operations per sample.

Example 1: Modeling of a Third-order Kernel

The third-order kernel considered in this example is obtained from a system modeling experiment in which a diode limiter circuit used in guitar distortion pedals [27, 28] is modeled using an LMS-adapted Volterra filter (with memory size N_a=10, sampling rate of f_s=44.1 kHz, and white input signal). The PC and the proposed reduced-rank approaches are used here for implementing the third-order kernel of the Volterra model obtained in such an experiment. As a result, the curves of normalized misalignment as a function of the number of operations per sample shown in Fig. 4 are obtained. A vertical dotted line pointing out the number of operations per sample required by the corresponding redundancy-removed Volterra implementation (around 495 operations) is also included in this figure aiming to establish a limit after which it is no longer interesting to use reduced-rank implementations. One can notice from Fig. 4 that the proposed approach outperforms the PC one for the case considered here; since, a smaller normalized misalignment is obtained for any given number of operations per sample. For instance, if a misalignment below −15 dB is desired, a PC-based implementation would require at least 280 operations per sample (i.e., a reduction of about 43% with respect to the number of coefficients of the redundancy-removed Volterra kernel), whereas an implementation based on the proposed approach would require only 139 operations per sample (a reduction of almost 72%).

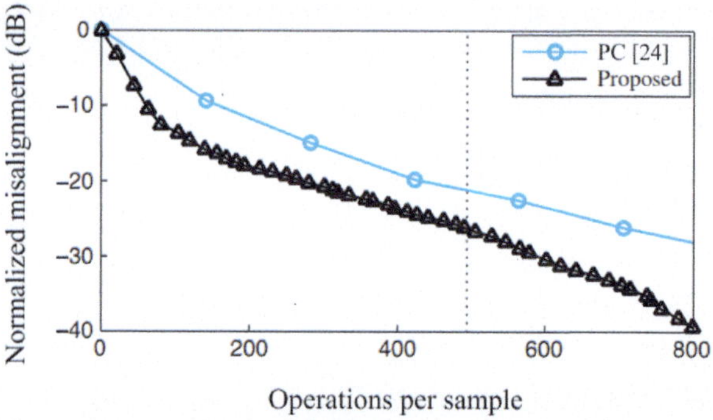

Figure 4: Number of coefficients required by different reduced-rank implementations of a third-order kernel.

Example 2: Modeling of a Fourth-order Kernel

The kernel considered in this example is similar to the fourth-order satellite-system model used in [24]. Such a model is obtained by using a cascade of a Butterworth low-pass filter with a memoryless fourth-power nonlinearity and a Chebyshev low-pass filter. For the sake of simplicity, we consider a version of this model in which the Butterworth and Chebyshev IIR filters are replaced by FIR versions obtained by truncating their impulse responses to 30 samples. Then, a fourth-order kernel with memory size 59 is obtained, which is again truncated to memory size 30. As a result, this kernel has a total of $\underline{N}_4 = 40,920$ coefficients in its redundancy-removed representation, requiring around 86,800 operations per sample for its implementation. It is important to mention that this kernel admits a symmetric representation for its coefficient matrix $\hat{\mathbf{H}}_{4,2,2}$. Consequently, the branches of PC-based structures will be composed of a single second-order kernel with its squared output multiplied by one of the singular values of $\hat{\mathbf{H}}_{4,2,2}$. The results obtained for this example are shown in Fig. 5. For a better visualization (due to the high density of the obtained points), we have used one marker for each 30 points in the curve of the proposed approach. From such results, one notices that the reduced-rank approaches are capable of modeling the considered fourth-order kernel with very good accuracy. For instance, −60 dB of misalignment is obtained through the proposed approach with around 12,520 operations per sample, which corresponds to almost 86% of complexity reduction. Moreover, Fig. 5 also shows that the proposed approach outperforms the PC approach in terms of trade-off between performance and complexity either for the range of computational cost from 0 to almost 27,000 operations per sample or for the range of misalignment from 0 to about −120 dB.

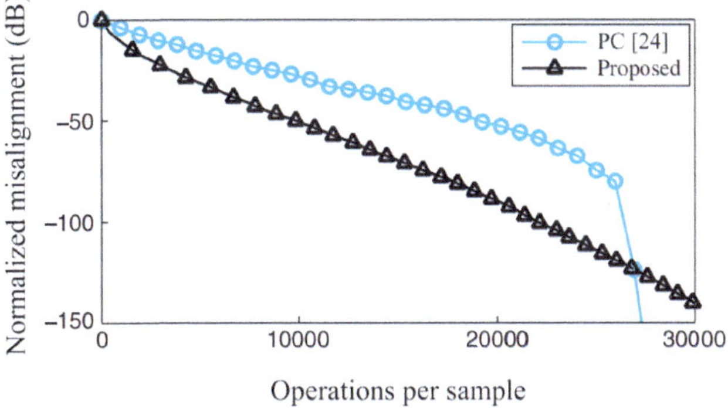

Figure 5: Number of coefficients required by different reduced-rank implementations of a fourth-order kernel.

CONCLUSIONS

In this paper, a novel reduced-rank approach for implementing higher-order Volterra filters was introduced. Such an approach is based on a new form of kernel implementation that allows converting any higher-order kernel into a structure composed exclusively of second-order kernels. Then, exploiting the singular value decomposition along with the coefficient matrices of these second-order kernels, a reduced-rank implementation for the higher-order Volterra filter can be achieved. Numerical simulation results were shown, confirming the effectiveness of the proposed approach in obtaining reduced-complexity implementations of Volterra filters.

ACKNOWLEDGEMENTS

The authors are thankful to Gabriel Celso Kulevicz da Silva for providing the experimental data used in Example 1 of Section 4. Thanks are also due to the editors and the anonymous reviewers for their valuable and constructive comments and suggestions, from which the revision of this paper has benefited significantly.

REFERENCES

1. VJ Mathews, GL Sicuranza, Polynomial Signal Processing (John Wiley & Sons, Inc., New York, 2000).
2. S Boyd, LO Chua, Fading memory and the problem of approximating nonlinear operators with Volterra series. IEEE Trans. Circ. Syst. CAS-32(11), 1150–1161 (1985).
3. IW Sandberg, R+ fading memory and extensions of input-output maps. IEEE Trans. Circ. Syst. I, Reg. Papers. 49(11), 1586–1591 (2002).
4. S Orcini, Improving the approximation ability of Volterra series identified with a cross-correlation method. Nonlinear Dyn. 78(4), 2861–2869 (2014).
5. A Carini, GL Sicuranza, in IEEE Int. Conf. Acoust. Speech Signal Process. Even mirror Fourier nonlinear filters (IEEEVancouver, 2013), pp. 5608–5612.
6. A Carini, GL Sicuranza, Fourier nonlinear filters. Signal Process. 94:, 183–194 (2014).
7. A Carini, S Cecchi, L Romoli, GL Sicuranza, Legendre nonlinear filters. Signal Process. 109:, 84–94 (2015).
8. A Carini, GL Sicuranza, A study about Chebyshev nonlinear filters. Signal Process. 122:, 24–32 (2016).
9. ELO Batista, R Seara, On the performance of adaptive pruned Volterra filters. Signal Process. 93(7), 1909–1920 (2013).
10. L Tan, J Jiang, An adaptive technique for modeling second-order Volterra systems with sparse kernels. IEEE Trans. Circ. Syst. II: Analog Digit. Sig. Proc. 45(12), 1610–1615 (1998).
11. F Kuech, W Kellermann, Orthogonalized power filters for nonlinear acoustic echo cancellation. Signal Process. 86(6), 1168–1181 (2006).
12. A Fermo, A Carini, GL Sicuranza, Low-complexity nonlinear adaptive filters for acoustic echo cancellation in GSM handset receivers. Eur. Trans. Telecommun. 14(2), 161–169 (2003).
13. M Zeller, W Kellermann, in Proc. Int. Work. Acoustic Echo and Noise Control (IWAENC). Coefficient pruning for higher-order diagonals of Volterra filters representing Wiener-Hammerstein models (Seattle, 2008).
14. M Zeller, LA Azpicueta-Ruiz, J Arenas-Garcia, W Kellermann, Adaptive Volterra filters with evolutionary quadratic kernels using

a combination scheme for memory control. IEEE Trans. Signal Process. 59(4), 1449–1464 (2011).

15. ELO Batista, OJ Tobias, R Seara, A sparse-interpolated scheme for implementing adaptive Volterra filters. IEEE Trans. Signal Process. 58(4), 2022–2035 (2010).
16. ELO Batista, R Seara, in Proc. 19th Eur. Signal Process. Conf. Adaptive NLMS diagonally-interpolated Volterra filters for network echo cancellation (IEEE/EURASIPBarcelona, 2011), pp. 1430–1434.
17. ELO Batista, R Seara, A fully LMS/NLMS adaptive scheme applied to sparse-interpolated Volterra filters with removed boundary effect. Signal Process. 92(10), 2381–2393 (2012).
18. MJ Reed, MOJ Hawksford, Efficient implementation of the Volterra filter. IEE Proc.-Vis. Image Signal Process. 147(2), 109–114 (2000).
19. F Kuech, W Kellerman, Partitioned block frequency-domain adaptive second-order Volterra filter. IEEE Trans. Signal Process. 53(2), 564–575 (2005).
20. H-H Chiang, CL Nikias, AN Venetsanopoulos, Efficient implementations of quadratic digital filters. Trans. Acoust., Speech, Signal Process. 34(6), 1511–1528 (1986).
21. Y Lou, CL Nikias, AN Venetsanopoulos, Efficient VLSI array processing structures for adaptive quadratic digital filters. Circuits, Syst. Signal Process. 7(2), 253–273 (1988).
22. S Marsi, GL Sicuranza, in Conf. Rec. 27th Asilomar Conf. Signals, Syst. Comput, vol. 2. On reduced-complexity approximations of quadratic filters (IEEEPacific Grove, 1993), pp. 1026–1030.
23. RD Nowak, BD Van Veen, Tensor product basis approximations for Volterra filters. IEEE Trans. Signal Process. 44(1), 36–50 (1996).
24. TM Panicker, VJ ews, GL Sicuranza, Adaptive parallel-cascade truncated Volterra filters. IEEE Trans. Signal Process. 46(10), 2664–2673 (1998).
25. G Favier, AY Kibangou, T Bouilloc, Nonlinear system modeling and identification using Volterra-PARAFAC models. Int. J. Adapt. Control Signal Process. 26(1), 30–53 (2012).
26. J Benesty, C Paleologu, S Ciochina, Sparse Adaptive Filters for Echo Cancellation (Morgan and Claypool Publishers, San Rafael, CA, 2010).
27. DT Yeh, JS Abel, JO Smith, in Proc. 10th Conf. on Digital Audio Effects (DAFx-07). Simulation of the diode limiter in guitar distortion circuits

by numerical solution of ordinary differential equations (Bordeaux, France, 2007).

28. GCK da Silva, Contribuições para Implementação Digital de um Pedal de Efeito de Áudio do Tipo Overdrive (in Portuguese). Master's thesis, Federal University of Santa Catarina, Florianópolis, Brazil, 2016.

Chapter 10

A Semi-smoothing Augmented Lagrange Multiplier Algorithm for Low-rank Toeplitz Matrix Completion

Ruiping Wen[1], Shuzhen Li[2] and Yonghong Duan[3]

[1]Key Laboratory of Engineering & Computing Science, Shanxi Provincial Department of Education/Department of Mathematics, Taiyuan Normal University, Jinzhong, P.R. China

[2]Department of Mathematics, Taiyuan Normal University, Jinzhong, P.R. China

[3]Department of Mathematics, Taiyuan University, Taiyuan, P.R. China

ABSTRACT

The smoothing augmented Lagrange multiplier (SALM) algorithm is a generalization of the augmented Lagrange multiplier algorithm for completing a Toeplitz matrix, which saves computational cost of the singular value decomposition (SVD) and approximates well the solution. However, the communication of numerous data is computationally demanding at each iteration step. In this paper, we propose an accelerated scheme to the SALM

Citation (APA): Wen, R., Li, S., & Duan, Y. (2019). A semi-smoothing augmented Lagrange multiplier algorithm for low-rank Toeplitz matrix completion. Journal of Inequalities and Applications, 2019(1). (16 pages), DOI: https://doi.org/10.1186/s13660-019-2033-7

algorithm for the Toeplitz matrix completion (TMC), which will reduce the extra load coming from data communication under reasonable smoothing. It has resulted in a semi-smoothing augmented Lagrange multiplier (SSALM) algorithm. Meanwhile, we demonstrate the convergence theory of the new algorithm. Finally, numerical experiments show that the new algorithm is more effective/economic than the original algorithm.

Keywords: Toeplitz matrix, Completion, Augmented Lagrange multiplier, Data communication

INTRODUCTION

Completing a low-rank matrix from a subset of its entries has been a hot problem recently, first introduced by [8], that has arisen in a wide variety of practical contexts across all disciplines of engineering and computational science such as model reduction [19], machine learning [1, 2], control [22], pattern recognition [12], imaging inpainting [3], video denoising [16], computer vision [28], and so on. Despite matrix completion (MC) requiring the global solution of a non-convex objective, there are many computational efficient algorithms which are effective for a broad class of matrices. The problem has received intensive research from both theoretical and algorithmic aspects, see, e.g., [4, 5, 6, 7, 8, 9, 10, 11, 13, 14, 15, 16, 17, 18, 21, 23, 26, 27, 29, 30, 31, 32, 34, 35], and the references therein for partial review. It is well known that the mathematical model of the MC problem is of the following form:

$$\min_{A\in\mathbb{R}^{m\times n}} \|A\|_*,$$

$$\text{subject to } \mathcal{P}_\Omega(A) = \mathcal{P}_\Omega(M), \qquad (1.1)$$

where the matrix $M \in \mathbb{R}^{m\times n}$ is an underlying matrix to be completed, Ω is a random subset of indices for the available entries, and P_Ω is the associated sampling orthogonal projection operator which acquires only the entries indexed by $\Omega\subset\{1,2,\dots,m\}\times\{1,2,\dots,n\}$.

In the current MC problems, the matrix M is of special structure in general. Therefore, much attention has been paid to the completion of Toeplitz and Hankel matrices in recent years [20, 24, 25, 30, 32]. Many scholars have conducted in-depth research on the special structure, property, and application of the Toeplitz and Hankel matrices; for example, nuclear norm minimization for the low-rank Hankel matrix reconstruction problem under the random Gaussian sampling model is investigated in [7]. In addition, Hankel matrix reconstitution in the sense of minimizing the nuclear norm with non-uniform sampling of entries is researched in [11]. To make full use of the special structure of a Toeplitz matrix, a mean value algorithm is presented in [30]; the modified Lagrange multiplier (MALM) algorithm [31] and the smoothing augmented Lagrange multiplier (SALM) algorithm [34] are also proposed. Therefore, the Toeplitz matrix completion (TMC) is one of the most important MC problems and has attracted a large amount of attention recently. As is well known, an n×n Toeplitz matrix is of the following form:

$$T = (t_{j-i})_{i,j=1}^{n} = \begin{pmatrix} t_0 & t_1 & \cdots & t_{n-2} & t_{n-1} \\ t_{-1} & t_0 & \cdots & t_{n-3} & t_{n-2} \\ \vdots & \vdots & \ddots & \vdots & \vdots \\ t_{-n+2} & t_{-n+3} & \cdots & t_0 & t_1 \\ t_{-n+1} & t_{-n+2} & \cdots & t_{-1} & t_0 \end{pmatrix} \in \mathbb{T}^{n\times n} \subset \mathbb{R}^{n\times n}, \quad (1.2)$$

which is determined by 2n−1 entries, say, the first row and the first column. Explicitly seeking the lowest rank Toeplitz matrix consistent with the known entries is mathematically considered as

$$\min_{A\in\mathbb{T}^{n\times n}} \|A\|_*,$$

$$\text{subject to } H \circ A = M, \quad (1.3)$$

where "∘" is the Hadamard product, $H = (H_{ij}) \in \mathbb{R}^{n\times n}$ is the weighted matrix with entries $H_{ij} = 1$ for $j - i \in \Omega \subset \{-n+1, \ldots, n-1\}$ and $H_{ij} = 0$ for any other (i,j), $M = (M_{ij}) \in \mathbb{T}^{n\times n}$ is the underlying Toeplitz matrix to be completed, namely $M_{ij} = 0$ for $j - i \notin \Omega$.

The SALM algorithm switches the iteration matrix into the Toeplitz structure at each iteration step by the smoothing operator, which saves computational cost of the singular value decomposition and approximates well the solution. Unfortunately, numerous data have to be shifted at each iteration step in the process of implementing this algorithm. However, there is a cost, sometimes relatively great, associated with the moving of data. The control of memory traffic is crucial to performance in many computers.

These factors motivated us to reduce the traffic jam of data, resulting in a semi-smoothing augmented Lagrange multiplier (SSALM) algorithm based on the selecting technique of the optimal parameter $\omega^{(k)}$ at each of the five iteration steps in [33]. Compared with the SALM algorithm, the new algorithm either saves computation cost or reduces data communication. Two aspects are taken into account, which results in more practical or economic implementation. The new algorithm not only overcomes the slowness of the SVD for the original ALM algorithm, but also reduces the greatness of the data communication for the ALM algorithm. We can see that the CPU of SSALM algorithm is reduced to 30.44% from the numerical experiments. The rest of this paper is organized as follows. Some preliminaries are provided in Sect. 2. Section 3 presents the semi-smoothing augmented Lagrange multiplier (SSALM) algorithm after giving an outline of the ALM algorithm, the dual approach, and the SALM algorithm. The convergence property of the SSALM algorithm is constructed in Sect. 4. We report the numerical results to indicate the effectiveness of the SSALM algorithm in Sect. 5. Finally, we end the paper with the concluding remarks in Sect. 6.

PRELIMINARIES

This section is devoted to some of the necessary notations and preliminaries. $\mathbb{R}^{m\times n}$ denotes the set of m×n real matrices, $\mathbb{T}^{n\times n}$ is the set of n×n real Toeplitz matrices. The nuclear norm of a matrix A is denoted by $\|A\|_*$, and the Frobenius norm $\|A\|_F$ is the maximum absolute value of the matrix entries of a matrix A. A^T is used to express the transpose of a matrix $A \in \mathbb{R}^{n\times n}$, rank(A) is equal to the rank of a matrix A, and tr(A)tr(A) represents the trace of A. The standard inner product between two matrices is denoted by $\langle X, Y\rangle = \mathrm{tr}(X^T Y)$. For $A = (a_{ij}) \in \mathbb{R}^{m\times n}$, $B = (b_{ij}) \in \mathbb{R}^{m\times n}$, their Hadamard product $A \circ B$ is an m×n matrix whose (i,j) entry is the $a_{ij}b_{ij}$, i.e., $A \circ B = (a_{ij}b_{ij}) \in \mathbb{R}^{m\times n}$.

The singular value decomposition (SVD) of a matrix $A \in \mathbb{R}^{m\times n}$ of r-rank is defined by

$$A = U\Sigma_r V^T, \qquad \Sigma_r = \mathrm{diag}(\sigma_1, \sigma_2, \ldots, \sigma_r),$$

where $U \in \mathbb{R}^{m\times r}$ and $V \in \mathbb{R}^{n\times r}$ are column orthonormal matrices, and $\sigma_1 \geq \sigma_2 \geq \cdots \geq \sigma_r > 0$.

Definition 2.1

(Singular value thresholding operator [6])

For each $\tau \geq 0$, the singular value thresholding operator D_τ is defined as follows:

$$\mathcal{D}_\tau(A) := U\mathcal{D}_\tau(\Sigma)V^T, \qquad \mathcal{D}_\tau(\Sigma) = \mathrm{diag}(\{\sigma_i - \tau\}_+),$$

where

$$A = U\Sigma_r V^T \in \mathbb{R}^{m\times n}, \qquad \{\sigma_i - \tau\}_+ = \begin{cases} \sigma_i - \tau, & \text{if } \sigma_i > \tau, \\ 0, & \text{if } \sigma_i \leq \tau. \end{cases}$$

$I_n = (e_1, e_2, \ldots, e_n) \in \mathbb{R}^{n\times n}$ denotes the n×n identity matrix and $Z_n = (e_2, e_3, \ldots, e_n, 0) \in \mathbb{R}^{n\times n}$ is called the shift matrix. It is clear that

$$Z_n^r = \begin{cases} \begin{pmatrix} O & O \\ I_{n-r} & O \end{pmatrix}, & 1 < r < n, \\ O, & r \geq n, \end{cases}$$

where "*O*" stands for a zero-matrix. Thus, a Toeplitz matrix $T \in \mathbb{T}^{n\times n}$, shown in (1.2), can be written as a linear combination of these shift matrices, that is,

$$T = \sum_{l=1}^{n-1} t_{-l} Z_n^l + \sum_{l=0}^{n-1} t_l \left(Z_n^T\right)^l.$$

$\Omega \subset \{-n+1, \ldots, n-1\}$ is an indices set of observed diagonals of a Toeplitz matrix M∈ $\mathbb{T}^{n\times n}$, $\bar{\Omega}$ is the complementary set of Ω. For any Toeplitz matrix $A \in \mathbb{T}^{n\times n}$, the vector vec(A,α) denotes the th diagonal of *A*, α=−n+1,−n+2,… ,n−1, that is to say,

$$\text{vec}(H \circ A, \alpha) = \begin{cases} \text{vec}(A, \alpha), & \alpha \in \Omega, \\ \mathbf{0}, & \alpha \notin \Omega, \end{cases} \quad (\mathbf{0} \text{ is a zero-vector}).$$

Definition 2.2

(Toeplitz structure smoothing operator [34])

For any matrix $A = (a_{ij}) \in \mathbb{R}^{n\times n}$, the Toeplitz structure smoothing operator T is defined as follows:

$$\mathcal{T}(A) := \sum_{l=1}^{n-1} \tilde{a}_{-l} Z_n^l + \sum_{l=0}^{n-1} \tilde{a}_l \left(Z_n^T\right)^l, \qquad (2.1)$$

where $\tilde{a}_\alpha = \frac{{}_\alpha A^{\min} + {}_\alpha A^{\max}}{2}, \alpha = -n+1, -n+2, \ldots, n-1$ with

$${}_\alpha A^{\min} = \min_{i,j \in \{1,2,\ldots,n\}} \{a_{ij}, i-j=\alpha\}, \quad \text{and} \quad {}_\alpha A^{\max} = \max_{i,j \in \{1,2,\ldots,n\}} \{a_{ij}, i-j=\alpha\}.$$

It is clear that T(A) is a Toeplitz matrix derived from the matrix *A*. Namely, any $A \in \mathbb{R}^{n\times n}$ can be changed into a Toeplitz structure via the smoothing operator T(·).

ALGORITHMS

First of all in this section, for completeness as well as for the purpose of comparison, we briefly review and summarize some relative algorithms for approximately minimizing the nuclear norm of a matrix under convex constraints.

Since the matrix completion problem is closely connected to the robust principal component analysis (RPCA) problem, then it can be formulated in the same way as RPCA, an equivalent problem of (1.1) can be considered as follows.

As E will compensate for the unknown entries of M, the unknown entries of M are simply set as zeros. Suppose that the given data are arranged as the columns of a large matrix $M \in \mathbb{R}^{m \times n}$. The mathematical model for estimating the low-dimensional subspace is to find a low-rank matrix $A \in \mathbb{R}^{m \times n}$ such that the discrepancy between A and M is minimized, leading to the following constrained optimization:

$$\min_{A,E \in \mathbb{R}^{m \times n}} \|A\|_*,$$

$$\text{subject to } A + E = M, \mathcal{P}_\Omega(E) = 0, \qquad (3.1)$$

where E will compensate for the unknown entries of M, the unknown entries of $M \in \mathbb{R}^{m \times n}$ are simply set as zeros. And $P\Omega$: $\mathbb{R}^{m \times n} \rightarrow \mathbb{R}^{m \times n}$ is a linear operator that keeps the entries in Ω unchanged and sets those outside Ω (say, in $\overline{\Omega}$) zeros.

The Dual Algorithm

The dual algorithm proposed in [10] tackles problem (3.1) via its dual. That is, one first solves the dual problem

$$\max_Y \langle M, Y \rangle,$$

$$\text{subject to } J(Y) \leq 1, \qquad (3.2)$$

for the optimal Lagrange multiplier Y, where

$$J(Y) = \max\left(\|Y\|_2, \lambda^{-1}\|Y\|_\infty\right). \qquad (3.3)$$

A steepest ascend algorithm constrained on the surface {Y|J(Y)=1} can be adopted to solve (3.2), where the constrained steepest ascend direction is obtained by projecting M onto the tangent cone of the convex body {Y|J(Y)≤1}. It turns out that the optimal solution to the primal problem (3.1) can be obtained during the process of finding the constrained steepest ascend direction.

The Augmented Lagrange Multiplier Algorithm

The augmented Lagrange multiplier (ALM) algorithm was proposed in [18] for solving a convex optimization (1.1). It should be described subsequently.

It is famous that the partial augmented Lagrangian function of problem (3.1) is

$$\mathcal{L}(A,E,Y,\mu) = \|A\|_* + \langle Y, M-A-E\rangle + \frac{\mu}{2}\|M-A-E\|_F^2. \tag{3.4}$$

Hence, the augmented Lagrange multiplier algorithm is designed as follows.

Algorithm 3.1

([18])

Given a sampled set Ω, a sampled matrix $D = \mathcal{P}_\Omega(M), \mu_0 > 0, \rho > 1$. Given also two initial matrices $Y_0 = 0, E_0 = 0. k := 0$.

1. Compute the SVD of the matrix $(D - E_k + \mu_k^{-1}Y_k)$:

$[U_k, \Sigma_k, V_k] = \mathrm{svd}\big(D - E_k + \mu_k^{-1}Y_k\big)$;

2. Set

$A_{k+1} = U_k \mathcal{D}_{\mu_k^{-1}}(\Sigma_k)V_k^T$,

solve $E_{k+1} = \arg\min_{\mathcal{P}_\Omega(E)=0} \mathcal{L}(A_{k+1}, E, Y_k, \mu_k)$,

$E_{k+1} = \mathcal{P}_{\bar{\Omega}}\big(D - A_{k+1} + \mu_k^{-1}Y_k\big)$;

3. If $\|D - A_{k+1} - E_{k+1}\|_F/\|D\|_F < \epsilon_1$ and $\mu_k\|E_{k+1} - E_k\|_F/\|D\|_F < \epsilon_2$, stop; otherwise, go to the next step;
4. Set $Y_{k+1} = Y_k + \mu_k(D - A_{k+1} - E_{k+1})$.

If $\mu_k\|E_{k+1} - E_k\|_F/\|D\|_F < \epsilon_2$, set $\mu_{k+1} = \rho\mu_k$; otherwise, go to Step 1.

Note that the ALM algorithm performs better both in theory and practice than other algorithms with a Q-linear convergence speed globally. It is of much better numerical behavior or higher accuracy. It is reported that the ALM algorithm has been applied to MC problem (1.1). The algorithm, however, has the disadvantage of the penalty function, namely the matrix sequences $\{A_k\}$ generated by the ALM algorithm are the accepted solutions but feasible ones.

The Smoothing Augmented Lagrange Multiplier Algorithm

In this subsection, we make mention of a stepped-up scheme for the TMC problem. The smoothing augmented Lagrange multiplier (SALM) approach employs the smoothing operator T (see (2.1)) to approximate a matrix generated in the *k*th iteration so that the current approximation is of a Toeplitz structure.

Then our problem can be expressed as the following convex programming:

$$\min_{A,E\in\mathbb{T}^{n\times n}} \|A\|_*,$$

$$\text{subject to} \quad A+E=\mathcal{P}_\Omega(M), \qquad \mathcal{P}_\Omega(E)=0, \tag{3.5}$$

where $M\in\mathbb{T}^{n\times n}$ is a real Toeplitz matrix, and $\Omega\subset\{-n+1,\ldots,n-1\}$. Let $D=\mathcal{P}_\Omega(M)$. Then the partial augmented Lagrangian function is

$$\mathcal{L}(A,E,Y,\mu)=\|A\|_*+\langle Y,D-A-E\rangle+\frac{\mu}{2}\|D-A-E\|_F^2, \tag{3.6}$$

where $Y\in\mathbb{R}^{n\times n}$.

Algorithm 3.2

([34])

Given a sampled set Ω, a sampled matrix D, $\mu_0>0$, $\rho>1$. Given also two initial matrices $Y_0=0$, $E_0=0$. $k:=0$.

1. Compute the SVD of the matrix $(D-E_k+\mu_k^{-1}Y_k)$ using the Lanczos method

$$[U_k,\Sigma_k,V_k]=\text{lansvd}\left(D-E_k+\mu_k^{-1}Y_k\right);$$

2. Set

$$X_{k+1}=U_k\mathcal{D}_{\mu_k^{-1}}(\Sigma_k)V_k^T,$$

compute for smoothing $\tilde{a}_\alpha=\frac{\alpha X_{k+1}^{\min}+\alpha X_{k+1}^{\max}}{2}$, $\alpha\in\{-n+1,-n+2,\ldots,n-1\}$, and

$$A_{k+1}=\mathcal{T}(X_{k+1})=\sum_{l=1}^{n-1}\tilde{a}_{-l}Z_n^l+\sum_{l=0}^{n-1}\tilde{a}_l(Z_n^T)^l,$$

$$E_{k+1}=\mathcal{P}_{\bar{\Omega}}\left(D-A_{k+1}+\mu_k^{-1}Y_k\right);$$

3. If $\|D-A_{k+1}-E_{k+1}\|_F/\|D\|_F<\epsilon_1$ and $\mu_k\|E_{k+1}-E_k\|_F/\|D\|_F<\epsilon_2$, stop; otherwise, go to the next step;
4. Set $Y_{k+1}=Y_k+\mu_k(D-A_{k+1}-E_{k+1})$.

If $\mu_k\|E_{k+1}-E_k\|_F/\|D\|_F<\epsilon_2$, set $\mu_{k+1}=\rho\mu_k$; otherwise, go to Step 1.

It is reported that the convergence speed of the SALM algorithm is greater than that of the ALM and APG algorithms. A merit of smoothing is that the fast SVD procedure can be utilized to reduce the computation.

As we know, the SVD time is saved at the expense of data communication. Sometimes, this is not worth the candle. This motivated us to come up with the following algorithm.

The Semi-smoothing Augmented Lagrange Multiplier Algorithm

In this subsection, we propose a semi-smoothing augmented Lagrange multiplier algorithm based on the ALM and SALM algorithms for the TMC problem. The new algorithm consists of two stages: one is $\ell-1$ iterations by the ALM scheme, which is free moving of data; another is the lth smoothing by the SALM procedure, which is keeping the iteration matrix as a Toeplitz structure.

Now, the semi-smoothing augmented Lagrange multiplier (SSALM) algorithm will be presented in the following.

Algorithm 3.3
(SSALM algorithm)
Input: A sampled set Ω, a sampled matrix $D, Y_{0,0} = 0, E_{0,0} = 0$; parameters $\mu_0 > 0, \rho > 1\ \ell > 1, \epsilon_1, \epsilon_2$.

Let $k := 0, q := 1, q = 1, 2, \ldots, \ell - 1$.

Repeat:

1. $\ell-1$ iterations.

(1) : Compute the SVD of the matrix $(D - E_{k,q} + \mu_{k,q}^{-1} Y_{k,q})$:

$$[U_{k,q}, \Sigma_{k,q}, V_{k,q}] = \text{lansvd}\left(D - E_{k,q} + \mu_{k,q}^{-1} Y_{k,q}\right);$$

(2) : Set

$$X_{k+1,q+1} = U_{k,q} \mathcal{D}_{\mu_{k,q}^{-1}}(\Sigma_{k,q}) V_{k,q}^T,$$

$$E_{k+1,q+1} = \mathcal{P}_{\bar{\Omega}}\left(D - X_{k+1,q+1} + \mu_{k,q}^{-1} Y_{k,q}\right);$$

(3) : If $\|D - X_{k+1,q+1} - E_{k+1,q+1}\|_F / \|D\|_F < \epsilon_1$ and $\mu_{k,q} \|E_{k+1,q+1} - E_{k,q}\|_F / \|D\|_F < \epsilon_2$, stop; otherwise, go to the next step;

(4) : Set $Y_{k+1,q+1} = Y_{k,q} + \mu_{k,q}(D - X_{k+1,q+1} - E_{k+1,q+1})$, $\mu_{k+1,q+1} = \rho \mu_{k,q}$; otherwise, go to step (1);

2. lth smoothing.

(1) : Compute the SVD of the matrix $(D - E_{k,\ell} + \mu_{k,\ell}^{-1} Y_{k,\ell})$:

$$[U_{k,\ell}, \Sigma_{k,\ell}, V_{k,\ell}] = \text{svd}\left(D - E_{k,\ell} + \mu_{k,\ell}^{-1} Y_{k,\ell}\right);$$

(2) : Set

$$X_{k+1,\ell} = U_{k,\ell}\mathcal{D}_{\mu_{k,\ell}^{-1}}(\Sigma_{k,\ell})V_{k,\ell}^T,$$

compute the factors of smoothing $\tilde{a}_\alpha = \frac{{}_\alpha X_{k+1,\ell}^{\min} + {}_\alpha X_{k+1,\ell}^{\max}}{2}$, $\alpha = -n+1, -n+2, \ldots,$ …,n−1,

and set

$$A_{k+1,\ell} = \mathcal{T}(X_{k+1,\ell}) = \sum_{l=1}^{n-1} \tilde{a}_{-l} Z_n^l + \sum_{l=0}^{n-1} \tilde{a}_l (Z_n^T)^l.$$

Update

$$E_{k+1,\ell} = \mathcal{P}_{\bar{\Omega}}\left(D - A_{k+1,\ell} + \mu_{k,\ell}^{-1} Y_{k+1,\ell}\right);$$

3. If $\|D - A_{k+1,\ell} - E_{k+1,\ell}\|_F / \|D\|_F < \epsilon_1$ and $\mu_{k,\ell}\|E_{k+1,\ell} - E_{k,\ell}\|_F / \|D\|_F < \epsilon_2$, stop; otherwise, go to the next step;

4. Set $Y_{k+1,q+1} = Y_{k,q} + \mu_{k,q}(D - A_{k+1,q+1} - E_{k+1,q+1})$.

If $\mu_{k,q}\|E_{k+1,q+1} - E_{k,q}\|_F / \|D\|_F < \epsilon_2$, set $\mu_{k+1,q+1} = \rho\mu_{k,q}$; otherwise, go to Step 1.

In fact, Algorithm 3.3 includes the above Algorithm 3.2 as a special case if ℓ=1. Obviously, the algorithm is an acceleration of the SALM algorithm in [34].

CONVERGENCE ANALYSIS

This section will analyze the convergence of Algorithm 3.3.

Let $(\ddot{A}, \ddot{E})$ be the solution of model (3.5) and $\ddot{Y}$ be that of the optimal problem (3.2). We provide some lemmas in the following firstly.

Lemma 4.1

([8])

Let $A \in \mathbb{R}^{m\times n}$ be an arbitrary matrix and $U\Sigma V^T$ be its SVD. Then the set of subgradients of the nuclear norm of A is given by

$$\partial\|A\|_* = \left\{UV^T + W : W \in \mathbb{R}^{m\times n}, U^T W = 0, WV = 0, \|W\|_2 \le 1\right\}.$$

Lemma 4.2

([18])

If μ_k is nondecreasing, then each entry of the following series is nonnegative and their sum is finite, i.e.,

$$\sum_{k=1}^{+\infty} \mu_k^{-1}\left(\langle Y_{k+1} - Y_k, E_{k+1} - E_k\rangle + \langle A_{k+1} - \ddot{A}, \hat{Y}_{k+1} - \ddot{Y}\rangle + \langle E_{k+1} - \ddot{E}, Y_{k+1} - \ddot{Y}\rangle\right)$$

$$< +\infty. \tag{4.1}$$

Lemma 4.3

([18])

The sequences $\{\ddot{Y}_k\}, \{Y_k\}$, *and* $\{\hat{Y}_k\}$ *are all bounded, where* $\hat{Y}_k = Y_{k+1} + \mu_{k-1}(D - A_k - E_{k-1})$.

Lemma 4.4

The sequence $\{Y_{k,q}\}$ generated by Algorithm 3.3 is bounded.

Proof

Let $B = \mu_{k,q}(D - E_{k,q} + \mu_{k,q}^{-1} Y_{k,q} - X_{k+1,q+1})$, $\mathcal{T}(B) = \sum_{l=1}^{n-1} \tilde{b}_{-l} Z_n^l + \sum_{l=0}^{n-1} \tilde{b}_l (Z_n^T)^l$, defined as (2.1).

First of all, we indicate that $Y_{k,q}, E_{k,q}, k = 1, 2, \ldots, q = 1, 2, \ldots, \ell - 1$, are all the Toeplitz matrices. Clearly, $Y_{0,0} = 0, E_{0,0} = 0$ are both smoothed into a Toeplitz structure. Suppose that $Y_{k,q}, E_{k,q}$ are both the Toeplitz matrices, so is $E_{k+1,q+1} = \mathcal{P}_{\bar{\Omega}}(D - X_{k+1,q+1} + \mu_{k,q}^{-1} Y_{k,q})$. Thus, $Y_{k+1,q+1}$ is a Toeplitz matrix also because of Step 4 in Algorithm 3.3.

$$\begin{aligned} Y_{k+1,q+1} &= Y_{k,q} + \mu_{k,q}(D - A_{k+1,q+1} - E_{k+1,q+1}) \\ &= Y_{k,q} + \mu_{k,q}(D - A_{k+1,q+1} - E_{k,q}) + \mu_{k,q}(E_{k,q} - E_{k+1,q+1}). \end{aligned}$$

Moreover,

$$\begin{aligned} \mu_{k,q}(E_{k,q} - E_{k+1,q+1}) &= \mu_{k,q}\mathcal{P}_{\bar{\Omega}}\left(E_{k,q} - (D - A_{k+1,q+1} + \mu_{k,q}^{-1} Y_{k,q})\right) \\ &= \mu_{k,q}\mathcal{P}_{\bar{\Omega}}\left(E_{k,q} - (D - \mathcal{T}(X_{k+1,q+1}) + \mu_{k,q}^{-1} Y_{k,q})\right) \\ &= \mu_{k,q}\mathcal{P}_{\bar{\Omega}}\left(E_{k,q} - \left(D - \mathcal{T}\left(D - E_{k,q} + \mu_{k,q}^{-1} Y_{k,q} - \mu_{k,q}^{-1} B\right) + \mu_{k,q}^{-1} Y_{k,q}\right)\right) \end{aligned}$$

$$= \mu_{k,q}\mathcal{P}_{\bar{\Omega}}\left(E_{k,q} - \left(D - (D - E_{k,q} + \mu_{k,q}^{-1} Y_{k,q} - \mu_{k,q}^{-1}\mathcal{T}(B)) + \mu_{k,q}^{-1} Y_{k,q}\right)\right)$$

$$= \mathcal{P}_{\bar{\Omega}}\mathcal{T}(B).$$

It is clear that by Steps 1–2 in Algorithm 3.3 we have

$$D - E_{k,q} + \mu_{k,q}^{-1} Y_{k,q} = \check{U}_{k,q} \check{\Sigma}_{k,q} \check{V}_{k,q}^T + \tilde{U}_{k,q} \tilde{\Sigma}_{k,q} \tilde{V}_{k,q}^T,$$

where $\check{U}_{k,q}, \check{V}_{k,q}$ are the singular vectors associated with singular values that are more than $\frac{1}{\mu_{k,q}}$ and $\tilde{U}_{k,q}, \tilde{V}_{k,q}$ are those associated with singular values that are less than or equal to $\frac{1}{\mu_{k,q}}$, the elements of the diagonal matrix $\check{\Sigma}_{k,q}$ are more than $\frac{1}{\mu_{k,q}}$, and those of the diagonal matrix $\tilde{\Sigma}_{k,q}$ are less than or equal to $\frac{1}{\mu_{k,q}}$. Hence, it is drawn that $A_{k+1,q+1} = \check{U}_{k,q}(\check{\Sigma}_{k,q} - \mu_{k,q}^{-1} I)\check{V}_{k,q}^T$, and

$$\begin{aligned}
\|B\|_F &= \left\|\mu_{k,q}(D - E_{k,q} + \mu_{k,q}^{-1} Y_{k,q} - A_{k+1,q+1})\right\|_F \\
&= \left\|\mu_{k,q}(\mu_{k,q}^{-1}\check{U}_{k,q}\check{V}_{k,q}^T + \tilde{U}_{k,q}\tilde{\Sigma}_{k,q}\tilde{V}_{k,q}^T)\right\|_F \\
&= \left\|\check{U}_{k,q}\check{V}_{k,q}^T + \mu_{k,q}\tilde{U}_{k,q}\tilde{\Sigma}_{k,q}\tilde{V}_{k,q}^T\right\|_F \\
&\le \sqrt{n}.
\end{aligned}$$

We can obtain hence that $Y_{k,q} + \mu_{k,q}(D - A_{k+1,q+1} - E_{k,q}) \in \partial\|A_{k+1,q+1}\|_*$ from Lemmas 4.2 and 4.3.

It is known that, for $A_{k+1,q+1} = U\Sigma V^T$, by Lemma 4.1,

$$\partial\|A_{k+1,q+1}\|_* = \{UV^T + W : W \in \mathbb{R}^{n\times n}, U^T W = 0, WV = 0, \|W\|_2 \le 1\}.$$

We also have

$$\|UV^T + W\|_F^2 = \mathrm{tr}((UV^T + W)^T (UV^T + W)) = \mathrm{tr}(VV^T + W^T W) \le n.$$

Therefore, the following inequalities can be obtained:

$$\left\|Y_{k,q} + \mu_{k,q}(D - A_{k+1,q+1} - E_{k,q})\right\|_F \le \sqrt{n}$$

and

$$\left\|\mathcal{P}_{\bar{\Omega}}(\mathcal{T}(B))\right\|_F \le \left\|\mathcal{T}(B))\right\|_F \le \|B\|_F \le \sqrt{n}.$$

It is evident that the sequence $\{Y_{k,q}\}$ is bounded. □

Theorem 4.1

Suppose that $\langle A_{k+1,q+1} - A_{k,q}, D - A_{k+1,q+1} - E_{k,q}\rangle \ge 0$*, then the sequence* $\{A_{k,q}\}$ *converges to the solution of* (3.5) *when* $\mu_{k,q} \to \infty$ *and* $\sum_{k=1}^{+\infty} \mu_{k,q}^{-1} = +\infty$.

Proof

It is true that

$$\lim_{k\to\infty} (D - A_{k+1} - E_{k+1}) = 0$$

since $\mu_{k,q}^{-1}(Y_{k+1,q+1} - Y_{k,q}) = D - A_{k+1,q+1} - E_{k+1,q+1}$ and Lemma 4.4.

Let $(\ddot{A}, \ddot{E})$ be the solution of (3.5). Then $A_{k+1,q+1}, Y_{k+1,q+1}, E_{k+1,q+1}$, $k = 1, 2, \ldots,$

are all Toeplitz matrices since $\ddot{A}+\ddot{E}=D$. We prove first that

$$\|E_{k+1,q+1}-\ddot{E}\|_F^2+\mu_{k,q}^{-2}\|Y_{k+1,q+1}-\ddot{Y}\|_F^2$$
$$=\|E_{k,q}-\ddot{E}\|_F^2-\|E_{k+1,q+1}-E_{k,q}\|_F^2+\mu_{k,q}^{-2}\|Y_{k,q}-\ddot{Y}\|_F^2$$
$$-\mu_{k,q}^{-2}\|Y_{k+1,q+1}-Y_{k,q}\|_F^2-2\mu_{k,q}^{-1}\langle A_{k+1,q+1}-\ddot{A},\hat{Y}_{k+1,q+1}-\ddot{Y}\rangle, \tag{4.2}$$

where $\hat{Y}_{k+1,q+1}=Y_{k,q}+\mu_{k,q}(D-A_{k+1,q+1}-E_{k,q})$, $\ddot{Y}$ is the optimal solution to the dual problem (3.2).

$$\|E_{k,q}-\ddot{E}\|_F^2=\left\|\mathcal{P}_{\bar{\Omega}}(E_{k,q}-\ddot{E})\right\|_F^2$$
$$=\left\|\mathcal{P}_{\bar{\Omega}}(E_{k+1,q+1}-\ddot{E}-E_{k+1,q+1}+E_{k,q})\right\|_F^2$$
$$=\left\|\mathcal{P}_{\bar{\Omega}}(E_{k+1,q+1}-\ddot{E})\right\|_F^2+\left\|\mathcal{P}_{\bar{\Omega}}(E_{k+1,q+1}-E_{k,q})\right\|_F^2$$
$$-2\left\langle\mathcal{P}_{\bar{\Omega}}(E_{k+1,q+1}-\ddot{E}),\mathcal{P}_{\bar{\Omega}}(E_{k+1,q+1}-E_{k,q})\right\rangle$$
$$=\|E_{k+1,q+1}-\ddot{E}\|_F^2+\|E_{k+1,q+1}-E_{k,q}\|_F^2$$
$$+2\mu_{k,q}^{-1}\left\langle\mathcal{P}_{\bar{\Omega}}(A_{k+1,q+1}-\ddot{A}),\hat{Y}_{k+1,q+1}-\ddot{Y}\right\rangle.$$

We obtain the following result with the same analysis:

$$\mu_{k,q}^{-2}\|Y_{k,q}-\ddot{Y}\|_F^2=\mu_{k,q}^{-2}\left\|\mathcal{P}_{\Omega}(Y_{k,q}-\ddot{Y})\right\|_F^2$$
$$=\mu_{k,q}^{-2}\left\|\mathcal{P}_{\Omega}(Y_{k+1,q+1}-\ddot{Y})\right\|_F^2+\mu_{k,q}^{-2}\left\|\mathcal{P}_{\Omega}(Y_{k+1,q+1}-Y_{k,q})\right\|_F^2$$
$$-2\mu_{k,q}^{-1}\left\langle\mathcal{P}_{\Omega}(Y_{k+1,q+1}-\ddot{Y}),\mathcal{P}_{\Omega}(Y_{k+1,q+1}-Y_{k,q})\right\rangle$$
$$=\mu_{k,q}^{-2}\|Y_{k+1,q+1}-\ddot{Y}\|_F^2+\mu_{k,q}^{-2}\|Y_{k+1,q+1}-Y_{k,q}\|_F^2$$
$$+2\mu_{k,q}^{-1}\left\langle\mathcal{P}_{\Omega}(A_{k+1,q+1}-\ddot{A}),\hat{Y}_{k+1,q+1}-\ddot{Y}\right\rangle.$$

Then

$$\min_{A\in\mathbb{T}^{m\times n}}\mu\|A\|_*+\frac{1}{2}\left\|\mathcal{P}_{\Omega}(A-M)\right\|_F^2 \tag{4.3}$$

holds true.

The sum $\sum_{k=1}^{\infty}\mu_{k,q}^{-1}\langle A_{k,q}-\ddot{A},\hat{Y}_{k,q}-\ddot{Y}\rangle<+\infty$ and $\|E_{k,q}-\ddot{E}\|^2+\mu_{k,q}^{-2}\|Y_{k,q}-\ddot{Y}\|^2$ is nonincreasing since $\|A\|_*$ is a convex function and $\hat{Y}_{k+1,q+1}\in\partial\|A\|_*$, $\langle A_{k+1,q+1}-\ddot{A},\hat{Y}_{k+1,q+1}-\ddot{Y}\rangle\geq 0$. On the other hand, the following are true by Algorithm 3.3:

$$\langle Y_{k,q},E_{k,q}-\ddot{E}\rangle=0 \quad\text{and}\quad \langle\ddot{Y},E_{k,q}-\ddot{E}\rangle=0.$$

Therefore, by the same idea of Theorem 2 in [18], it is obtained that $\ddot{A}$ is the solution of (3.5). □

Theorem 4.2

Let $X = (x_{ij}) \in \mathbb{R}^{n\times n}, \mathcal{T}(X) = (\tilde{x}_{ij}) \in \mathbb{T}^{n\times n}$ *be the Toeplitz matrix derived fromX, introduced in* (2.1). *Then, for any Toeplitz matrix* $Y = (y_{ij}) \in \mathbb{T}^{n\times n}$, $\langle X - \mathcal{T}(X), Y\rangle = 0$.

Proof

By the definition of $\mathcal{T}(X)$, we have $\sum_{i,j}(x_{ij} - \tilde{x}_{ij}) = 0,\ i,j = 1,2,\ldots,n.$ Since Y is a Toeplitz matrix and $y_\alpha = y_{ij},\ \alpha = i - j,\ i,j = 1,2,\ldots,n$, then

$$\begin{aligned}\langle X - \mathcal{T}(X), Y\rangle &= \sum_{i=1}^{n}\sum_{j=1}^{n}(x_{ij} - \tilde{x}_{ij})y_{ji}\\ &= \sum_{l=1}^{n-1} y_{-l}(x_{ij} - \tilde{x}_{ij}) + \sum_{l=0}^{n-1} y_l(x_{ij} - \tilde{x}_{ij})\\ &= 0.\end{aligned}$$

Theorem 4.3

In Algorithm 3.3, $A_{k,q}$ *is a Toeplitz matrix generated by* $X_{k,q}$, *then*

$$\|A_{k,q} - \ddot{A}\|_F < \|X_{k,q} - \ddot{A}\|_F$$

is true with $\ddot{A}$ *being the solution of* (3.5).

Proof

$$\begin{aligned}\|X_{k,q} - \ddot{A}\|_F^2 &= \|X_{k,q} - A_{k,q} + A_{k,q} - \ddot{A}\|_F^2\\ &= \langle X_{k,q} - A_{k,q}, X_{k,q} - A_{k,q}\rangle + 2\langle X_{k,q} - A_{k,q}, A_{k,q} - \ddot{A}\rangle\\ &\quad + \langle A_{k,q} - \ddot{A}, A_{k,q} - \ddot{A}\rangle\\ &= \|X_{k,q} - A_{k,q}\|_F^2 + \|A_{k,q} - \ddot{A}\|_F^2.\end{aligned}$$

Thus, $\|A_{k,q} - \ddot{A}\|_F < \|X_{k,q} - \ddot{A}\|_F$ holds true.

NUMERICAL EXPERIMENTS

In this section we report some original numerical results of two algorithms (SALM, SSALM) for some n×n matrices with different ranks. All the experiments are conducted on the same workstation with an Intel(R) Core(TM) i7-6700 CPU @ 3.40 GHz that has 16 GB memory and 16-bit operating system, running Windows 7 and Matlab (vision 2016a). We analyze and compare iteration numbers (IT), computing time in seconds (time(s)), deviation (error 1, error 2), and ratio which are defined in the following. It can be seen that the SSALM algorithm proposed in this study is highly effective compared with the SALM algorithm.

$$\text{ERROR 1} = \frac{\|A+E-D\|_F}{\|D\|_F}, \qquad \text{ERROR 2} = \frac{\|A-M\|_F}{\|M\|_F},$$

$$\text{RATIO} = \frac{\text{the CPU of the SSALM algorithm}}{\text{the CPU of the SALM algorithm}} \times 100\%.$$

In the experiments, $M \in \mathbb{T}^{n\times n}$ represents an uncompleted Toeplitz matrix, the sampling density p=m/(2n−1), where m is the number of the known diagonal entries. By the way, p∈{0.3,0.4,0.5,0.6} here. Due to the special structure of a Toeplitz matrix, we have 0≤m≤2n−1. The SALM and SSALM algorithms share the same values of all parameters, say, $\tau_0 = 1/\|D\|_2, \delta = 1.2172 + 1.8588m/n^2, \epsilon_1 = 10^{-9}, \epsilon_2 = 5\times 10^{-6}$. In the test of the SSALM algorithm, ℓ=2 or ℓ=3 as a rule of semi-smoothing.

The experimental results of two algorithms are presented in Tables 1–6. From the tables, two algorithms can successfully compute an approximate solution of the prescriptive stop condition for all the test matrix *M*. And computing time of our SSALM algorithm in far less than that of the SALM algorithm. In particular, compared with the cost of the SALM algorithm, we can find that the cost of the SSALM algorithm is decreased up to 30.44%. The "ratio" in Tables 5–6 can show this effectiveness.

Table 1: Comparison between SALM and SSALM for p=0.6

n	rank(*M*)	Algorithm	IT	time (s)	error 1	error 2
500	10	SALM	44	2.6563	8.2372e−10	1.0363e−07
		SSALM	42	1.8853	9.2608e−10	1.9658e−09
800	10	SALM	54	5.9456	8.1138e−10	8.2779e−09
		SSALM	52	5.0976	5.6384e−10	7.6367e−09
1000	10	SALM	57	8.2474	8.9636e−10	4.1947e−09
		SSALM	58	6.3207	7.9333e−10	1.9217e−09
1500	10	SALM	64	16.7198	9.5636e−10	2.1154e−09
		SSALM	64	15.0399	9.1550e−10	1.0551e−09
2000	10	SALM	71	31.7924	9.4056e−10	7.5879e−08
		SSALM	68	26.8256	8.1847e−10	2.3201e−08
2500	10	SALM	71	47.3399	7.0194e−10	1.5048e−08
		SSALM	72	42.5214	9.8809e−10	2.6892e−09
3000	10	SALM	77	70.2107	6.7658e−10	2.4626e−09
		SSALM	76	60.6145	6.9232e−10	1.1348e−09

4000	20	SALM	74	146.3660	6.5154e−10	5.1684e−05
		SSALM	72	129.2216	5.3346e−10	1.4304e−09
5000	20	SALM	74	220.6386	9.9475e−10	2.1943e−08
		SSALM	74	198.3210	5.5006e−10	1.8351e−09
8000	25	SALM	78	550.5427	9.2197e−10	3.6735e−09
		SSALM	80	471.4695	5.5647e−10	1.5932e−09

Table 2: Comparison between SALM and SSALM for p=0.5

n	**rank(*M*)**	**Algorithm**	**IT**	**time (s)**	**error 1**	**error 2**
500	10	SALM	44	2.8955	9.5801e−10	2.2756e−08
		SSALM	44	2.1933	4.4278e−10	2.3334e−09
800	10	SALM	56	8.6741	7.0000e−10	2.6659e−06
		SSALM	52	5.4772	9.2740e−10	3.3480e−09
1000	10	SALM	60	11.0450	8.2993e−10	6.6858e−08
		SSALM	56	6.7779	8.6651e−10	1.8478e−09
1500	10	SALM	68	20.6328	8.45398e−10	9.6939e−07
		SSALM	64	15.6997	9.3584e−10	1.5261e−09
2000	10	SALM	71	44.8591	9.9033e−10	7.2479e−08
		SSALM	68	27.9845	9.6632e−10	2.8679e−09
2500	10	SALM	78	67.4042	7.3828e−10	1.2312e−07
		SSALM	74	45.3268	7.1109e−10	2.5905e−09
3000	10	SALM	77	74.2323	7.2763e−10	9.9884e−09
		SSALM	76	64.4300	8.2104e−10	1.2376e−08
4000	20	SALM	73	155.9955	9.4452e−10	1.8879e−06
		SSALM	72	128.0443	7.1470e−10	1.8278e−09
5000	20	SALM	74	235.1514	9.1421e−10	7.9369e−06
		SSALM	76	200.7122	6.2859e−10	3.5799e−09
8000	25	SALM	80	622.5518	9.5065e−10	5.5408e−06
		SSALM	80	498.8602	5.1021e−10	3.0270e−08

Table 3: Comparison between SALM and SSALM for p=0.4

n	**rank(*M*)**	**Algorithm**	**IT**	**time (s)**	**error 1**	**error 2**
500	10	SALM	45	7.8497	6.8884e−10	5.3595e−06
		SSALM	44	2.3892	5.3103e−10	2.3267e−09

800	10	SALM	54	7.4948	9.3427e–10	2.1464e–06
		SSALM	54	5.9618	8.9128e–10	1.5780e–08
1000	10	SALM	58	9.9528	9.7372e–10	3.2663e–05
		SSALM	56	7.2641	9.7968e–10	2.4662e–09
1500	10	SALM	67	25.3226	9.8741e–10	2.7106e–06
		SSALM	64	16.1484	9.0982e–10	1.5210e–09
2000	10	SALM	70	34.6083	8.4557e–10	1.8757e–07
		SSALM	70	29.7298	6.4346e–10	1.7874e–09
2500	10	SALM	76	54.5748	7.1288e–10	8.4514e–08
		SSALM	74	46.5716	4.9981e–10	1.6792e–09
3000	10	SALM	78	84.9121	5.3019e–10	2.2285e–06
		SSALM	76	66.0761	9.2876e–10	1.1697e–09
4000	20	SALM	73	164.7655	8.6946e–10	7.3598e–06
		SSALM	72	144.5329	5.1358e–10	1.4439e–09
5000	20	SALM	75	251.7641	8.9821e–10	7.9543e–08
		SSALM	76	203.3167	8.3168e–10	3.4049e–08
8000	25	SALM	79	598.8095	9.5617e–10	3.7230e–08
		SSALM	80	530.8533	4.9436e–10	1.2710e–09

Table 4: Comparison between SALM and SSALM for p=0.3

n	rank(M)	Algorithm	IT	time (s)	error 1	error 2
500	10	SALM	46	11.3753	9.3029e–10	3.2509e–06
		SSALM	46	5.1926	6.5512e–10	1.8892e–08
800	10	SALM	57	17.3212	9.0562e–10	1.0264e–04
		SSALM	58	13.5300	5.7672e–10	1.9937e–05
1000	10	SALM	59	12.2900	9.8849e–10	2.9149e–07
		SSALM	58	7.9075	8.0278e–10	6.0819e–09
1500	10	SALM	69	38.0422	8.9932e–10	6.4286e–05
		SSALM	66	17.6928	5.5536e–10	4.7554e–08
2000	10	SALM	72	39.9752	9.1752e–10	3.8681e–07
		SSALM	70	30.0543	8.0322e–10	2.2090e–09
2500	10	SALM	78	92.3154	9.8020e–10	6.9561e–06
		SSALM	74	47.8242	9.2726e–10	6.3568e–09
3000	10	SALM	80	119.9347	8.6992e–10	6.7548e–06
		SSALM	80	84.1827	7.6381e–10	1.6034e–07
4000	20	SALM	74	187.0096	7.3863e–10	2.0127e–06
		SSALM	72	142.3911	8.5136e–10	3.0778e–08
5000	20	SALM	74	246.3862	9.4517e–10	4.4710e–07
		SSALM	78	219.6853	6.8201e–10	2.8025e–06
8000	25	SALM	83	1.4409e+03	9.9003e–10	2.7735e–04
		SSALM	84	1.1312e+03	8.0522e–10	3.7995e–04

Table 5: The values of ratio for ℓ=2

p=0.6	n	500	800	1000	1500	2000	2500	3000	4000	5000	8000
	rank(M)	10	10	10	10	10	10	10	20	20	25
	ratio(%)	70.97	85.74	76.64	89.95	84.38	89.82	86.33	88.29	89.88	85.64
p=0.5	n	500	800	1000	1500	2000	2500	3000	4000	5000	8000
	rank(M)	10	10	10	10	10	10	10	20	20	25
	ratio(%)	75.75	63.41	61.37	76.09	62.38	67.25	86.80	82.08	85.35	80.13
p=0.4	n	500	800	1000	1500	2000	2500	3000	4000	5000	8000
	rank(M)	10	10	10	10	10	10	10	20	20	25
	ratio(%)	30.44	79.55	72.99	63.77	85.90	85.34	77.82	87.72	80.76	88.65
p=0.3	n	500	800	1000	1500	2000	2500	3000	4000	5000	8000
	rank(M)	10	10	10	10	10	10	10	20	20	25
	ratio(%)	45.65	78.11	64.34	46.51	75.18	51.81	70.19	76.14	89.16	78.51

Table 6: Comparison between SALM and SSALM for ℓ=3

n	rank(M)	p	Algorithm	IT	time (s)	error 1	error 2	RATIO (%)
500	10	0.4	SALM	47	7.5689	5.2053e–10	3.9970e–06	71.71
			SSALM	47	5.4276	3.8663e–10	3.2564e–06	
1000	10	0.4	SALM	60	13.9741	8.8726e–10	2.6235e–07	61.58
			SSALM	59	8.6048	9.8580e–10	9.1642e–09	
1500	10	0.5	SALM	68	29.6545	9.1634e–10	4.5666e–05	63.99
			SSALM	68	18.9748	8.6885e–10	4.3619e–09	
1500	10	0.3	SALM	69	47.1327	9.3547e–10	6.8306e–06	65.78
			SSALM	68	31.0025	9.0834e–10	2.2110e–06	
2000	10	0.5	SALM	73	50.5371	8.4794e–10	3.1914e–05	59.00
			SSALM	71	29.8151	9.8594e–10	8.2954e–09	
2000	10	0.4	SALM	72	38.5308	6.6310e–10	9.5342e–07	75.51
			SSALM	69	29.0941	9.3350e–10	3.5384e–09	
3000	10	0.5	SALM	79	83.3712	9.9293e–10	1.2756e–05	76.68
			SSALM	78	63.9294	6.5972e–10	2.8985e–09	
3000	10	0.4	SALM	77	80.7144	8.9678e–10	2.0862e–07	83.38
			SSALM	78	67.2993	9.9493e–10	7.3612e–09	
3000	10	0.3	SALM	79	118.3675	9.8435e–10	9.6348e–05	81.24
			SSALM	80	96.1642	7.0585e–10	1.4254e–06	
4000	20	0.6	SALM	73	152.4618	8.2869e–10	1.7109e–06	75.89
			SSALM	72	115.7083	8.2721e–10	2.4255e–09	
4000	20	0.5	SALM	72	166.5827	9.3233e–10	1.0321e–07	74.92
			SSALM	72	124.8054	7.5040e–10	2.4066e–09	
4000	20	0.4	SALM	75	186.7383	9.0624e–10	7.5345e–07	73.03
			SSALM	74	136.3765	8.7658e–10	1.0216e–07	

5000	20	0.5	SALM	75	232.0879	9.8661e–10	5.7423e–08	80.86
			SSALM	75	187.6559	9.6715e–10	2.2527e–08	
5000	20	0.4	SALM	76	250.9495	8.5572e–10	1.1830e–06	83.20
			SSALM	77	208.7798	9.7333e–10	7.8349e–07	
5000	20	0.3	SALM	76	273.7735	9.4519e–10	6.4327e–06	86.76
			SSALM	80	237.5358	5.0062e–10	8.2082e–06	
8000	25	0.5	SALM	80	575.2382	9.1308e–10	5.0447e–05	82.81
			SSALM	81	476.3530	7.4764e–10	4.6774e–08	
8000	25	0.4	SALM	80	512.3396	9.7395e–10	6.3150e–06	97.00
			SSALM	81	496.9932	7.2501e–10	9.3571e–09	
8000	25	0.3	SALM	83	579.2973	7.1372e–10	1.3349e–06	90.08
			SSALM	80	521.8350	9.4433e–10	6.5484e–09	

CONCLUDING REMARKS

As is known to all, matrix completion usually means to reconstruct a matrix from a subset of the items of a matrix by taking advantage of low-rank structure matrix interdependencies between the entries. It is well known but NP-hard in general. In recent years, the Toeplitz matrix completion has attracted widespread attention and has become one of the most important completion problems. In order to solve such problems, we proposed an accelerated technique of the SALM algorithm in this study, namely the SSALM algorithm, and the theory of the SSALM algorithm convergence is established. Theoretical analysis and numerical results have shown that the SSALM scheme is an effective algorithm for solving the TMC problem. In particular, the CPU of the SSALM algorithm is consistently reduced up to 30.44% in all cases. The SSALM algorithm overcomes the original ALM algorithm both complexity of the singular value decomposition and surmounts the property of the extra load of the SALM algorithm. The reason is that data communication congestion is far more expensive than computing in many computers. Therefore, the SSALM algorithm has better numerical behavior for solving the TMC problem than the SALM algorithm (Tables 1–6).

ACKNOWLEDGEMENTS

The authors are very much indebted to the editor and anonymous referees for their helpful comments and suggestions which greatly improved the original manuscript of this paper. The authors also are thankful for the support from the NSF of Shanxi Province (201601D011004) and the SYYJSKC-1803.

AUTHORS' CONTRIBUTIONS

All authors contributed equally to the writing of this paper. All authors read and approved the final manuscript.

REFERENCES

1. Amit, Y., Fink, M., Srebro, N., Ullman, S.: Uncovering shared structures in multiclass classification. In: Proceeding of the 24th International Conference on Machine Learning, pp. 17–24. ACM, New York (2007)
2. Argyriou, A., Evgeniou, T., Pontil, M.: Multi-task feature learning. Adv. Neural Inf. Process. Syst. 19, 41–48 (2007)
3. Bertalmio, M., Sapiro, G., Caselles, V., Ballester, C.: Multi-task feature learning, image inpainting. Comput. Graph. 34, 417–424 (2000)
4. Blanchard, J., Tanner, J., Wei, K.: CGIHT: conjugate gradient iterative hard thresholding for compressed sensing and matrix completion. In: Numerical Analysis Group (2014) Preprint 14/08
5. Boumal, N., Absil, P.A.: RTRMC: a Riemannian trust-region method for low-rank matrix completion. In: Shawe-Taylor, J., Zemel, R.S., Bartlett, P., Pereira, F.C.N., Weinberger, K.Q. (eds.) Advances in Neural Inf. Processing Systems, NIPS, vol. 24, pp. 406–414 (2011)
6. Cai, J.F., Candès, E.J., Shen, Z.: A singular value thresholding method for matrix completion. SIAM J. Optim. 20(4), 1956–1982 (2010)
7. Cai, J.F., Qu, X., Xu, W., Ye, G.B.: Robust recovery of complex exponential signals from random Gaussian projections via low rank Hankel matrix reconstruction. Appl. Comput. Harmon. Anal. 41(2), 470–490 (2016)
8. Candès, E.J., Recht, B.: Exact matrix completion via convex optimization. Found. Comput. Math. 9(6), 717–772 (2009)
9. Chen, C., He, B., Yuan, X.: Matrix completion via an alternating direction method. IMA J. Numer. Anal. 32, 227–245 (2012)
10. Chen, M., Ganesh, A., Lin, Z., Ma, Y., Wright, J., Wu, L.: Fast convex optimization algorithms for exact recovery of a corrupted low-rank matrix. J. Mar. Biol. Assoc. UK 56(3), 707–722 (2009)
11. Chen, Y., Chi, Y.: Robust spectral compressed sensing via structured matrix completion. IEEE Trans. Inf. Theory 60(10), 6576–6601 (2014)
12. Eldén, L.: Matrix Methods in Data Mining and Pattern Recognition. Society for Industrial and Applied Mathematics, Philadelphia (2007)
13. Hu, Y., Zhang, D.B., Liu, J., Ye, J.P., He, X.F.: Accelerated singular value thresholding for matrix completion. In: KDD'12, Beijing, August 12-16, 2012 (2012)
14. Jain, P., Meka, R., Dhillon, I.: Guaranteed rank minimization via

singular value projection. In: Proceeding of the Neural Information Processing Systems Conf., NIPS, pp. 937–945 (2010)

15. Jain, P., Netrapalli, P., Sanghavi, S.: Low-rank matrix completion using alternating minimization. In: Proceedings of the 45th Annual ACM Symposium on Theory of Computing (STOC), pp. 665–674 (2013)
16. Ji, H., Liu, C., Shen, Z., Xu, Y.: Robust video denoising using low rank matrix completion. In Proceedings of IEEE Computer Society Conference on Computer Vision and Pattern Recognition CVPR (2010)
17. Kyrillidis, A., Cevher, V.: Matrix recipes for hard thresholding methods. J. Math. Imaging Vis. 48(2), 235–265 (2013)
18. Lin, Z., Chen, M., Wu, L., Ma, Y.: The augmented Lagrange multiplier method for exact recovery of corrupted low-rank matrices. UIUC Technicial Report UIUL-ENG-09-2214, (2010)
19. Liu, Z., Vandenberghe, L.: Interior-point method for nuclear norm approximation with application to system identification. SIAM J. Matrix Anal. Appl. 31(3), 1235–1256 (2009)
20. Luk, F.T., Qiao, S.: A fast singular value algorithm for Hankel matrices. In: Olshevsky, V. (ed.) Fast Algorithms for Structured Matrices: Theory and Applications. Contemporary Mathematics, vol. 323. Am. Math. Soc., Providence (2003)
21. Ma, S., Goldfarb, D., Chen, L.: Fixed point and Bregman iterative methods for matrix rank minimization. Math. Program. 128, 321–353 (2011)
22. Mesbahi, M., Papavassilopoulos, G.P.: On the rank minimization problem over a positive semidefinite linear matrix inequality. IEEE Trans. Autom. Control 42(2), 239–243 (1997)
23. Mishra, B., Apuroop, K.A., Sepulchre, R.: A Riemannian geometry for low-rank matrix completion (2013) arXiv:1306.2672
24. Sebert, F., Zou, Y.M., Ying, L.: Toeplitz block matrices in compressed sensing and their applications in imaging. IEEE, Inf. Tech. Appl. in Biomedicine, 47–50 (2008)
25. Shaw, A.K., Pokala, S., Kumaresan, R.: Toeplitz and Hankel approximation using structured approach. Acoustics. Speech and signal processing. Proc. IEEE Int. Conf. 2349, 12–15 (1998)
26. Tanner, J., Wei, K.: Low rank matrix completion by alternating steepest descent methods. Appl. Comput. Harmon. Anal. 40, 417–429 (2016)
27. Toh, K.C., Yun, S.: An accelerated proximal gradient algorithm for

nuclear norm regularized linear least squares problems. Pac. J. Optim. 6, 615–640 (2010)

28. Tomasi, C., Kanade, T.: Shape and motion from image streams under orthography: a factorization method. Int. J. Comput. Vis. 9(2), 137–154 (1992)
29. Vandereycken, B.: Low rank matrix completion by Riemannian optimization. SIAM J. Optim. 23(2), 1214–1236 (2013)
30. Wang, C.L., Li, C.: A mean value algorithm for Toeplitz matrix completion. Appl. Math. Lett. 41, 35–40 (2015)
31. Wang, C.L., Li, C., Wang, J.: A modified augmented Lagrange multiplier algorithm for Toeplitz matrix completion. Adv. Comput. Math. 42, 1209–1224 (2016)
32. Wang, C.L., Li, C., Wang, J.: Comparisons of several algorithms for Toeplitz matrix recovery. Comput. Math. Appl. 71(1), 133–146 (2016)
33. Wen, R.P., Li, S.D., Meng, G.Y.: SOR-like methods with optimization model for augmented linear systems. East Asian J. Appl. Math. 7(1), 101–115 (2017)
34. Wen, R.P., Li, S.Z., Zhou, F.: Toeplitz matrix completion via smoothing augmented Lagrange multiplier algorithm. Appl. Math. Comput. 355, 299–310 (2019)
35. Wen, R.P., Yan, X.H.: A new gradient projection method for matrix completion. Appl. Math. Comput. 258, 537–544 (2015)

Chapter 11

Singular Spectrum-based Matrix Completion for Time Series Recovery and Prediction

Grigorios Tsagkatakis[1], Baltasar Beferull-Lozano[2] and Panagiotis Tsakalides[1,3]

[1]Institute of Computer Science, Foundation for Research & Technology - Hellas (FORTH), Crete, Greece

[2]Lab Intelligent Signal Processing & Wireless Networks (WISENET), Department of Information and Communication Technologies, University of Agder, Grimstad, Norway

[3]Department of Computer Science, University of Crete, Crete, Greece

ABSTRACT

Big data, characterized by huge volumes of continuously varying streams of information, present formidable challenges in terms of acquisition, processing, and transmission, especially when one considers novel technology platforms such as the Internet-of-Things and Wireless Sensor

Citation (APA): Tsagkatakis, G., Beferull-Lozano, B., & Tsakalides, P. (2016). Singular spectrum-based matrix completion for time series recovery and prediction. EURASIP Journal on Advances in Signal Processing, 2016(1), 66. (14 pages). DOI: https://doi.org/10.1186/s13634-016-0360-0

Networks. Either by design or by physical limitations, a large number of measurements never reach the central processing stations, making the task of data analytics even more problematic. In this work, we propose Singular Spectrum Matrix Completion (SS-MC), a novel approach for the simultaneous recovery of missing data and the prediction of future behavior in the absence of complete measurement sets. The goal is achieved via the solution of an efficient minimization problem which exploits the low rank representation of the associated trajectory matrices when expressed in terms of appropriately designed dictionaries obtained by leveraging the theory of Singular Spectrum Analysis. Experimental results in real datasets demonstrate that the proposed scheme is well suited for the recovery and prediction of multiple time series, achieving lower estimation error compared to state-of-the-art schemes.

Keywords: Compressed Sensing, Reconstruction Error, Singular Spectrum Analysis, Matrix Completion, Nuclear Norm

INTRODUCTION

The dynamic nature of Big Data, a feature termed velocity, is a critical aspect of massive data streams from a signal processing viewpoint [1]. Due to the high velocity of the input streams, measurements may be missing with a high probability. This phenomenon can be attributed to three factors, namely: (a) intentionally collecting a subset of the measurements for efficiency purposes; (b) unintentional subsampling due to desynchronization; and (c) missing measurements due to communications errors including packet drops, outages, and congestion. To elaborate on these factors, we consider data streams associated with the Internet-of-Things (IoT) paradigm and we focus on Wireless Sensor Networks (WSNs) since WSNs can serve as an enabling platform for IoT applications [2, 3]. In the context of IoT/WSNs, one source of missing measurements is attributed to intentional subsampling, a scenario where the designer/operator reduces the sampling rate of the sensing infrastructure in order to increase the lifetime of the network. The relationship between sampling rate and lifetime is governed by the limited energy availability that typically characterizes WSNs. While efficient compression and aggregation schemes can be employed to reduce power consumption, reducing the number of measurements is the most efficient approach to achieve this goal [4]. Even when a specific sampling rate is selected, desynchronization between nodes inevitably leads to a reduction of the network-wide sampling rate, since nodes that were supposed to sample at the same time instance end up acquiring measurements at different instances [5]. This issue is also closely related to the quantization of the sampling time, as measurements that were collected in succession can be mapped to different sampling instances, introducing missing measurements for particular time slots. In addition to energy consumption and desynchronization, missing measurements can also be attributed to network outages and packet losses, which are frequent in WSNs deployed in harsh and cluttered environments, causing a large number of packets to fail in reaching their destination.

In this work, we investigate a novel paradigm in distributed data acquisition and centralized reconstruction and forecasting. The proposed sampling, reconstruction, and prediction scheme assumes that only a small number of randomly selected nodes acquire measurements during each sampling instance, while nodes that are not in the sampling group enter a low-power state. Because of the sampling scheme, in addition to missing data due to packet losses, the base station only observes a subset of the entire collection of measurements. To address this issue, we propose the so-called Singular Spectrum Matrix Completion (SS-MC) scheme, a formal approach for the recovery of missing values and the forecasting of future ones from a single or multiple time series measurements. The proposed SS-MC scheme builds upon the recently proposed framework of Matrix Completion (MC) [6, 7] for the recovery of low-rank matrices from a minimal set of measurements by extending the low-rank matrix recovery framework to the estimation of missing measurements from appropriately generated trajectory matrices and combines it with the Singular Spectrum Analysis framework for exploiting the information encoded in training data. Figure 1 presents a visual overview of the proposed reconstruction scheme, where incomplete trajectory matrices are recovered, providing accurate estimations of past and future measurements. In short, the key novelties of this work include the following:

- A novel efficient paradigm for estimating missing measurements which extents the recently developed framework of low-rank matrix recovery by exploiting inherent correlations without the need for explicit models.

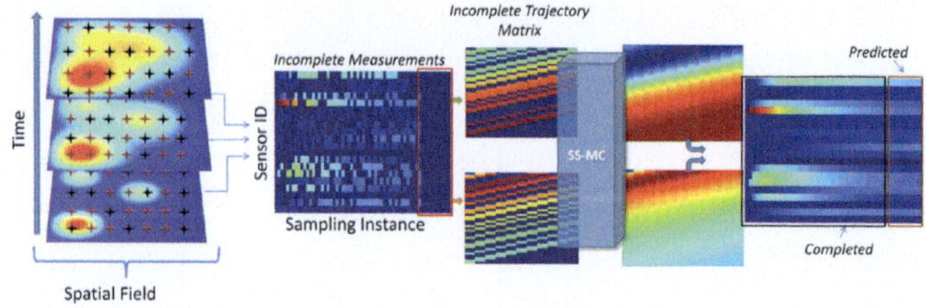

Figure 1: Overview of the proposed sampling, recovery, and prediction scheme. On the left, the three images correspond to the spatial field at three different time instances, where the star symbols indicate the sensing nodes. During each sampling instance, red stars indicate sampling sensors while black stars indicate non-sampling sensors. The figure in the center depicts the incomplete measurement matrix where rows correspond to measurements from a specific sensor and columns to different sampling instances. The red square over the right part of the matrix highlights that in addition to missing value estimation, our system can generate a number of instances (columns) corresponding to future predictions. Individual sensor measurements are transformed to trajectory matrices that are introduced to the proposed SS-MC framework. The SS-MC algorithm produces completed trajectory matrices that can be joined to generate a fully completed (past and future) measurement matrix.

- The proposed SS-MC scheme is an integrated approach for accurately predicting future values even when only a limited number of past measurements is available. This is radical departure from traditional time series forecasting schemes which assume the full availability of historical data.
- The proposed scheme can naturally handle a single or multiple time series sources extending traditional estimation approaches that operate strictly on either single or multi-source data.
- The performance of the proposed method against state-of-the-art techniques is evaluated on real data acquired by a distributed sensor network, which serves as an illustrative example of a Big Data application.

The rest of the paper is organized as follows: Section 2 presents an overview of state-of-the-art methods for energy-efficient data collection. Sections 3 and 4 provide the description of the two theoretical models we consider in this work, namely time series modeling via Singular Spectrum Analysis and missing measurement estimation via the Matrix Completion framework. Section 5 introduces SS-MC, our proposed recovery and prediction method, including the mathematical formulation as well as an efficient optimization approach based on Augmented Lagrange Multipliers. The performance of the proposed scheme is experimentally validated against state-of-the-art methods in Section 6 and the paper concludes in Section 7.

RELATED WORK

Designing efficient techniques for minimizing the cost of continuous data collection by exploiting data correlations has been extensively studied from multiple aspects and different perspectives in the context of WSNs [8]. Jindal and Psounis [8] presented a method for inferring the spatial correlation of WSN data and for generating synthetic data using a statistical tool called variagriam. Estimating the sampling field at a given location, based on the available sensor data at other additional locations is a common approach for energy efficient sampling. Data imputation and interpolation techniques, such as Nearest Neighbors Imputationand Kriging, are two very efficient schemes for estimating unavailable data [9]. While in interpolation, one seeks the value of the field in a location where no sensors are present, imputation approaches try to estimate the value at the sensor location at a time instance where sampling did not take place. Kriging relies on the semi-variogram, a statistical tool developed by geo-statisticians [10] in order to estimate the value of a field at a specific location, given prior knowledge about the inherent correlations of data from neighboring nodes. In k-Nearest Neighbors, this objective is reached by using a weighted nearest neighbor interpolation, where the weight corresponding to each sample is based on statistical information indicating the degree of spatial dependence in the field [11].

Another line of work for data imputation exploits probabilistic models for estimating the missing entries. In [12], an Expectation Maximization (EM) algorithm is presented which estimates the parameters of the probability distribution of the data by iteratively maximizing the likelihood of the available data as a function of these parameters. In order to increase the robustness of the process, the authors proposed the regularized EM (RegEM) where a regularization term is added during the inversion of the correlation matrix in order to increase the robustness of the algorithm when more variables are present than data records. RegEM is currently one of the state-of-the-art data imputation techniques, and its performance is compared against the proposed and other schemes in the experimental section.

Data compression has also been extensively explored in the context of energy-efficient data collection in WSNs, based on the premise that data processing is less demanding in terms of energy consumption compared to transmission; hence, energy reduction can be achieved. For example, the recently proposed framework of Compressed Sensing (CS), a state-of-the-art signal sampling and compression scheme, was investigated for WSN data acquisition and aggregation [13, 14] exploiting the sparsity of the sampled data when expressed in an appropriate basis [15]. Distributed compression schemes such as Distributed Source Coding [16] have also been proposed for compressing WSN measurements in densely deployed networks, since utilizing side information from neighboring nodes can dramatically reduce communication cost. The sparse characteristics of correlated datasets have also been recently considered for transmission of EEG signals [17, 18]. Although sparsity and CS-based methods can have a dramatic reduction in transmission power, typically in these scenarios, the signals are first fully sampled and then compressed.

While the CS framework requires a particular form of sampling (incoherent sampling), the related paradigm of low-rank matrix recovery (MC) assumes a random sampling of the matrix entries. Due to the intuitive sampling, the MC framework has been considered for a variety of signal recovery problems including collaborative spectrum sensing [19], sensor localization [20, 21], and image reconstruction problems [22, 23] among others. MC has been recently explored as a sampling scheme for WSNs [24, 25, 26, 27]. In [24], the authors investigated the scenario where sensors lie on a uniform rectangular grid and random sub-sampling is taking place by each sensor. Our work bares some similarities with this line of work; however, we do not pose specific deployment constraints and we allow the sensors to occupy any location in the sensed region. Furthermore, our work differs significantly in the exploitation of prior knowledge in the form of a dictionary, which is utilized during the reconstruction stage. The utilization of the singular spectrum dictionary allows for the incorporation of prior knowledge regarding the data generation process which can significantly improve the reconstruction performance [25]. Furthermore, the proposed scheme is able to predict future measurements in addition to estimating missing past ones.

Low-rank recovery was also recently considered in [28] where the authors employ MC for the recovery of undersampled correlated EEG signals. Our work in this paper investigates different extensions of MC-based recovery by considering trajectory matrices and singular spectrum dictionaries. We develop a generative model where the sampled data can be jointly represented as a low-rank linear combination of dictionary elements, spanning the subspace where data is lying. A similar situation was recently explored, leading to the low rank representations (LRR) framework [29] where the objective is to identify a low rank matrix which can accurately represent the source data. LRR has been considered for subspace clustering problems [30]; however, only fully populated matrices were considered.

In the context of Big Data, matrix and tensor data recovery via an online rank minimization process [31] was recently proposed for scalable imputation of missing data. This was achieved by low-dimensional subspace tracking through the minimization of a weighted least squares regression, regularized with a nuclear norm. While this work bares resemblance to our work, our generative model does not require a fixed bilinear factorization due to a pre-specified rank, while it exploits the subspace identified by the SSA for simultaneous missing past measurement imputation and future predictions.

ANALYSIS OF TIME SERIES DATA

Singular Spectrum Analysis (SSA) is a model-free method for time series analysis and forecasting which has been widely exploited in the analysis of environmental, economical, and computer network data [32, 33]. The basic assumption underlying SSA is that one can approximate a time series $\mathbb{M}_i$ of length K from L lagged samples, by considering the spectral analysis of specialized matrices, called trajectory matrices. Embedding at sampling instance T, the first step of SSA, involves the process of generating a trajectory matrix $\mathbf{M}_i = \{\mathbf{m}_{i,t} | t = T - L : T\} \in \mathbb{R}^{K \times L}$ of lag L measurement vectors, where each vector $\mathbf{m}_{i,t'} = \{m_{i,t'} | t' = t - K : t\}$ encodes the measurements corresponding to a sampling window of length K for sensor i. The length K of the time window and the lag L are two critical parameters encoding important aspects of the underlying data.

In SSA, once the trajectory matrix of the time series has been generated, the subsequent step involves the spectral analysis of the lag-covariance matrix. Formally, given the matrix $\mathbf{M}_i$, the lag-covariance matrix defined as $\mathbf{C}_i = \mathbf{M}_i \mathbf{M}_i^T$ can be used for extracting the eigenvectors of $\mathbf{C}$ which define an L-dimensional subspace where the time series $\mathbb{M}_i$ resides, while the associated eigenvalues encode the variance along the direction of the associated eigenvector. Alternatively, one can apply the SVD decomposition to the original trajectory matrix $\mathbf{M}_i$ in which case the outputs are two matrices containing the right and left singular vectors **U** and **V** and a diagonal matrix Σ containing the singular values. Given the SVD decomposition, the trajectory matrix $\mathbf{M}_i$ can be expressed as the sum of rank-1 matrices given by $\mathbf{M}_j = \sum_j \sqrt{\lambda_j}\mathbf{u}_j\mathbf{v}_j^T$, where each collection (λ_j,$\mathbf{u}_j$,$\mathbf{v}_j$) is called eigentriple. Given the eigenvectors extracted via the SSA, one can project and reconstruct the time series or perform prediction by employing two steps, eigentriple grouping and diagonal averaging. Eigentriple grouping aims at arranging the eigentripes in sets in order to separate additive components that are exactly or approximate separable, facilitating the analysis of the eigenvectors. Diagonal averaging aims at translating the recovered trajectory matrix into a time series according to

$$\hat{m}[k] = \begin{cases} \frac{1}{k}\sum_{m=1}^{K} m^*[m, k-m+1], \\ \qquad\qquad \text{for } T-L-K \leq k < L \\ \frac{1}{L}\sum_{m=1}^{L} m^*[m, k-m+1], \quad \text{for } L \leq k < K \\ \frac{1}{T-K+1}\sum_{m=k-K+1}^{T-K+1} m^*[m, k-m+1], \\ \qquad\qquad \text{for } K \leq k < T \end{cases} \tag{1}$$

where m *[i,j]=m[i,j] for L<K and m *[i,j]=m[j,i] otherwise.

It is worth noting that SSA has also been considered in situations when a number of measurements are missing. A straightforward approach, also employed here, is to estimate the eigenvectors and eigenvalues using only the available measurements during the lag-covariance matrix generation [34]. SSA has also been considered when missing measurements are present [35, 36]; however, the proposed methods differ from our work in that we exploit prior knowledge in the form of a dictionary. Furthermore, the proposed scheme is able to perform missing value estimation, either past or future, while there is no constraint associated with the structure of the missing measurements.

In addition to the analysis of time series, SSA can also be used as a forecasting mechanism. In recurrent forecasting SSA, the time series of known measurements and unknown components is transformed to its Hankel form and the linear recurrent relation coefficients are utilized for forecasting the future values. While typical SSA considers the trajectory matrices associated with a single time series, the Multivariate Singular Spectrum Analysis (MSSA) method has been proposed for handling multiple time series [37, 38, 39]. In this work, we consider a simple extension of SSA where instead of analyzing a single trajectory matrix, we consider a compound trajectory matrix generated by the concatenation of S individual matrices, i.e., $\mathbf{M} = [\mathbf{M}_1, \mathbf{M}_2, \ldots, \mathbf{M}_S] \in \mathbb{R}^{S(K\times L)}$. Introducing multiple sources of data can have a dramatic impact in performance as will be shown in the experimental results, with at most linear increase in computational complexity.

LOW-RANK MATRIX COMPLETION

The low-rank approximation of a given matrix is a frequent problem in data analysis [40]. The rank of the matrix indicates the number of linearly independent columns (or rows), and thus it is a indicator of the degree of linear correlation that exists within the data. There are multiple reasons that justify the need for such an analysis. For example, prior knowledge regarding the linear correlation of the data may suggest that the requested matrix is low rank. In other situations, noise in the data artificially increases the rank of the matrix, so reducing the rank effectively amounts to a denoising process. Assuming without loss of generality that S=1, given a noisy (K×L) matrix **M**, the objective of low-rank approximation is to identify a matrix **X** such that:

$$\underset{\mathbf{X}}{\text{minimize}} \ \text{rank}(\mathbf{X}) \quad \text{subject to} \ \|\mathbf{X} - \mathbf{M}\|_F < \epsilon \qquad (2)$$

where ε is the approximation error, related to the noise power. By utilizing the SVD decomposition **M**=**U** **S** **V** ^{T}, a low-rank approximation matrix **X** can be found by X= $\mathbf{U}\mathcal{T}(\mathbf{S})\mathbf{V}^T$, where $\mathcal{D}_\tau(\mathbf{S}) = diag([\sigma_i(\mathbf{S}) - \tau]_+)$ is a thresholding operator that selects only the elements with values greater than τ from the diagonal matrix **S**and sets the rest to zero. The effect of this process is that only a small number of singular values are kept for the low-rank approximation **X** of **M**.

The rank of the matrix is a key property in the recently proposed framework of Matrix Completion (MC) where one tries to estimate the (K×L) entries of the matrix **M** from a smaller number of q entries, where q≪(K×L). According to MC, such a recovery is possible provided the matrix is characterized by a small rank (compared to its dimensions) and enough randomly selected entries of the matrix are acquired [6, 41]. More specifically, one can recover an accurate approximation **X** of the matrix **M** from a small number of entries by solving the minimization problem:

$$\underset{\mathbf{X}}{\text{minimize}}\ \text{rank}(\mathbf{X})$$
$$\text{subject to } \mathcal{P}_{\Omega}(\mathbf{X}) = \mathcal{P}_{\Omega}(\mathbf{M}) \tag{3}$$

where P_{Ω} is a random sampling operator which records only a small number of entries from the matrix **M**, i.e.,

$$\mathcal{P}_{\Omega}(\mathbf{M}) = \begin{cases} m_{ij}, & \text{if } ij \in \Omega \\ 0, & \text{otherwise} \end{cases} \tag{4}$$

where Ω is the sampling set. In the context of WSN for example, the set Ω specifies the collection of sensors that are active at each specific sampling instance. In general, to solve the MC problem, the sampling operator P must satisfy the modified restricted isometry property, which is the case when uniform random sparse sampling is employed in both rows and columns of matrix **M** [42]. The incoherence of sampling introduced by P with respect to **M** guarantees that recovery is possible from a limited number of measurements.

Although solving the above problem will generate a low-rank matrix consistent with the observations, rank minimization is an NP-hard problem. Fortunately, a relaxation of the above problem was shown to produce very accurate approximations, by replacing the rank constraint by the tractable nuclear norm, which represents the convex envelope of the rank [6]. The minimization in Eq. (4) can then be reformulated as:

$$\underset{\mathbf{X}}{\text{minimize}}\ \|\mathbf{X}\|_*$$
$$\text{subject to } \mathcal{P}_{\Omega}(\mathbf{X}) = \mathcal{P}_{\Omega}(\mathbf{M}) \tag{5}$$

where the nuclear norm is defined as $\|\mathbf{X}\|_* = \sum \|\sigma_i\|_1$, i.e., the sum of absolute values of the singular values. Candès and Tao showed that under certain conditions the nuclear norm minimization in Eq. (5) can estimate the same matrix as the rank minimization in Eq. (3) with high probability provided $q \geq C K^{6/5} r \log(K)$ randomly selected entries of the rank r matrix are acquired [7] (assuming $K \geq L$).

To solve the nuclear norm minimization problem, various approaches have been proposed including Singular Value Thresholding [43] and the Augmented Lagrange Multipliers [44], among others. We review the technique based on the ALM due to its exceptional performance in terms of both processing complexity and reconstruction accuracy and since it is used as a basis for the extended scheme we discuss next.

To express the MC problem in Eq. (5) in the ALM form, we reformulate it as:

$$\begin{aligned}&\underset{\mathbf{X},\mathbf{E}}{\text{minimize}}\ \|\mathbf{X}\|_*\\&\text{subject to}\ \mathbf{X}+\mathbf{E}=\mathbf{M}\\&\qquad\mathcal{P}_\Omega(\mathbf{E})=0\end{aligned}\tag{6}$$

The additional variable **E** is introduced in order to encode the unknown values in the trajectory matrix **M**, by restricting the estimation error on the recorded values only. The optimization encoded in Eq. (6) can be expressed in an augmented Lagrangian form by defining the Lagrangian function:

$$\begin{aligned}\mathcal{L}(\mathbf{X},\mathbf{E},\mathbf{Y},\mu)&=\|\mathbf{X}\|_*+tr(\mathbf{Y}^T(\mathbf{X}-\mathbf{M}+\mathbf{E}))\\&\quad+\frac{\mu}{2}\|\mathbf{X}-\mathbf{M}+\mathbf{E}\|_F^2\end{aligned}\tag{7}$$

where **Y** is the Lagrange multiplier matrix associated to the first equality constraint and μ is the penalty parameter. Minimization of the problem in Eq. (7) involves an iterative process, where a sequential minimization over all variables, i.e., **X**,**E**, and **Y**, takes place at each iteration. This method of iteratively minimizing over each variable is refereed to as the Alternating Directions Method of Multipliers (ADMM) [45, 46].

One of the key characteristics of MC is the minimal conditions that are imposed for successful recovery, namely the incoherence of sampling and the low rank of the recovered matrix. While a minimal set of requirements is beneficial in situations where limited prior information is available, when such information exists introducing additional constraints can lead to a significantly better recovery. In this section, we exploit the temporal dynamic that time series exhibit in order to enhance typical MC with an additional dictionary which encodes past behavior in a proposed SS-MC framework.

THE SS-MC ALGORITHM

We consider the truncated trajectory matrices **M** formed by concatenating the individual trajectory matrices according to the MSSA approach. The objective of this work is to consider a generative model that produces the time series Hankel matrices **M** according to the factorization **M** = **D L** where **M** may correspond to a single or multiple sources. In both cases, our key assumption is that given a full rank dictionary matrix **D** obtained through training data, the coefficient matrix **L** is approximately low rank, i.e., the number of significant singular vectors is much smaller than the ambient dimensions of the matrix. To apply the low-rank representation scheme on matrices with missing data, the introduction of the random sub-sampling operator is necessary. Our proposed sampling scheme is a combination of MC and reduced rank multivariate linear regression and it seeks a low-rank presentation coefficient matrix **L** from a small number of measurements $P_{\Omega}(M)$. Based on this generative model, our proposed Singular Spectrum Matrix Completion (SS-MC) formulation is given by:

$$\begin{aligned} &\underset{\mathbf{L}}{\text{minimize}}\ \mathit{rank}(\mathbf{L}) \\ &\text{subject to}\ \mathcal{P}_{\Omega}(\mathbf{M}) = \mathcal{P}_{\Omega}(\mathbf{DL}) \end{aligned} \tag{8}$$

where **D** is a dictionary of elementary atoms that span a low-rank data-induced subspace. Figure 2 presents an example of a real trajectory matrix (left), the representations coefficients **L** (center), and the singular value distribution of the coefficients (right).

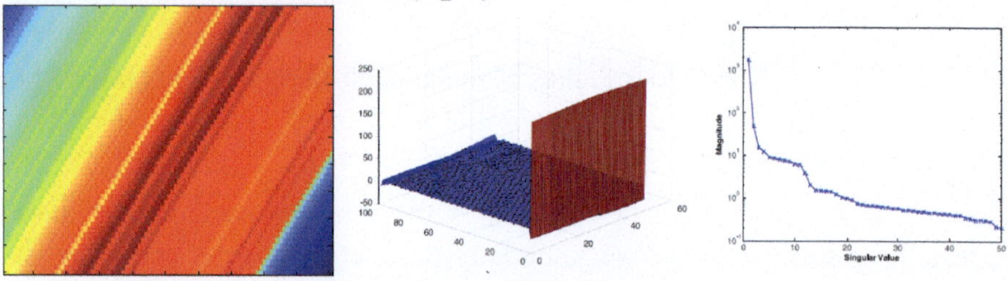

Figure 2: Example of the generative process of a real trajectory matrix from the Intel-Berkeley dataset. The matrix on the left is the Hankel matrix generated by a sensor for a given set of window and lag values. By utilizing the SSA-based dictionary, the mapping of the Hankel matrix results in an extremely low rank representation matrix shown in the middle, where a small number of singular values capture most of the signal energy, as shown in the right figure.

Efficient Optimization

Similarly to MC optimization, the problem in Eq. (8) is NP-hard due to the rank in the objective function and thus it cannot be solved efficiently for reasonably sized data. A remedy to this problem is to replace the rank constraint with the nuclear norm constraint, thus solving:

$$\begin{aligned} &\underset{\mathbf{L}}{\text{minimize}}\ \|\mathbf{L}\|_{*} \\ &\text{subject to}\ \mathcal{P}_{\Omega}(\mathbf{M}) = \mathcal{P}_{\Omega}(\mathbf{DL}) \end{aligned} \tag{9}$$

A key novelty of our work is that in addition to the low rank of the matrix,

during the recovery, we employ a dictionary for modeling the generative process that produces the sensed data, as it can be seen in Eq. (9).
The problem in Eq. (9) can be transformed to a semidefinite programming problem and solved using interior point methods [47, 48]. However, utilizing such off-the-shelf solvers introduces a very high algorithmic complexity which renders them impractical, even for moderately sized scenarios. Motivated by the requirements for a data collection mechanism that is both accurate and efficient, we reformulate the SS-MC problem in an Augmented Lagrangian form. By utilizing the ALM formulation for SS-MC, we can achieve efficient recovery, tailored to the specific properties of the problem. Introducing the intermediate dummy variables **Z** and **E**, Eq. (9) can be written as:

$$\begin{aligned} &\underset{\mathbf{L},\mathbf{Z},\mathbf{E}}{\text{minimize}} \ \|\mathbf{L}\|_* \\ &\text{subject to } \mathbf{M} = \mathbf{DZ} + \mathbf{E} \\ &\qquad\qquad \mathbf{Z} = \mathbf{L} \\ &\qquad\qquad \mathcal{P}_\Omega(\mathbf{E}) = 0 \end{aligned} \tag{10}$$

where **L** , **Z**, and **E** are the minimization variables. The extra variable **Z** is introduced in order to decouple the minimization variables by separating the **L** variable in the objective function with the **Z** variable in the first constraint. Similar to the ALM formulation for MC in Eq. (7), **E** is introduced in order to account for the missing entries in **M**. More specifically, the constraint on the error matrix **E** is applied only on the available data via the sampling operator PP. The ALM form of Eq. (10) is an unconstrained minimization given by:

$$\begin{aligned} \mathcal{L}(\mathbf{L},\mathbf{Z},\mathbf{E},\mathbf{Y}_1,\mathbf{Y}_2,\mu) = \|\mathbf{L}\|_* &+ tr\left(\mathbf{Y}_1^T(\mathcal{P}_\Omega(\mathbf{M}-\mathbf{DZ}))\right) \\ &+ tr\left(\mathbf{Y}_2^T(\mathbf{Z}-\mathbf{L})\right) \\ &+ \frac{\mu}{2}(\|\mathcal{P}_\Omega(\mathbf{M}-\mathbf{DZ})\|_F^2 + \|\mathbf{Z}-\mathbf{L}\|_{F^2}) \end{aligned} \tag{11}$$

where $\mathbf{Y}_1$ and $\mathbf{Y}_2$ are Lagrange multiplier matrices. The solution can be found by iteratively minimizing Eq. (11) with respect to each of the variables via an ADMM approach. Formally, the minimization problem with respect to **L** is given by:

$$\begin{aligned} \mathbf{L}^{(k+1)} &= \min_{\mathbf{L}} \mathcal{L}\left(\mathbf{L}^{(k)},\mathbf{Z}^{(k)},\mathbf{E}^{(k)},\mathbf{Y}_1^{(k)},\mathbf{Y}_2^{(k)},\mu^{(k)}\right) \\ &= \min \|\mathbf{L}\|_* + tr\left(\mathbf{Y}_2^{\mathbf{T}}(\mathbf{Z}-\mathbf{L})\right) + \frac{\mu}{2}\left(\|\mathbf{Z}-\mathbf{L}\|_F^2\right) \end{aligned}$$

$$= \min \frac{1}{\mu}\|\mathbf{L}\|_* + \frac{1}{2}\|\mathbf{L} - (\mathbf{Z} + \mathbf{Y}_2/\mu)\|_F^2 . \tag{12}$$

The sub-problem in Eq. (12) is a nuclear norm minimization problem and can be solved very efficiently by the Singular Value Thresholding operator [43]. The minimization with respect to **Z** is given by:

$$\begin{aligned}\mathbf{Z}^{(k+1)} &= \min_{\mathbf{Z}} \mathcal{L}\left(\mathbf{L}^{(k+1)}, \mathbf{Z}^{(k)}, \mathbf{E}^{(k)}, \mathbf{Y}_1^{(k)}, \mathbf{Y}_2^{(k)}, \mu^{(k)}\right) \\ &= \min tr\left(\mathbf{Y}_1^T(\mathcal{P}_\Omega(\mathbf{M}) - \mathcal{P}_\Omega(\mathbf{DZ}))\right) + tr\left(\mathbf{Y}_2^T(\mathbf{Z} - \mathbf{L})\right) \\ &+ \frac{\mu}{2}\left(\|\mathcal{P}_\Omega(\mathbf{M} - \mathbf{DZ})\|_F^2 + \|\mathbf{Z} - \mathbf{L}\|_F^2\right) .\end{aligned} \tag{13}$$

Calculating the gradient of the expression in Eq. (13), we obtain:

$$\frac{\partial \mathcal{L}}{\partial \mathbf{Z}} = \mathbf{D}^T\mathbf{Y}_1 - \mathbf{Y}_2 + \mu\left(\mathbf{D}^T(\mathbf{M} - \mathbf{E} - \mathbf{DZ}) - \mathbf{Z} + \mathbf{L}\right) \tag{14}$$

which after setting it equal to zero provides the update equation for Eq. (14) given by:

$$\begin{aligned}\mathbf{Z}^{(k+1)} &= \left(\mathbf{I} + \mathbf{D}^T\mathbf{D}\right)^{-1}\left(\mathbf{D}^T\left(\mathbf{M}^{(k)} - \mathbf{E}^{(k)}\right) + \mathbf{L}^{(k)}\right. \\ &\left. + \left(\mathbf{D}^T\mathbf{Y}_1^{(k)} - \mathbf{Y}_2^{(k)}\right)/\mu^{(k)}\right) .\end{aligned} \tag{15}$$

Furthermore, the augmented Lagrangian in Eq. (11) has to be minimized with respect to **E**, i.e.,

$$\begin{aligned}\mathbf{E}^{(k+1)} &= \min_{\mathcal{P}_\Omega(\mathbf{M})=0} \mathcal{L}\left(\mathbf{L}^{(k+1)}, \mathbf{Z}^{(k+1)}, \mathbf{E}^{(k)}, Y_1^{(k)}, Y_2^{(k)}, \mu^{(k)}\right) \\ &= \min_{\mathcal{P}_\Omega(M)=0} \mathbf{Y}_1 + \mu(\mathbf{E} - \mathbf{M} + \mathbf{DZ})\end{aligned} \tag{16}$$

which provides the update equation for Eq. (16) that is given by:

$$\mathbf{E}^{(k+1)} = \mathcal{P}_{\bar{\Omega}}\left(\mathbf{M} - \mathbf{DZ}^{(k+1)} + \frac{1}{\mu^{(k)}}\mathbf{Y}_1^k\right) \tag{17}$$

where the notation $\mathcal{P}_{\bar{\Omega}}$ is used to restrict the error estimation only on the measurements that do not belong to the sampling set. Last, we perform updates on the two Lagrange multipliers $\mathbf{Y}_1$ and $\mathbf{Y}_2$. The steps at each iteration of the optimization are shown in Algorithm 1.

Algorithm 1: Singular Spectrum Matrix Completion (SS-MC).

Input: The subsampled trajectory matrix
$\mathbf{M}_{ij}, (i, j) \in \Omega$,
The dictionary of examples **D**,
The error tolerance *threshold,*
The maximum number of iterations *limit.*
Output: The representation coefficients matrix **L** and the estimated matrix $\mathbf{X} = \mathbf{DL}$.

1: **initialization**
$\mathbf{L}^0 = \mathbf{0}\,,\ \mathbf{E}^{(0)} = \mathbf{0}\,,\ \mathbf{Z}^{(0)} = \mathbf{0}\,,\ k = 0,\ \alpha = 1.1$
2: **while** *error* $\geq$ *threshold* or *iterations* $\leq$ *limit* **do**
3: Minimize with respect to **L** to obtain $\mathbf{L}^{(\mathbf{k+1})}$

$$(\mathbf{U}, \mathbf{S}, \mathbf{V}) = SVD(\mathbf{Z} + \mathbf{Y}_2/\mu)$$
$$\mathbf{L}^{(\mathbf{k+1})} = U\mathcal{D}_\tau(\mathbf{S})\mathbf{V}^\mathbf{T}$$

4: Minimize with respect to **Z** to obtain $\mathbf{Z}^{(\mathbf{k+1})}$

$$\mathbf{Z}^{(\mathbf{k+1})} = (\mathbf{I} + \mathbf{D}^\mathbf{T}\mathbf{D})^{-1}(\mathbf{D}^\mathbf{T}(\mathbf{M} - \mathbf{E}) + \mathbf{L} + (\mathbf{D}^\mathbf{T}\mathbf{Y}_1 - \mathbf{Y}_2)/\mu)$$

5: Minimize with respect to **E** to obtain $\mathbf{E}^{(\mathbf{k+1})}$

$$\mathbf{E}^{(k+1)} = \mathcal{P}_{\bar{\Omega}}\left(\mathbf{M} - \mathbf{DZ} + \frac{1}{\mu^{(k)}}\mathbf{Y}_1\right)$$

6: Update the Lagrangian multipliers

$$\mathbf{Y}_\mathbf{1}^{(\mathbf{k+1})} = \mathbf{Y}_\mathbf{1}^{(\mathbf{k})} + \mu^{(\mathbf{k})}(\mathbf{M} - \mathbf{DZ}^{(\mathbf{k+1})} - \mathbf{E}^{(\mathbf{k+1})})$$
$$\mathbf{Y}_\mathbf{2}^{(\mathbf{k+1})} = \mathbf{Y}_\mathbf{2}^{(\mathbf{k})} + \mu^{(\mathbf{k})}(\mathbf{Z}^{(\mathbf{k+1})} - \mathbf{L}^{(\mathbf{k+1})})$$

set $k \longleftarrow k + 1$
7: **end while**

Due to its numerous applications, the ADMM method has been extensively studied in the literature for the case of two variables [45, 46] where it has been shown that under mild conditions regarding the convexity of the cost functions, the two-variables ADMM converges at a rate O(1/r) [49]. Although extending the convergence properties to a larger number of variables has not been shown in general, recently the convergence properties of ADMM for a sum of two or more non-smooth convex separable functions subject to linear constraints were examined [50].

The proposed minimization scheme in Eq. (11) satisfies a large number of the constraints suggested in [50] such as the convexity of each sub-problem, the strict convexity and continuous differentiability of the nuclear norm, the full rank of the dictionary, and the size of the step for the dual update α, while empirical evidence suggests that the closed form solution of each sub-problem allows the SS-MC algorithm to converge to an accurate solution in a small number of iterations.

Singular Spectrum Dictionary

In this work, we investigate the utilization of prior knowledge for the efficient reconstruction of severely under-sampled time series data. To model the data, we follow a generative scheme where the full collection of acquired measurements is encoded in the trajectory matrix M$\in \mathbb{R}^{K\times L}$. **M** is assumed to be generated from a combination of a dictionary D$\in \mathbb{R}^{K\times K}$ and a coefficient matrix L$\in \mathbb{R}^{K\times L}$according to **M** = **D L**, where we assume that K≤L. This particular factorization is related to SVD by **M** = **D L** = **U** (**S V** ^{T}) where the orthonormal matrix **D** = **U** is a basis for the subspace associated with the column space of **M**, while **L** = **S V** ^{T}is a low-rank representation matrix encoding the projection of the trajectory matrix onto this subspace.

This particular choice of dictionary **D** implies a specific relationship between the spectral characteristics of the trajectory matrix **M** and the low-rank representation matrix **L**. To understand this relationship, we consider the spectral decomposition of each individual matrix in the form $\mathbf{D}=\mathbf{U}\,\mathbf{G}_1\,\mathbf{R}^{-1}$ and $\mathbf{L}=\mathbf{R}\,\mathbf{G}_2\,\mathbf{V}^*$ The matrices **U** ,**R** and **V** are unitary while $\mathbf{G}_1$ and $\mathbf{G}_2$ are diagonal matrices containing the singular values of the **D** and **L**, respectively. The particular factorization permits us to utilize the product SVD [51, 52] and expresses the singular value decomposition of the product according to the expression $\mathbf{D}\,\mathbf{L}=\mathbf{U}(\mathbf{G}_1\,\mathbf{G}_2)\mathbf{V}^*$, where the singular values of the matrix product are given by the product of the singular values of the corresponding matrices.

In this work, we consider orthogonal dictionaries, as opposed to overcomplete ones. Orthogonality of the dictionary guarantees that the vectors encoded in the dictionary span the low-dimensional subspace and therefore the representation of the measurements is possible. Furthermore, an orthonormal dictionary, such as the one considered in this work, is characterized by $\mathbf{G}_1=\mathbf{I}$, leaving $\mathbf{G}_2$ responsible for the representation. We target exactly $\mathbf{G}_2$ in our problem formulation by seeking a low-rank representation matrix **L**.

In our experimental results, we consider sets of training data associated with fully sampled time series from the first days of each experiment for generating the dictionaries. The subspace identified by the fully sampled data is used for the subsequent recovery of past measurements and prediction of future ones. Alternatively, the dictionary could be updated during the course of the SS-MC application via an incremental subspace learning method [53, 54]. We opted out from an incremental subspace learning since although it can potentially lead to better estimation, it is also associated with increased computational load and the higher probability of estimation drift and lower performance.

Networking aspects of SS-MC

In the context of IoT applications utilizing WSN infrastructures, communication can take place among nodes, but most typically between the nodes and the base station where data analytics are extracted. This communication can be supported (a) by a direct wireless link between the nodes and the sink/base station; (b) via appropriate paths that allow multi-hop communications; or (c) via more powerful cluster heads what forward the measurements to the base station.

For the multi-hop scheme, equal weight of each sample (democratic sampling) implies that no complicated processing needs to take place by the resource limited forwarding nodes. Furthermore, for high-performance WSNs, where point-to-point communication between nodes is available and processing capabilities are sufficient, nodes could perform reconstruction of a local neighborhood thus offering advantages similar to other distributed estimation schemes [55].

From a practical point-of-view, we argue that recovery and prediction of measurements from low sampling rates offer numerous advantages. First, it saves energy by reducing the number of samples that have to be acquired, processed, and communicated thus increasing the lifetime of the network. The proposed sampling scheme also reduces the frequency of sensor re-calibrations for sensors that perform complex signal acquisition, including chemical and biological sampling. As a result, higher quality measurements and therefore more reliable estimation of the field samples can be achieved. Furthermore, the method increases robustness to communication errors by estimating measurements included in lost or dropped packets, without the need for retransmission. Last, our scheme does not require explicit knowledge of node locations for the estimation of the missing measurements, since the incomplete measurement matrices and the corresponding trajectory matrices are indexed by the sensor id, thus allowing greater flexibility during deployment.

EXPERIMENTAL RESULTS

To evaluate the performance of the proposed low-rank reconstruction and prediction scheme, we consider real data from the Intel Berkeley Research Lab dataset[1] [56] and the SensorScope Grand St-Bernard dataset[2] [57]. The former dataset contains the recordings of 54 multimodal sensors located in an indoor environment over a 1-month period, while the latter contains multimodal measurements from 23 stations deployed at the Grand-St-Bernard pass between Switzerland and Italy.

In both cases, we analyze temperature measurements as an exemplary modality, while we exclude failed sensors from the recovery process. Unless stated otherwise, in all cases, we fix the SSA parameters, K=50 and L=100, and we train using a single day's worth of data while testing on the five consecutive ones. The threshold τ for the singular value thresholding operator is set to preserve 90 % of the signals' energy, while the parameter μ was set to 0.01 through a validation process, although the specific value had a minimal impact in performance.

To evaluate the performance, we consider three state-of-the-art methods and we compare them to the proposed SS-MC. More specifically, we evaluate the performance of the ADMM version of MC [44], the Knn-imputation [58], and the RegEM [12]. The reconstruction error is measured by the normalized mean squared error between the true **M** and the estimated **X** trajectory matrices given by $\frac{\sum \|\mathbf{M}-\mathbf{X}\|^2}{\sum \|\mathbf{M}\|^2}$.

Recovery with Respect to Measurement Availability

The objective of this subsection is to present the recovery capabilities of the proposed SS-MC and state-of-the-art methods with respect to the availability of measurements, i.e., the sampling rate.

The two plots shown in Fig. 3 present the reconstruction error for the Intel-Berkeley data at 20 % (top) and 50 % (bottom) sampling rates, averaged over all sensing nodes. Naturally, one can see that increasing the sampling rate has a positive effect on all methods. Nevertheless, we also observe that not all sampling instances are equally difficult to estimate and that the reconstruction error exhibits a periodic trend across sampling instances. These variations are attributed to the significant changes in the environmental conditions due to the transition from nighttime to daytime.

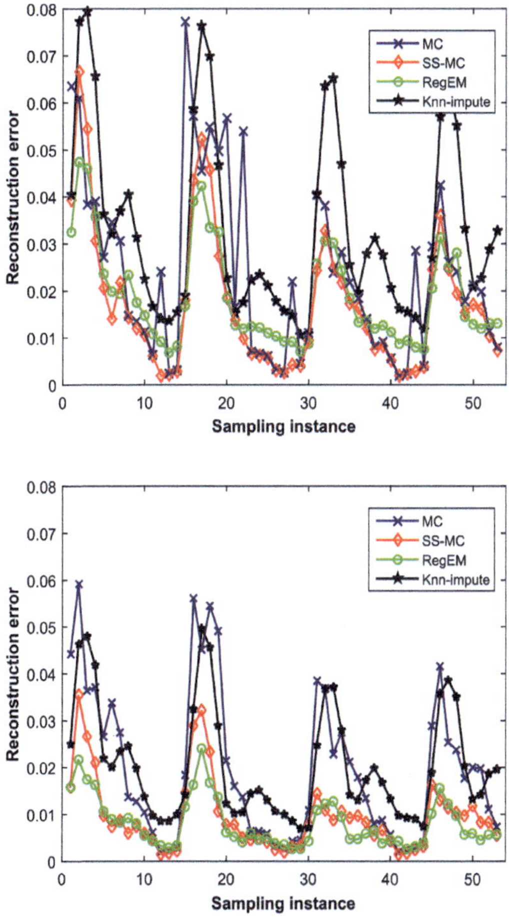

Figure 3: Reconstruction error at 20 % (top) and 50 % (bottom) sampling rates for the Intel-Berkeley dataset.

Comparing the four methods, we observe that under all measurement availability scenarios, the proposed SS-MC scheme typically achieves the lowest reconstruction error and exhibits the most stable performance. The performance of SS-MC is closely followed, especially in low sampling rates, by RegEM which also exhibits a very stable performance, while on the other hand, MC and Knn-impute are more sensitive to the sampling instance, exhibiting a more erratic behavior.

To further illuminate the behavior of each method, we consider a large set of sampling instances and present the averaged recovery performance as a function of the sampling rate in Fig. 4 for Intel-Berkeley (top) and

SensorScope (bottom) data. Regarding the performance on the Intel-Berkeley dataset, we observe that the proposed SS-MC and RegEM achieve comparable performance, much better than typical MC and Knn-impute. An interesting observation is that while SS-MC, RegEM, and Knn-impute all exhibit a monotonic reduction in reconstruction error at higher sampling rates, MC reaches a performance plateau around a 25 % sampling rate. This phenomenon is attributed to the rank constrains of MC leading to a low rank estimation which causes an incorrect estimation of missing measurements.

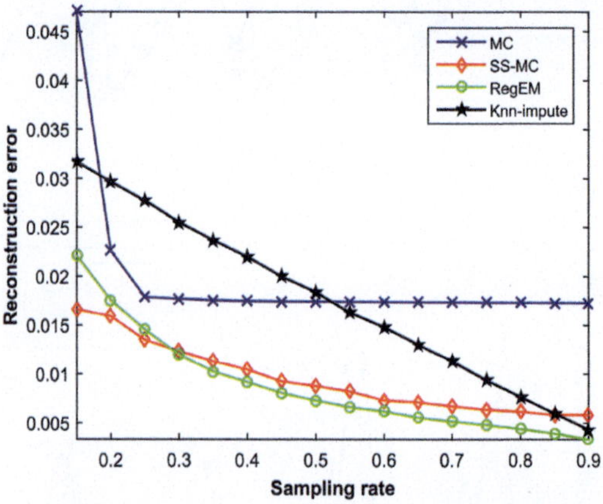

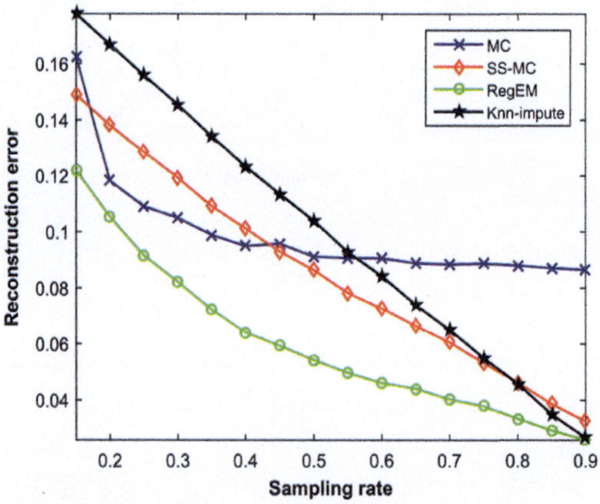

Figure 4: Average reconstruction error for the Intel-Berkeley (top) and the SensorScope (bottom) platforms.

Regarding the performance on the SensorScope data, one can observe that in this case RegEM achieves a significantly better performance compared to the other methods, followed by MC at low sampling rates and SS-MC at large ones. Similar to the behavior observed for the Intel-Berkeley data, MC again reaches a performance plateau while the other methods achieve a monotonically reducing reconstruction error. Note that although RegEM achieves the lowest reconstruction error, it is also the most computationally demanding of the four methods.

Recovery from Multiple Sources

In this subsection, we investigate the recovery capabilities of the SS-MC and state-of-the-art method as a function of the number of sensors/sources that are simultaneously considered. Figure 5 presents the reconstruction error for the multiple source/sensor cases, where 2 (top), and 5 (bottom) sources from the Intel-Berkeley dataset are simultaneously considered. Comparing these results with the results shown Fig. 4 (top), one can observe that increasing the number of sources that a method considers simultaneously can have a different effect for each method, although no method appears to be able to exploit the additional sources of data.

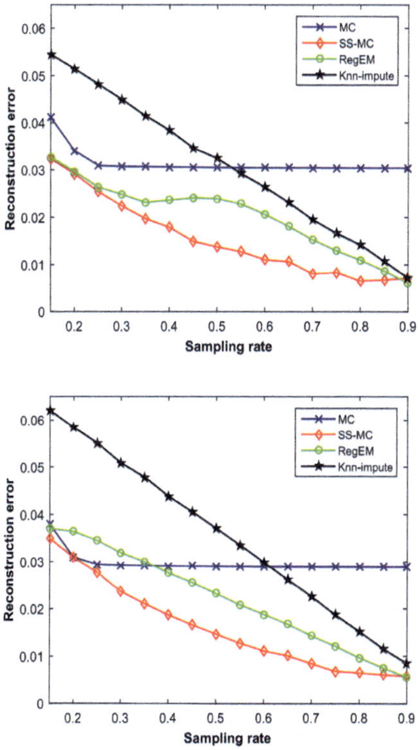

Figure 5: Reconstruction error using 2 (top), and 5 (bottom) sources of the Intel-Berkeley dataset.

State-of-the-art methods, like Knn-impute and RegEM, not only appear to be unable to exploit the additional sources of data, but introducing the additional sources leads to an increase in reconstruction error for a given sampling rate. On the other hand, typical MC is unaffected by the different scenarios, exhibiting the same plateau in behavior regardless of the number of sources under consideration. Unlike the other methods, the proposed SS-MC is able to better handle the additional data. Although applying SS-MC with multiple sources of data does not lead to better performance, the proposed method is better in handling such complex data streams, offering the lowest reconstruction error among all methods considered.

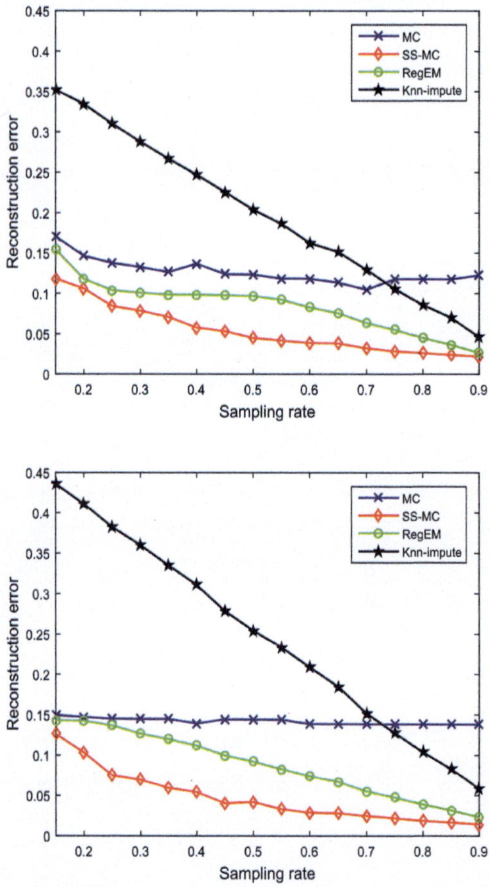

Figure 6: Reconstruction error using 2 (top) and 5 (bottom) sources of the SensorScope dataset.

The situation differs however for the SensorScope data shown in Fig. 6 for 2 (top) and 5 (bottom) sources, respectively. In this case, Knn-input appears to suffer a significant reduction in reconstruction quality due to the additional data sources, leading to a notable increase in reconstruction error compared to the single stream case. RegEM and typical MC also do not appear to benefit from the additional sources. In contrast to these methods, the proposed SS-MC achieves a more robust behavior leading to a significantly better behavior compared to the single source case. The improvement is more dramatic when moving from the single to two sources; however, introducing additional sources has a positive effect on recovery performance.

In general, for the state-of-the-art methods we consider, experimental results suggest that introducing multiple correlated sources does not necessarily aid in the recovery performance, while under different scenarios, the aggregation of multiple sources may also introduce prohibitively large communication overheads. On the other hand, the proposed SS-MC can smoothly transition from the single sensor/source case to multiple sensors/ sources achieving compelling gains in certain scenarios.

Joint Recovery and Prediction

In this set of results, we consider the more challenging scenario where the method must simultaneously recover and predict future measurements. The results for the SensorScope data shown in Fig. 7 demonstrate the competitive performance of the proposed SS-MC method compared to state-of-the-art methods for both 10 (top) and 20 (bottom) look-ahead steps. The benefits of our method are more clearly shown for the short-term prediction (top) while for the long term, we observe a similar behavior for all methods. Naturally, the performance is significantly better for the short term compared to the long term; however, we observe that both the MC and the SS-MC approaches achieve a very stable performance in both cases, suggesting that the low-rank regularization can provide strong benefits in this challenging scenario.

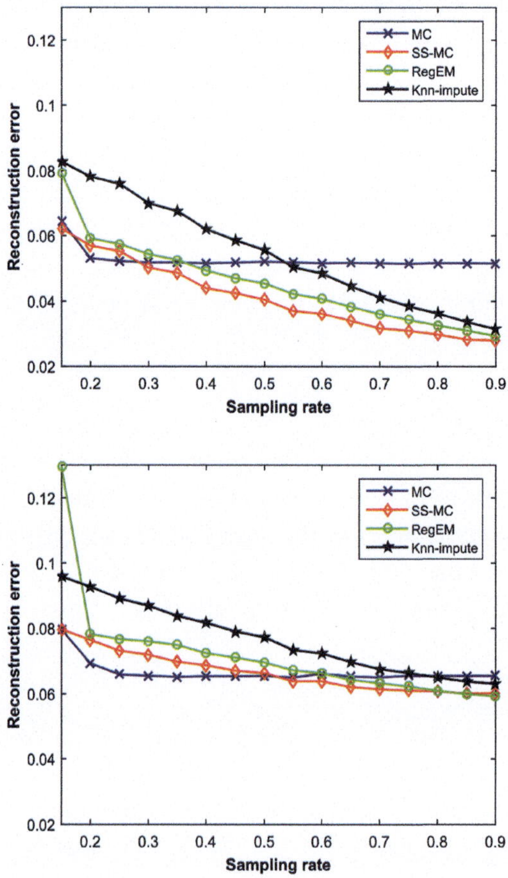

Figure 7: Joint recovery and prediction for 10 (top) and 20 (bottom) look-ahead steps on SensorScope data.

Figure 8 illustrates the recovery/estimation performance on the Intel-Berkeley data where we observe that the proposed SS-MC method achieves a dramatic reduction in reconstruction error, clearly surpassing the other methods in both short-term and long-term predictions. Similar to the SensorScope data, both MC and SS-MC achieve a very stable performance while SS-MC is much less affected by the increase in prediction horizon. Considering the results for both cases, we can conclude that SS-MC is an excellent choice for the challenging problem, achieving a very low prediction error even when only a small subset of measurements is available.

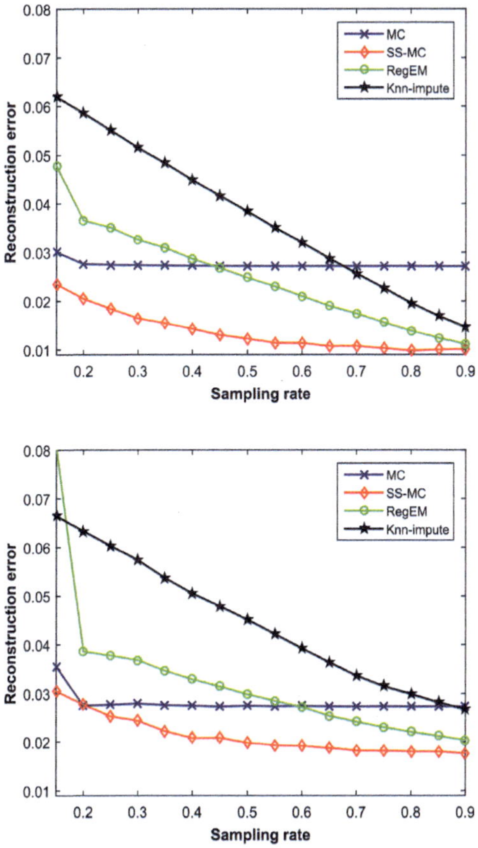

Figure 8: Joint recovery and prediction for 10 (top) and 20 (bottom) look-ahead steps on Intel-Berkeley data.

Performance with Respect to Computational Resources

The results reported in the previous subsections assume that a single day's worth of data is utilized during the training phase where the dictionary **D** is obtained. Here, we investigate the recovery capability of the proposed SS-MC method as a function of the amount of training data, i.e., the number of days used for training.

Figure 9 presents the reconstruction error for the Intel-Berkeley data using 1, 2, and 3 days of training data. The results clearly indicate that introducing more data from training has limited impact on the reconstruction performance. When one considers that the process of collecting fully sampled data can have a dramatic impact on the lifetime of the network, we can conclude that given a limited set of representative data suffices for SS-MC.

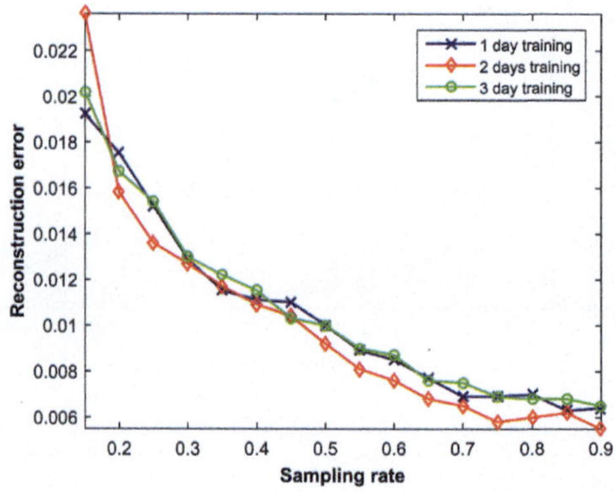

Figure 9: Reconstruction error for different training set sizes on the Intel-Berkeley dataset.

This aspect is critical since we assume that the training data is fully populated without any missing measurements. To achieve the acquisition of such training data requires extra care in terms of communication robustness as well as a larger energy consumption due to full sampling.

In addition to the amount of the training data that is required for a given performance, we also investigated the SS-MC recovery as a function of the number of iterations and the sampling rate. The results shown in Fig. 10 demonstrate that the quality of the recovery is affected by the availability of the measurements where for larger sampling rates, a smaller number of iterations is required. Despite this relationship, however, we also observe that there is a clear limit on the performance gain above 50 iterations. This is the number of iterations we have assumed in our experiments unless the approximation error drops below 10^{-4}.

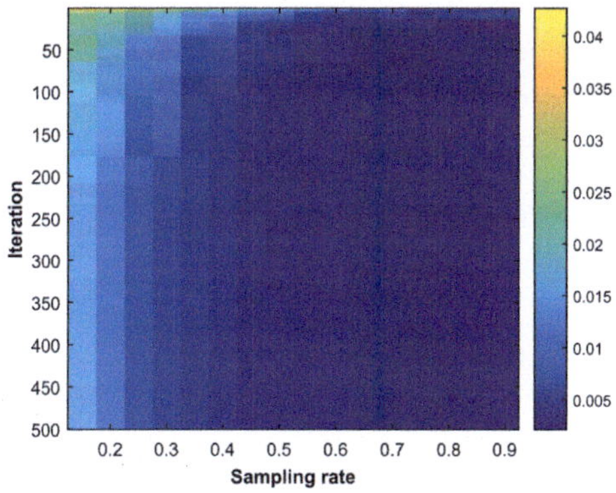

Figure 10: Reconstruction error as a function of sampling rate and iteration number.

The requirements of Big Data processing mandate algorithm that can achieve high quality performance with minimal processing requirements. To better illustrate the computational requirements for each method, Table 1 presents the processing time (in seconds) for the proposed (SS-MC) and the three state-of-the-art methods under different sampling rates when considering a single (1) or multiple (5) sources.

Table 1: Computational time for different number of sensors and measurement availability

	25 %		**50 %**		**75 %**	
	1	**5**	**1**	**5**	**1**	**5**
SS-MC	0.188	0.950	0.137	0.719	0.087	0.358
MC	0.101	0.140	0.101	0.146	0.103	0.152
RegEM	0.092	0.137	0.098	0.407	0.154	1.194
Knn	0.153	0.866	0.102	0.632	0.051	0.275

Table 1 clearly demonstrates the relationships of each method with respect to the sampling rate where we observe that for the proposed SS-MC method, increasing the sampling rate leads to lower processing time for both the single and the multiple source cases. On the other hand, MC requires a fixed processing time independently of the number of available measurements, while the effect of the number of sources is minimal. RegEM's processing time is increasing as the number of available measurements increase due to the inner mechanics of the algorithm which require multiple regression to take place. Last, the Knn-impute method exhibits a decrease in processing time with respect to the measurement availability and an increase associated with multiple sources. Overall, the proposed SS-MC exhibits a stable and predicable performance, achieving a very good trade-off between processing requirements and reconstruction quality.

CONCLUSIONS

Acquiring, transmitting, and processing Big Data presents numerous challenges due to the complexity and volume issues among others. The situation becomes even more complicated when one considers data sources associated with the Internet-of-Things paradigm, where component and architecture limitations, including processing capabilities, energy availability, and communication failures, must also be considered. In this work, we proposed a distributed sampling-centralized recovery scheme where due to various design choices and physical constraints, only a small subset of the entire set of measurements is collected during each sampling instance. The proposed SS-MC approach exploits the low-rank representation of appropriately generated trajectory matrices, when expressed in the subspace associated with dictionaries learned using training data, in order to recover missing measurements as well as predict future values. The recovery and prediction procedures are implemented via an efficient optimization based on the augmented Lagrange multipliers method. Experimental results on real data from the Intel-Berkeley and the SensorScope datasets validate the merits of the proposed scheme compared to state-of-the-art methods like typical matrix completion, RegEM, and Knn-imputation, both in terms of pure reconstruction as well as in the demanding case of simultaneous recovery and prediction.

Endnotes

[1] http://db.csail.mit.edu/labdata/labdata.html.

[2] http://lcav.epfl.ch/page-86035-en.html.

ACKNOWLEDGEMENTS

This work was founded by the DEDALE (contract no. 665044) within the H2020 Framework Program) of the EC. This work was also supported by the PETROMAKS Smart-Rig (grant 244205 /E30), SFI Offshore Mechatronics (grant 237896/O30), both from the Research Council of Norway, and the RFF Agder UiA CIEMCoE grant.

REFERENCES

1. K Slavakis, G Giannakis, G Mateos. IEEE Signal Proc. Mag. 31(5), 18 (2014).
2. J Gubbi, R Buyya, S Marusic, M Palaniswami. Futur. Gener. Comput. Syst. 29(7), 1645 (2013).
3. DB Rawat, JJ Rodrigues, I Stojmenovic, Cyber-Physical Systems: From Theory to Practice (CRC Press, Boca Raton, 2015).
4. G Tzagkarakis, G Tsagkatakis, D Alonso, E Celada, C Asensio, A Panousopoulou, P Tsakalides, B Beferull-Lozano, in: Cyber Physical Systems: From Theory to Practice. (DB Rawat, J Rodrigues, I Stojmenovic, eds.) (CRC Press, USA, 2015).
5. F Sivrikaya, B Yener. IEEE Netw. 18(4), 45 (2004).
6. B EJ Candès, Recht, Found. Comput. Math. 9(6), 717 (2009).
7. T EJ Candès, IEEE Tao, Trans. Inf. Theory. 56(5), 2053 (2010).
8. A Jindal, K Psounis. ACM Trans. Sens. Netw. (TOSN). 2(4), 466 (2006).
9. GE Batista, MC Monard. Appl. Artif. Intell. 17(5–6), 519 (2003).
10. N Cressie. Terra Nova. 4(5), 613 (1992).
11. J Li, A Heap. Ecol. Informa. 6(3), 228 (2011).
12. T Schneider. J. Clim. 14(5), 853 (2001).
13. C Luo, F Wu, J Sun, C Chen, in International conference on Mobile computing and networking (ACM, Beijing China, 2009).
14. C Luo, F Wu, J Sun, CW Chen, IEEE Trans. Wirel. Commun. 9(12), 3728 (2010).
15. A Fragkiadakis, I Askoxylakis, E Tragos, in: International Workshop on Computer Aided Modeling and Design of Communication Links and Networks (IEEE, 2013).
16. Z Xiong, A Liveris, S Cheng, IEEE Signal Proc. Mag. 21(5), 80 (2004).
17. A Majumdar, RK Ward, Biomed. Signal Process. Control. 13:, 142 (2014).
18. A Shukla, A Majumdar, Biomed. Signal Process. Control. 18:, 174 (2015).
19. JJ Meng, W Yin, H Li, E Houssain, Z Han, in: Acoustics Speech and Signal Processing (ICASSP) 2010 IEEE International Conference on (IEEE, Dallas, 2010).

20. S Nikitaki, G Tsagkatakis, P Tsakalides, IEEE Trans. Mob. Comput. 14(11), 2244 (2015).
21. S Nikitaki, G Tsagkatakis, P Tsakalides, in: Signal Processing Conference (EUSIPCO) 2012 Proceedings of the 20th European (IEEE, Bucharest, 2012).
22. PJ Shin, PE Larson, MA Ohliger, M Elad, JM Pauly, DB Vigneron, M Lustig, Magn. Reson. Med. 72(4), 959 (2014).
23. G Tsagkatakis, P Tsakalides, in: Machine Learning for Signal Processing (MLSP) 2012 IEEE International Workshop on (IEEE, Santander, 2012).
24. A Majumdar, R Ward, in: Data Compression Conference, 2010 (IEEE, Snowbird, 2010).
25. G Tsagkatakis, P Tsakalides, in: Sensor Array and Multichannel Signal Processing Workshop (SAM) (IEEE, Hoboken, 2012).
26. F Fazel, M Fazel, M Stojanovic, in: Information Theory and Applications Workshop (ITA) (IEEE, San Diego, 2012).
27. S Savvaki, G Tsagkatakis, P Tsakalides, in ACM International Workshop on Cyber-Physical Systems for Smart Water Networks (ACM, New York, 2015).
28. A Majumdar, A Gogna, Sensors. 14(9), 15729 (2014).
29. G Liu, Z Lin, S Yan, J Sun, Y Yu, Y Ma, IEEE Trans. Pattern Anal. Mach. Intell.35(1), 171 (2013).
30. E Elhamifar, R Vidal, IEEE Trans. Pattern Anal. Mach. Intell. 35(11), 2765 (2013).
31. M Mardani, G Mateos, GB Giannakis, IEEE Trans. Signal Process. 63(10), 2663 (2015).
32. N Golyandina. Singular Spectrum Analysis for time series (Springer Science & Business MediaNew York, 2013).
33. G Tzagkarakis, M Papadopouli, P Tsakalides, in: ACM Symposium on Modeling, analysis, and simulation of wireless and mobile systems (ACM, Chania, 2007).
34. DH Schoellhamer, Geophys. Res. Lett. 28(16), 3187 (2001).
35. D Kondrashov, M Ghil, Nonlinear Process. Geophys. 13(2), 151 (2006).
36. N Golyandina, E Osipov, J. Stat. Plan. Infer. 137(8), 2642 (2007).
37. K Patterson, H Hassani, S Heravi, A Zhigljavsky, J. App. Stat. 38(10),

2183 (2011).

38. N Golyandina, D Stepanov, in: 5th St. Petersburg workshop on simulation, vol. 293 (St. Petersburg State University, St. Petersburg, 2005).
39. N Golyandina, A Korobeynikov, A Shlemov, K Usevich. J. Stat. Softw. 67(1), 1 (2015).
40. I Markovsky. Low rank approximation: algorithms, implementation, applications (Springer Science & Business MediaNew York, 2011).
41. Y E Candès, Plan, Proc. IEEE. 98(6), 925 (2010).
42. B Recht, M Fazel, P Parrilo, SIAM Rev. 52(3), 471 (2010).
43. JF Cai, EJ Candès, Z Shen, SIAM J. Optim. 20(4), 1956 (2010).
44. Z Lin, M Chen, Y Ma, arXiv preprint 1009.5055, (2010). http://arxiv.org/abs/1009.5055.
45. DP Bertsekas. 1st edn. Constrained Optimization and Lagrange Multiplier Methods (Optimization and Neural Computation Series) (Athena ScientificNashua, 1996).
46. S Boyd, N Parikh, E Chu, B Peleato, J Eckstein, Found. Trends Mach Learn. 3(1), 1 (2011).
47. Z Liu, L Vandenberghe, SIAM J. Matrix Anal. Appl. 31(3), 1235 (2009).
48. M Grant, S Boyd, Y Ye (2008). Online accessiable: http://stanford.edu/~boyd/cvx. Accessed 1 Jan 2014.
49. S Boyd, N Parikh, E Chu, B Peleato, J Eckstein, Distributed optimization and statistical learning via the alternating direction method of multipliers. Found. Trends®; Mach. Learn. 3(1), 1–122 (2011). Now Publishers Inc.
50. Luo ZQ, arXiv preprint arXiv:1208.3922 (2012). http://arxiv.org/abs/1208.3922.
51. KV Fernando, S Hammarling. Linear algebra in signals, systems, and control (Boston, MA, 1986), pp. 128–140(1988).
52. B De Moor, Signal Process. 25(2), 135 (1991).
53. DA Ross, J Lim, RS Lin, MH Yang, Int. J. Comput. Vis. 77(1–3), 125 (2008).
54. Y Li, Pattern Recogn. 37(7), 1509 (2004).
55. ID Schizas, GB Giannakis, ZQ Luo, IEEE Trans. Signal Process. 55(8), 4284 (2007).

56. S Madden, Intel lab data, 2004, (2012). http://db.csail.mit.edu/labdata/labdata.html.
57. F Ingelrest, G Barrenetxea, G Schaefer, M Vetterli, O Couach, M Parlange, ACM Trans. Sens. Netw. (TOSN). 6(2), 1 (2010).
58. O Troyanskaya, M Cantor, G Sherlock, P Brown, T Hastie, R Tibshirani, D Botstein, R Altman, Bioinformatics. 17(6), 520 (2001).

Chapter 12

An Effective Numerical Method to Solve a Class of Nonlinear Singular Boundary Value Problems using improved Differential Transform Method

Liejun Xie, Cailian Zhou and Song Xu

Department of Mathematics, Ningbo University, Fenghua Road 818, Jiangbei District, Ningbo City 315211, Zhejiang Province, People's Republic of China

ABSTRACT

In this work, an effective numerical method is developed to solve a class of singular boundary value problems arising in various physical models by using the improved differential transform method (IDTM). The IDTM applies the Adomian polynomials to handle the differential transforms of the nonlinearities arising in the given differential equation. The relation between the Adomian polynomials of those nonlinear functions and the coefficients of unknown truncated series solution is given by a simple formula, through

Citation (APA): Xie, L. J., Zhou, C. L., & Xu, S. (2016). An effective numerical method to solve a class of nonlinear singular boundary value problems using improved differential transform method. SpringerPlus, 5(1), 1066. (19 pages), DOI: https://doi.org/10.1186/s40064-016-2753-9.

which one can easily deduce the approximate solution which takes the form of a convergent series. An upper bound for the estimation of approximate error is presented. Several physical problems are discussed as illustrative examples to testify the validity and applicability of the proposed method. Comparisons are made between the present method and the other existing methods.

Keywords: Singular boundary value problem, Differential transform method, Adomian polynomials, Improved differential transform method, Approximate series solutions

BACKGROUND

Singular boundary value problems (SBVPs) is an important class of boundary value problems, and arises frequently in the modeling of many actual problems related to physics and engineering areas such as in the study of electro hydrodynamics, theory of thermal explosions, boundary layer theory, the study of astrophysics, three layer beam, electromagnetic waves or gravity driven flows, inelastic flows, the theory of elastic stability and so on. In general, SBVPs is difficult to solve analytically. Therefore, various numerical techniques have been proposed to treat it by many researchers. However, the solution of SBVPs is numerically challenging due to the singularity behavior at the origin.

In this work, we are interested again in the following SBVPs arising frequently in applied science and engineering:

$$u''(x) + \frac{\alpha}{x}u'(x) = f(x,u), \quad 0 < x \le 1, \quad \alpha \ge 1, \tag{1}$$

subject to the boundary value conditions

$$u'(0) = 0 \tag{2}$$

and

$$au(1) + bu'(1) = c, \tag{3}$$

where a, b and c are any finite real constants. If α=1, (1) becomes a cylindrical problem, and it becomes a spherical problem when α=2. It is assumed that f(x, u) is continuous, $\frac{\partial f}{\partial u}$ exists and is continuous and $\frac{\partial f}{\partial u} \ge 0$ for any $0<x\le 1$ such that Eq. (1) has a unique solution (Russell and Shampine 1975). The SBVPs (1–3) with different $\alpha\alpha$ arise in the study of various scientific problems for certain linear or nonlinear functions f(x, u). The common cases

related to the actual problems are summarized as follows. The first case for α=2 and

$$f(x,u) = f(u) = \frac{\delta u(x)}{u(x) + \mu} \tag{4}$$

emerges from the modeling of steady state oxygen diffusion in a spherical cell with Michaelis–Menten uptake kinetics (Lin 1976; McElwain 1978). In this case, u(x) represents the oxygen tension; δ and μ are positive constants involving the reaction rate and the Michaelis constant. Hiltmann and Lory (1983) proposed the existence and uniqueness of the solution for b=1 and a=c. Analytical bounding functions were given in Anderson and Arthurs (1985). The numerical methods to solve the SBVPs for this case have attracted a reasonable amount of research works, such as the finite difference method (FDM) (Pandey 1997), the cubic spline method (CSM) (Rashidinia et al. 2007; Ravi and Bhattacharya 2006), the Sinc-Galerkin method (SGM) (Babolian et al. 2015), the Adomian decomposition method (ADM) and its modified methods (Khuri and Sayfy 2010; Wazwaz et al. 2013; Singh and Kumar 2014), the variational iteration method (VIM) (Ravi and Aruna 2010; Wazwaz 2011), the series expansion technique (SEM) (Turkyilmazoglu 2013) and the B-spline method (BSM) (Çağlar et al. 2009).

The second case arises in the study of the distribution of heat sources in the human head (Flesch 1975; Gray 1980; Duggan and Goodman 1986), in which α=2 and

$$f(x,u) = f(u) = -le^{-lku(x)}, \quad l > 0,\ k > 0. \tag{5}$$

In Duggan and Goodman (1986), point-wise bounds and uniqueness results were presented for the SBVPs with the nonlinear function f(x, u) of the forms given by (4) and (5). Quite a little amount of works by using different approaches, including the FDM (Pandey 1997), the CSM (Rashidinia et al. 2007; Ravi and Bhattacharya 2006) and the SGM (Babolian et al. 2015), have been proposed to obtain the approximate solutions of this case.

The third important case of physical significance is when α=1,2 and

$$f(x,u) = f(u) = \nu e^{u(x)}, \tag{6}$$

which arises in studying the theory of thermal explosions (Khuri and Sayfy 2010; Kumar and Singh 2010; Chang 2014) and the electric double layer in a salt-free solution (Chang 2012). A variety of numerical methods have been applied to handle such SBVPs, for example, the fourth order finite difference method (FFDM) (Chawla et al. 1988), the modified Adomian decomposition method (Khuri and Sayfy 2010; Singh and Kumar 2014; Kumar and Singh 2010), the Taylor series method (TSM) (Chang 2014) and the BSM

(Çağlar et al. 2009).

Besides, Chandrasekhar (1939) derived another case for $\alpha=2, b=0$ and

$$f(x,u) = f(u) = -u^{\gamma}(x), \tag{7}$$

which γ is a physical constant. This case is in connection with the equilibrium of thermal gas thermal (Ames 1968). The numerical solution of this kind of equation for $\gamma=5$ was considered by using various methods, such as the FFDM (Chawla et al. 1988), the VIM (Ravi and Aruna 2010), the SEM (Turkyilmazoglu 2013) and the modified Adomian decomposition method (Singh and Kumar 2014).

All the aforementioned methods can yield a satisfied result. However, each of these methods has its own weaknesses. For example, the VIM (Ravi and Aruna 2010; Wazwaz 2011) has an inherent inaccuracy in identifying the Lagrange multiplier, and fails to solve the equation when the nonlinear function f(x, u) is of the forms (5) and (6). Those methods such as the FDM (Pandey 1997; Chawla et al. 1988), the SEM (Turkyilmazoglu 2013), the SGM (Babolian et al. 2015) and the spline method (Rashidinia et al. 2007; Ravi and Bhattacharya 2006; Çağlar et al. 2009) require a tedious process and huge volume of computations in dealing with the linearization or discretization of variables. The ADM (Wazwaz et al. 2013) needs to obtain the corresponding Volterra integral form of the given equation, via which one can overcome the difficulty of singular behavior at $x=0$. The modified ADM (Khuri and Sayfy 2010; Kumar and Singh 2010) needs to introduce a twofold indefinite integral operator to give better and accurate results; moreover, the success of method in (Singh and Kumar 2014) relies on constructing Green's function before establishing the recursive relation for applying the ADM to derive the solution components. All those manners are at the expense of computation budgets. Besides, none of above methods is applied to handle the equations with all forms of nonlinearities (4–7).

In recent years, a lot of attentions have been devoted to the applications of differential transform method (DTM) and its modifications. The DTM proposed by Pukhov (1980, 1982, 1986) at the beginning of 1980s. However, his work passed unnoticed. In 1986, Zhou (1986) reintroduced the DTM to solve the linear and nonlinear equations in electrical circuit problems. The DTM is a semi-numerical-analytic method that generates a Taylor series solution in the different manner. In the past forty years, the DTM has been successfully applied to solve a wide variety of functional equations; see Xie et al. (2016) and the references therein. Although being powerful, there still exist some difficulties in solving various of equations by the classical

DTM. Some researchers have devoted to deal with these obstacles so as to extend the applications of the DTM. For example, in view of the DTM numerical solution cannot exhibit the real behaviors of the problem, Odibat et al. (2010) proposed a multi-step DTM to accelerate the convergence of the series solution over a large region and applied successfully to handle the Lotka-Volterra, Chen and Lorenz systems. In Gökdoğan et al. (2012), Momani and Ertürk (2008) suggested an alternative scheme to overcome the difficulty of capturing the periodic behavior of the solution by combining the DTM, Laplace transform and Padé approximants. Another difficulty is to compute the differential transforms of the nonlinear components in a simple and effective way. By using the traditional approach of the DTM, the computational difficulties will inevitably arise in determining the transformed function of an infinity series. Compared to the traditional method, Chang and Chang (2008) proposed a relatively effective algorithm for calculating the differential transform through a derived recursive relation. Yet, by using their method, it is inevitable to increase the computational budget, especially in dealing with those differential equations which have two or more nonlinear terms being investigated. Recently, the authors Elsaid (2012), Fatoorehchi and Abolghasemi (2013) disclosed the relation between the Adomian polynomials and the differential transform of nonlinearities, and developed an inspiring approach to handle the nonlinear functions in the given functional equation. Meanwhile, the problem of tedious calculations in dealing with nonlinear problems by using the ADM has also been improved considerably by Duan (2010a, b, 2011). All of these effective works make it possible to broaden the applicability and popularity of the DTM considerably.

The aim of this work is to develop an efficient approach to solve the SBVPs (1–3) with those nonlinear terms (4–7). This scheme is mainly based on the improved differential transform method (IDTM), which is the improved version of the classical DTM by using the Adomian polynomials to handle the differential transforms of those nonlinear functions (4–7). No specific technique is required in dealing with the singular behavior at the origin. Meanwhile, unlike some existing approaches, the proposed method tackles the problem in a straightforward manner without any discretization, linearization or perturbation. The numerical solution obtained by the proposed method takes the form of a convergent series with those easily computable coefficients through the Adomian polynomials of those nonlinear functions as the forms of (4–7). The rest of the paper is organized as follows. In the next section, the concepts of DTM and Adomian polynomials are introduced. Algorithm for solving the problem (1–3) and an upper bound for the estimation of approximate error are presented in Sect. 3. Sect.4 shows some numerical examples to testify the validity and applicability of the proposed method. In Sect. 5, we end this paper with a brief conclusion.

ADOMIAN POLYNOMIAL AND DIFFERENTIAL TRANSFORM

Adomian Polynomial

In the Adomian decomposition method (ADM), a key notion is the Adomian polynomials, which are tailored to the particular nonlinearity to easily and systematically solve nonlinear differential equations. The interested readers are referred to Adomian (1990, 1994) for the details of the ADM.

For the applications of decomposition method, the solution of the given equation in a series form is usually expressed by

$$u = \sum_{m=0}^{\infty} u_m, \tag{8}$$

and the infinite series of polynomials

$$f(u) = f\left(\sum_{m=0}^{\infty} u_m\right) = \sum_{m=0}^{\infty} A_m \tag{9}$$

for the nonlinear term f(u), where A_m is called the Adomian polynomials, and depends on the solution components $u_0, u_1, \ldots, u_m$. The traditional algorithm for evaluating the Adomoan polynomials A_n was first provided in Adomian and Rach (1983) by the formula

$$A_n = \frac{1}{n!} \frac{d^n}{d\lambda^n} f\left(\sum_{m=0}^{\infty} u_m \lambda^m\right)\Bigg|_{\lambda=0}. \tag{10}$$

A large amount of works (Duan 2010b, b, 2011; Adomian and Rach 1983; Rach 2008, 1984; Wazwaz 2000; Abbaoui et al. 1995; Abdelwahid 2003; Azreg-AÏnou 2009) have been applied to give the more effective computational method for the Adomian polynomials. For fast computer generation, we favor Duan's Corollary 3 algorithm (Duan 2011) among all of these methods, as it merely involves the analytic operations of addition and multiplication without the differentiation operator, which is eminently convenient for symbolic implementation by computer algebraic systems such as Maple and Mathematics. The method to generate the Adomian polynomials in Duan (2011) is described as follows:

$$C_n^1 = u_n, \quad n \geq 1,$$
$$C_n^k = \frac{1}{n}\sum_{j=0}^{n-k}(j+1)u_{j+1}C_{n-1-j}^{k-1}, \quad 2 \leq k \leq n, \tag{11}$$

such that

$$yA_0 = f(u_0),$$
$$A_n = \sum_{k=1}^{n} C_n^k f^{(k)}(u_0), \quad n \geq 1. \tag{12}$$

It is worth mentioning that Duan's algorithm involving (11) and (12) has been testified to be one of the fastest subroutines on record (Duan 2011), including the fast generation method given by Adomian and Rach (1983).

Differential transform

The differential transform of the kthkth differentiable function u(x) at x=0 is defined by

$$U(k) = \frac{1}{k!}\left[\frac{d^k u(x)}{dx^k}\right]_{x=0}, \tag{13}$$

and the differential inverse transform of U(k) is described as

$$u(x) = \sum_{k=0}^{\infty} U(k)x^k, \tag{14}$$

where u(x) is the original function and U(k) is the transformed function. For the practical applications, the function u(x) is expressed by a truncated series and Eq. (14) can be written as

$$u(x) \approx u_N(x) = \sum_{k=0}^{N} U(k)x^k. \tag{15}$$

It is not difficult to deduce the transformed functions of the fundamental operations listed in Table 1.

Table 1: The fundamental operations of the DTM

Original function	Transformed function
$w(x) = \alpha u(x) \pm \beta v(x)$	$W(k) = \alpha U(k) \pm \beta V(k)$
$w(x) = u(x)v(x)$	$W(k) = \sum_{m=0}^{k} U(m)V(k-m)$
$w(x) = d^m u(x)/dx^m$	$W(k) = \frac{(k+m)!}{k!} U(k+m)$
$w(x) = x^m$	$W(k) = \delta(k-m) = \begin{cases} 1, & \text{if } k = m, \\ 0, & \text{if } k \neq m. \end{cases}$
$w(x) = \exp(x)$	$W(k) = 1/k!$
$w(x) = \sin(\alpha x + \beta)$	$W(k) = \alpha^k/k! \sin(k\pi/2 + \beta)$
$w(x) = \cos(\alpha x + \beta)$	$W(k) = \alpha^k/k! \cos(k\pi/2 + \beta)$

Note that α,β are constants and m is a nonnegative integer

METHOD OF SOLUTION OF SBVPS (1–3)

We want to find the approximate solution of the problem (1–3) with the type:

$$u_N(x) = \sum_{k=0}^{N} U(k)x^k, \tag{16}$$

where the coefficients U(0),U(1),…,U(N) are determined using the following steps:

- According to the definition (13) of the differential transform and the boundary value condition (2), we have

$$U(1) = 0. \tag{17}$$

Suppose that

$$U(0) = \beta, \tag{18}$$

where β is a real parameter to be determined.

- Multiplying both sides of Eq. (1) by variable x, we have

$$xu'(x) + \alpha u'(x) = xf(x, u). \tag{19}$$

Applying the differential transform (13) to Eq. (19), we get the following recurrence relation:

$$U(k+1) = \frac{F(k-1)}{(k+1)(k+\alpha)}, \quad k = 1, 2, \ldots, N-1, \tag{20}$$

where F(k) is the differential transform of the nonlinear function f(x,u)=f(u).

- Using Lemma 3.1 in Fatoorehchi and Abolghasemi (2013), we compute F(k) through the Adomian polynomials A_k:

$$F(k) = A_k, \quad k = 0, 1, 2, \ldots, N. \tag{21}$$

Remark 1

Lemma 3.1 in Fatoorehchi and Abolghasemi (2013) indicates that the differential transforms and the Adomian polynomials of nonlinear functions have the same mathematical structure such that we can derive the differential transforms of any nonlinear functions by merely calculating the relevant Adomian polynomials but with constants instead of variable components.

Remark 2

As mentioned before, we use Duan's Corollary 3 algorithm (Duan 2011) (11–12) to generate the Adomian polynomials.

- Substituting (21) into (20), and then combining the relations (16–18), we obtain the truncated series solution of the problem (1–3) as follows:

$$u_N(x) = \beta + \sum_{k=1}^{N-1} \frac{A_{k-1}}{(k+1)(k+\alpha)} x^{k+1}. \tag{22}$$

- Imposing the truncated series solution (22) on the boundary condition (3), we obtain a nonlinear algebraic equation with unknown parameter β:

$$g(\beta) = 0. \tag{23}$$

Solving Eq. (23), and substituting the value of β into (22), we obtain the final result.

An upper bound for the estimation of approximate error is presented in the following lemma.

Lemma 1

Suppose that $u(x) \in C^{N+1}$ [0,1] is the exact solution of the problem (1–3), $u_N(x) = \sum_{k=0}^{N} U(k)x^k$ is the truncated series solution with degree N, it holds that

$$||u(x) - u_N(x)||_\infty \le \frac{M}{(N+1)!} + \max_{0\le k\le N} |c_k|, \tag{24}$$

where $M = \max_{0\le x\le 1} |u^{(N+1)}(x)|, c_k = \frac{u^{(k)}(0)}{k!} - U(k)$.

Proof

Obviously, we have

$$||u(x) - u_N(x)||_\infty \le ||u(x) - \widetilde{u}_N(x)||_\infty + ||\widetilde{u}_N(x) - u_N(x)||_\infty, \tag{25}$$

where $\widetilde{u}_N(x) = \sum_{k=0}^{N} \frac{u^{(k)}(0)}{k!} x^k$ is the Taylor polynomial of the unknown function u(x) at x=0.

Since $u(x) \in C^{N+1}$ [0,1], it follows that

$$u(x) = \widetilde{u}_N(x) + R_N(x) = \widetilde{u}_N(x) + \frac{u^{(N+1)}(\xi)}{(N+1)!} x^{N+1}, \quad \xi \in (0,1),$$

where $R_N(x)$ is the remainder of Taylor polynomial $\tilde{u}_N(x)$. Therefore

$$\left|u(x)-\tilde{u}_N(x)\right| = |R_N(x)| = \left|\frac{u^{(N+1)}(\xi)}{(N+1)!}x^{N+1}\right| \le \frac{1}{(N+1)!}\max_{0\le x\le 1}\left|u^{(N+1)}(x)\right|. \tag{26}$$

Let

$$\mathbf{C} = (c_0, c_1, \ldots, c_N), \quad \Theta = (x^0, x^1, \ldots, x^N)^T,$$

where

$$c_k = \frac{u^{(k)}(0)}{k!} - U(k), \quad k = 0, 1, \ldots, N.$$

We then have

$$\left|\tilde{u}_N(x) - u_N(x)\right| = \left|\sum_{k=0}^{N}\left(\frac{u^{(k)}(0)}{k!} - U(k)\right)x^k\right| = |\mathbf{C}\cdot\Theta| \le ||\mathbf{C}||_\infty \cdot ||\Theta||_\infty \tag{27}$$

Combining the relations (25–27), it follows that

$$||u(x) - u_N(x)||_\infty \le \frac{1}{(N+1)!}\max_{0\le x\le 1}\left|u^{(N+1)}(x)\right| + ||\mathbf{C}||_\infty \cdot ||\Theta||_\infty$$
$$\le \frac{M}{(N+1)!} + \max_{0\le k\le N}|c_k|. \tag{28}$$

Thus, the proof is completed.

NUMERICAL EXAMPLES

In this section, based on the discussion in Sect. 3, we report numerical tests of five classical examples discussed frequently to testify the validity and applicability of the proposed method. All the numerical computations were performed using Maple and Matlab on personal computer. For comparison, we computed the absolute error defined by

$$\mathrm{E}_N(x) = |u(x) - u_N(x)| \tag{29}$$

and the maximal absolute error by

$$\mathrm{ME}_N = \max_{0\le x\le 1}|u(x) - u_N(x)|, \tag{30}$$

where u(x) is the exact solution and $u_N(x)$ is the truncated series solution with degree N.

Example 1

Consider the following nonlinear SBVP in the study of isothermal gas sphere (Singh and Kumar 2014; Ravi and Aruna 2010; Chawla et al. 1988):

$$u''(x) + \frac{2}{x}u'(x) = -u^5(x), \tag{31}$$

subject to the boundary conditions

$$u'(0) = 0, \quad u(1) = \frac{\sqrt{3}}{2}. \tag{32}$$

The exact solution of this problem is given by $u(x) = \sqrt{\frac{3}{3+x^2}}$. It is also known as the Emden-Fowler equation of the first kind. In what follows, we shall solve it with the proposed algorithm.

Firstly, we set

$$U(0) = \beta, \quad U(1) = 0.$$

The Adomian polynomials of nonlinear term $f(x,u) = -u^5(x)$ in this problem are computed as

$$\begin{aligned}
A_0 &= -U^5(0),\\
A_1 &= -5U^4(0)U(1),\\
A_2 &= -10U^3(0)U^2(1) - 5U^4(0)U(2),\\
A_3 &= -10U^2(0)U^3(1) - 20U^3(0)U(1)U(2) - 5U^4(0)U(3),\\
&\vdots
\end{aligned}$$

Furthermore, according to the relations (20) and (21), we obtain the differential transforms U(k) of the unknown function u(x)

$$\begin{aligned}
U(2) &= \frac{1}{2\cdot 3}A_0 = -\frac{1}{6}\beta^5,\\
U(4) &= \frac{1}{4\cdot 5}A_3 = \frac{1}{24}\beta^9,\\
U(6) &= \frac{1}{6\cdot 7}A_5 = -\frac{5}{432}\beta^{13},\\
&\vdots\\
U(k) &= 0, \quad \text{if } k \text{ is odd and } k \geq 3.
\end{aligned}$$

By using Eq. (22), we obtain the truncated series solution for N=10 as follows:

$$u_{10}(x) = \beta - \frac{1}{6}\beta^5x^2 + \frac{1}{24}\beta^9x^4 - \frac{5}{432}\beta^{13}x^6 + \frac{35}{10368}\beta^{17}x^8 - \frac{7}{6912}\beta^{21}x^{10}. \tag{33}$$

Secondly, imposing the truncated series solution (33) on the boundary conditions $u(1) = \sqrt{3}/2$, we get a nonlinear algebraic equation. By solving it, the unknown parameter β is computed as

$$\beta = 1.000553890. \tag{34}$$

Finally, substituting (34) into (33), we get the approximate solution with degree 10

$$u_{10}(x) = 1.000553890 - 0.1671287533x^2 + (0.4187483621e - 1)x^4 - (0.1165769154e - 1)x^6 + (0.3407699551e - 2)x^8 - (0.1024576736e - 2)x^{10}.$$

In Table 2, we compare the absolute errors (29) of numerical results obtained by the present method, the VIM (Ravi and Aruna 2010) and the modified ADM using Green functions (GIDM) (Singh and Kumar 2014) for N=12. Table 3 lists the theoretical estimate errors (24) and the maximal absolute errors (30) of the approximate solutions for changing approximation levels, and shows a comparison of the maximal absolute errors with the GIDM (Singh and Kumar 2014) and the FFDM (Chawla et al. 1988). We can see from Table 3 that the accuracy of our computational results is getting better as the approximation level is increasing. Moreover, our numerical solution $u_{10}(x)$ has an accuracy of $O(10^{-4})$, whereas the GIDM (Singh and Kumar 2014) needs to employ 14 terms to archive this goal as shown in Table 1of Singh and Kumar (2014); numerical solution with even 64 terms obtained by the FFDM (Chawla et al. 1988) still hovers at this level. In summary, Tables 2 and 3indicate that the results of our proposed method have higher accuracy than of the GIDM (Singh and Kumar 2014), the FFDM (Chawla et al. 1988) and the VIM (Ravi and Aruna 2010).

Table 2: Comparison of the absolute error $E_{12}(x)$ for Example 1

x	**GIDM (Singh and Kumar 2014)**	**VIM (Ravi and Aruna 2010)**	**Present method**
0.0	3.1880e−03	6.3220e−03	1.6776e−04
0.1	3.1209e−03	6.2702e−03	1.6637e−04
0.2	2.9269e−03	6.1173e−03	1.6227e−04
0.3	2.6263e−03	5.8687e−03	1.5568e−04
0.4	2.2489e−03	5.5281e−03	1.4691e−04
0.5	1.8284e−03	5.0903e−03	1.3639e−04
0.6	1.3978e−03	4.5347e−03	1.2450e−04
0.7	9.8413e−04	3.8201e−03	1.1132e−04
0.8	6.0707e−04	2.8837e−03	9.5269e−05
0.9	2.7774e−04	1.6426e−03	6.8180e−05
1.0	3.52e−08	1.00e−10	0

Table 3: The theoretical estimate errors TE_N and comparison of the maximal absolute errors ME_N of present method and of other methods for Example 1

N	TE_N	ME_N	*N*	TE_N	ME_N	*N*	**in Singh and Kumar (2014)**	*N*	**in Chawla et al. (1988)**
6	1.83e–02	6.80e–03	12	4.7721e–04	1.6776e-04	12	1.3978e–03	16	3.64e–04
8	5.10e–03	1.70e–03	16	4.6453e–05	1.6521e-05	16	2.4654e–04	32	2.49e–04
10	1.5666e–03	5.5389e–04	20	4.6453e–06	1.6614e-06	20	4.8643e–05	64	1.60e–04

Example 2

Consider the following nonlinear SBVP (Khuri and Sayfy 2010; Singh and Kumar 2014; Çağlar et al. 2009; Chawla et al. 1988):

$$u''(x) + \frac{1}{x}u'(x) = -e^{u(x)}, \tag{35}$$

subject to the boundary conditions

$$u'(0) = 0, \quad u(1) = 0. \tag{36}$$

The exact solution is given by $u(x) = 2\ln\frac{C+1}{Cx^2+1}$, where $C = 3 - 2\sqrt{2}$.

The Adomian polynomials of nonlinear term $f(x,u) = -e^{u(x)}$ in this problem are computed as

$$A_0 = -e^{U(0)},$$

$$A_1 = -U(1)e^{U(0)},$$

$$A_2 = -U(2)e^{U(0)} - \frac{1}{2}U^2(1)e^{U(0)},$$

$$A_3 = -U(3)e^{U(0)} - U(1)U(2)e^{U(0)} - \frac{1}{6}U^3(1)e^{U(0)},$$

$$\vdots$$

A comparison of the absolute errors (29) of the numerical solutions for N=10,20,40 obtained by the present method and the modified decomposition method (BSDM) (Khuri and Sayfy 2010) is described in Table 4. Table 5 lists the maximal absolute errors (30) of those numerical results derived from the proposed method, the BSM (Çağlar et al. 2009) and the FFDM (Chawla et al. 1988). And also, we list the theoretical estimate errors (24) in Table 5 for comparison. It can be seen from Tables 4 and 5 that one can obtain the better approximate solution by using the present method compared to the other mentioned methods, even if we take the relative smaller N. Moreover, the theoretical estimate errors, the absolute errors and the maximal absolute errors all decrease as the increase of N. Therefore, evaluation of more components of the numerical solution will reasonably improve the accuracy.

Table 4: Comparison of the absolute errors $E_N(x)$ for Example 2

x	**BSDM Khuri and Sayfy (2010)**			**Present method**		
	$E_{10}(x)$	$E_{20}(x)$	E40(x)E40(x)	$E_{10}(x)$	$E_{20}(x)$	$E_{40}(x)$
0.0	1.05e−05	1.05e−05	1.05e−05	1.05e−05	2.2e−09	1.4e−09
0.1	1.05e−05	1.05e−05	1.05e−05	1.05e−05	1.2e−09	4.0e−10
0.2	1.03e−05	1.03e−05	1.03e−05	1.03e−05	1.4e−09	6.0e−10
0.3	1.02e−05	1.02e−05	1.02e−05	1.02e−05	1.4e−09	6.0e−10
0.4	9.93e−06	9.93e−06	9.93e−06	9.93e−06	1.5e−09	8.0e−10
0.5	9.62e−06	9.62e−06	9.62e−06	9.62e−06	2.6e−09	1.8e−09
0.6	2.73e−06	6.07e−06	6.93e−06	9.25e−06	1.9e−09	1.2e−09
0.7	6.67e−07	3.65e−06	4.75e−06	8.75e−06	1.4e−09	7.0e−10
0.8	1.58e−06	2.02e−06	2.93e−06	7.88e−06	9.0e−10	3.0e−10
0.9	1.08e−06	8.76e−07	1.37e−06	5.78e−06	5.5e−10	1.1e−09
1.0	0	0	0	1.10e−10	2.74e−11	3.6e−11

Table 5: The theoretical estimate errors TE_N and comparison of the maximal absolute errors ME_N of present method and of other methods for Example 2

N	TE_N	ME_N	N	TE_N	ME_N	N	**in Çağlar et al. (2009)**	N	**in Chawla et al. (1988)**
10	6.9957e−05	1.0488e−05	16	2.2413e−07	3.5041e−08	20	3.1607e−05	16	2.52e−03
12	1.0042e−05	1.5380e−06	18	3.2730e−08	5.4593e−09	40	7.8742e−06	32	1.83e−04
14	1.4795e−06	2.3036e−07	20	6.6210e−09	8.4075e−10	60	3.5011e−06	64	1.28e−05

Example 3

Consider the following nonlinear SBVP in the study of steady-state oxygen diffusion in a spherical cell (Babolian et al. 2015; Khuri and Sayfy 2010; Wazwaz 2011; Çağlar et al. 2009):

$$u''(x) + \frac{\alpha}{x}u'(x) = \frac{\delta u(x)}{u(x)+\mu}, \quad \delta > 0, \quad \mu > 0, \tag{37}$$

subject to the boundary conditions

$$u'(0) = 0, \quad 5u(1) + u'(1) = 5, \tag{38}$$

where δ and μ are often taken as 0.76129 and 0.03119, respectively. We take the value of α as 1, 2 and 3.

The Adomian polynomials of nonlinear term $f(x,u) = \frac{\delta u(x)}{u(x)+\mu}$ in this problem are computes as

$$A_0 = \frac{\delta}{U(0)+\mu}U(0),$$

$$A_1 = \frac{\delta\mu}{(U(0)+\mu)^2}U(1),$$

$$A_2 = \frac{\delta\mu}{(U(0)+\mu)^2}U(2) - \frac{\delta\mu}{(U(0)+\mu)^3}U^2(1),$$

$$A_3 = \frac{\delta\mu}{(U(0)+\mu)^2}U(3) - \frac{2\delta\mu}{(U(0)+\mu)^3}U(1)U(2) + \frac{\delta\mu}{(U(0)+\mu)^4}U^3(1),$$

$$\vdots$$

Proceeding as before, we compute the approximate solution $u_{12,2}(x)$ for N=12 and α=2, and show a comparison of the numerical results compared to the other existing methods in Table 6, from which one can see that the results of our computations are in good agreement with those ones obtained by the SGM (Babolian et al. 2015), the BSDM (Khuri and Sayfy 2010), the VIM (Wazwaz 2011) and the BSM (Çağlar et al. 2009).

Moreover, since there is no exact solution of this problem, we instead investigate the absolute residual error functions and the maximal error remainder parameters, which are the measures of how well the numerical solution satisfies the original problem (37–38). The absolute residual error functions are

$$\left|\mathrm{ER}_{N,\alpha}(x)\right| = \left|u''_{N,\alpha}(x) + \frac{\alpha}{x}u'_{N,\alpha}(x) - \frac{\delta u_{N,\alpha}(x)}{\mu + u_{N,\alpha}(x)}\right|, \quad 0 < x \leq 1,$$

and the maximal error remainder parameters are

$$\mathrm{MER}_{N,\alpha} = \max_{0<x\leq 1}\left|\mathrm{ER}_{N,\alpha}(x)\right|.$$

In Fig. 1, we plot the absolute residual error functions $|\mathrm{ER}_{N,2}(x)|$ for N=2 through 12 by step 2. Besides, the maximal error remainder parameters $\mathrm{MER}_{N,\alpha}$ for the same N and α=1,2,3 are listed in Table 7, from which it is interesting to point out that for a given N the accuracy of our approximate solutions increases with the increase of α. Moreover, Fig. 1 and Table 7 show clearly that the accuracy of our method is getting better as the approximation level is increasing for a fixed α. The logarithm plots of the value of $\mathrm{MER}_{2,\alpha}$ through $\mathrm{MER}_{12,\alpha}$ for α=1,2,3 are displayed in Fig. 2, which demonstrates an approximately exponential rate of convergence for the obtained truncated series solutions and thus the presented method converges rapidly to the exact solution.

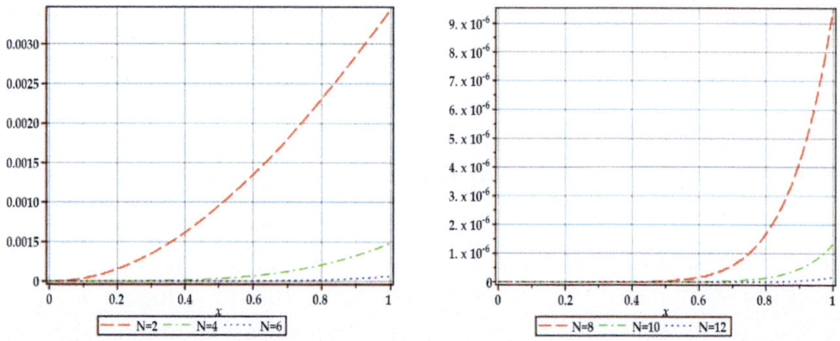

Figure 1: The absolute residual error functions $|ER_{N,2}(x)|$for N=2,4,6 (left) and 8, 10, 12 (right) of Example 3.

Table 6: Comparison of the approximate solutions for Example 3

x	**BSDM (Khuri and Sayfy 2010)**	**BSM (Çağlar et al. 2009)**	**VIM (Wazwaz 2011)**	**SGM (Babolian et al. 2015)**	**Present method**
0.0	0.8284832948	0.8284832729	0.8284832761	0.8284832912	0.8284832870
0.1	0.8297060968	0.8297060752	0.8297060781	0.8297060933	0.8297060890
0.2	0.8333747380	0.8333747169	0.8333747193	0.8333747345	0.8333747303
0.3	0.8394899183	0.8394898981	0.8394898996	0.8394899148	0.8394899106
0.4	0.8480527887	0.8480527703	0.8480527701	0.8480527859	0.8480527816
0.5	0.8590649275	0.8590649139	0.8590649108	0.8590649281	0.8590649239
0.6	0.8725283156	0.8725283084	0.8725282997	0.8725283208	0.8725283166
0.7	0.8884452994	0.8884452958	0.8884452781	0.8884453065	0.8884453023
0.8	0.9068185417	0.9068185402	0.9068185095	0.9068185490	0.9068185448
0.9	0.9276509830	0.9276509825	0.9276509392	0.9276509893	0.9276509853
1.0	0.9509457948	0.9509457946	0.9509457539	0.9509457994	0.9509457960

Table 7: The maximal error remainder parameters $MER_{N,\alpha}$ for Example 3

α	$MER_{2,\alpha}$	$MER_{4,\alpha}$	$MER_{6,\alpha}$	$MER_{8,\alpha}$	$MER_{10,\alpha}$	$MER_{12,\alpha}$
1	5.8000e−03	1.4000e−03	3.1751e−04	7.3547e−05	1.7000e−05	3.9243e−06
2	3.4000e−03	4.8431e−04	6.7761e−05	9.4474e−06	1.3142e−06	1.8267e−07
3	2.4000e−03	2.4481e−04	2.4485e−05	2.4388e−06	2.4240e−07	2.4065e−08

Example 4

Consider the following nonlinear SBVP which arises in the study of the distribution of heat sources in the human head (Pandey 1997; Rashidinia et al. 2007; Ravi and Bhattacharya 2006; Babolian et al. 2015; Khuri and Sayfy 2010; Singh and Kumar 2014; Çağlar et al. 2009; Duggan and Goodman 1986):

$$u''(x) + \frac{2}{x}u'(x) = -e^{-u(x)}, \qquad (39)$$

subject to the boundary conditions

$$u'(0) = 0, \quad au(1) + bu'(1) = 0. \qquad (40)$$

We consider the following two cases:

Case one: a=b=1.

Case two: a=0.1,b=1.

The Adomian polynomials of nonlinear term $f(x,u) = -e^{-u(x)}$ in this problem are computed as

$$A_0 = -e^{-U(0)},$$

$$A_1 = U(1)e^{-U(0)},$$

$$A_2 = U(2)e^{-U(0)} - \frac{1}{2}U^2(1)e^{-U(0)},$$

$$A_3 = U(3)e^{-U(0)} - U(1)U(2)e^{-U(0)} + \frac{1}{6}U^3(1)e^{-U(0)},$$

$$\vdots$$

Again no exact solution exists for this equation, hence it was handled numerically. Table 8 describes the numerical results of the first case obtained by the proposed method at the order of approximation N=12 and the other existing methods, including the FDM (Pandey 1997), the non-polynomial cubic spline method (NPCSM) (Rashidinia et al. 2007), the CSM (Ravi and Bhattacharya 2006) and the SGM (Babolian et al. 2015). Meanwhile, a comparison for the approximate solutions of the second case obtained by the present method with the same approximation level as the first case and the previous existing methods which include the CSM (Ravi and Bhattacharya 2006), the SGM (Babolian et al. 2015), the BSDM (Khuri and Sayfy 2010) and the BSM (Çağlar et al. 2009) is presented in Table 9. One can seen from two Tables that our computations are in good line with the results obtained by the other approaches compared. In fact, at the approximation level for N=12, the maximal absolute error is found to be order of magnitude $O(10^{-7})$ for the first case, and $O(10^{-9})$ for the second case.

Table 8: Comparison of the numerical results for the first case of Example 4

x	FDM (Pandey 1997)	NPCSM (Rashidinia et al. 2007)	CSM (Ravi and Bhattacharya 2006)	SGM (Babolian et al. 2015)	Present method
0.0	0.3675169710	0.3675181074	0.3675179806	0.3675168124	0.3675167997
0.1	0.3663623697	0.3663637561	0.3663634922	0.3663623265	0.3663623137
0.2	0.3628941066	0.3628959378	0.3628952219	0.3628940634	0.3628940507
0.3	0.3570975862	0.3570991429	0.3570986892	0.3570975430	0.3570975301
0.4	0.3489484612	0.3489499903	0.3489495462	0.3489484178	0.3489484049
0.5	0.3384121893	0.3384136581	0.3384132502	0.3384121459	0.3384121330
0.6	0.3254435631	0.3254450019	0.3254445925	0.3254435196	0.3254435063
0.7	0.3099860810	0.3099878567	0.3099870705	0.3099860373	0.3099860240
0.8	0.2919711440	0.2919789654	0.2919720836	0.2919711001	0.2919710864
0.9	0.2713170512	0.2713185637	0.2713179289	0.2713170072	0.2713169936
1.0	0.2479277646	0.2479292837	0.2479285659	0.2479277203	0.2479277073

Table 9: Comparison of the numerical results for the second case of Example 4

x	CSM (Ravi and Bhattacharya 2006)	BSM (Çağlar et al. 2009)	BIDM (Khuri and Sayfy 2010)	SGM (Babolian et al. 2015)	Present method
0.0	1.147041084	1.147039937	1.147040795	1.147039016	1.147039019
0.1	1.146511706	1.146510559	1.146511419	1.146509639	1.146509642
0.2	1.144922563	1.144921418	1.144922282	1.144920499	1.144920502
0.3	1.142270622	1.142269478	1.142270348	1.142268560	1.142268563
0.4	1.138550801	1.138549661	1.138550539	1.138548745	1.138548748
0.5	1.133755950	1.133754813	1.133755703	1.133753900	1.133753904
0.6	1.127876795	1.127875663	1.127876562	1.127874754	1.127874756
0.7	1.120901889	1.120900762	1.120901665	1.120899858	1.120899860
0.8	1.112817535	1.112816416	1.112817317	1.112815517	1.112815520
0.9	1.103607704	1.103606593	1.103607490	1.103605701	1.103605704
1.0	1.093253927	1.093252826	1.093253716	1.093251942	1.093251944

Example 5

Consider the following SBVP with nonlinear term different from the forms (4–7) which arises in the radial stress on a rotationally symmetric shallow membrane cap (Singh and Kumar 2014; Ravi and Aruna 2010):

$$u''(x) + \frac{3}{x}u'(x) = \frac{1}{2} - \frac{1}{8u^2(x)}, \qquad (41)$$

subject to the boundary conditions

$$u'(0) = 0, \quad u(1) = 1. \qquad (42)$$

The Adomian polynomials of nonlinear term $f(x,u) = \frac{1}{2} - \frac{1}{8u^2(x)}$ in this problem are computed as

$$A_0 = \frac{1}{2} - \frac{1}{8U^2(0)},$$

$$A_1 = \frac{1}{4}\frac{U(1)}{U^3(0)},$$

$$A_2 = -\frac{3}{8}\frac{U^2(1)}{U^4(0)} + \frac{1}{4}\frac{U(2)}{U^3(0)},$$

$$A_3 = \frac{1}{2}\frac{U^3(1)}{U^5(0)} - \frac{3}{4}\frac{U(1)U(2)}{U^4(0)} + \frac{1}{4}\frac{U(3)}{U^3(0)},$$

$$\vdots$$

Like the previous problems 3 and 4, a closed-form solution to this equation can not be written down. So we instead investigate the absolute residual error functions and the maximal error remainder parameters to examine the accuracy and the reliability of our numerical results. Here, the absolute residual error functions are

$$|\mathrm{ER}_N(x)| = \left| u_N''(x) + \frac{3}{x}u_N'(x) - \frac{1}{2} + \frac{1}{8u_N^2(x)} \right|, \quad 0 < x \le 1,$$

and the maximal error remainder parameters are

$$\mathrm{MER}_N = \max_{0<x\le 1} |\mathrm{ER}_N(x)|.$$

In Fig. 3, we plot the absolute residual error functions $|\mathrm{ER}_N(x)|$ for N=4 through 14 by step 2. The logarithm plot for the maximal error remainder parameters MER_N for the same N is shown in Fig. 4, which demonstrates an approximately exponential rate of convergence of the obtained truncated series solutions and thus the presented method converges rapidly to the exact solution. Even though there is no exact solution for this problem, the following 10th order approximation has an accuracy of $O(10^{-8})$ and can be used for practical applications

$$u_{10}(x) = 0.9541353070 + (0.4533672772e-1)x^2 + (0.5436871104e-3)x^4 - (0.1611538997e-4)x^6 + (0.3997114810e-6)x^8 - (0.6144814593e-8)x^{10}.$$

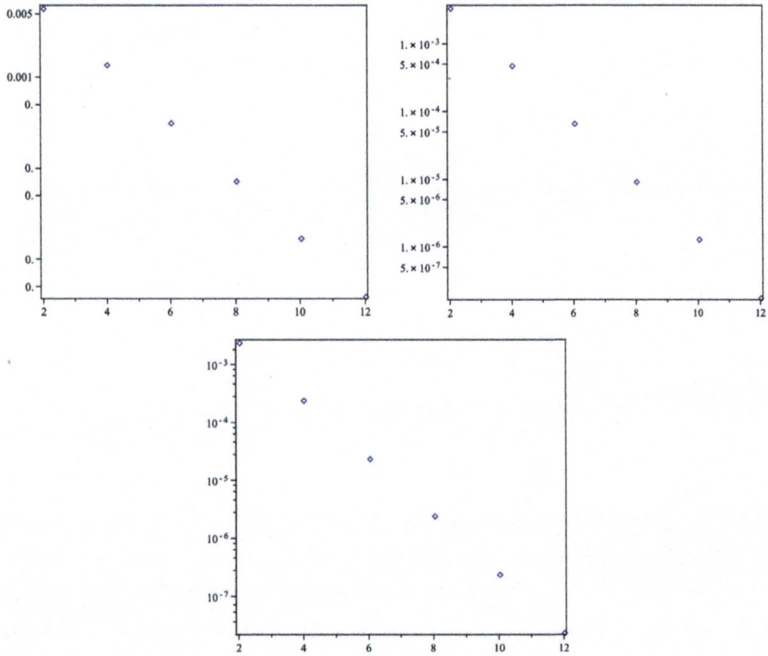

Figure 2: The logarithmic plots for the maximal error remainder parameters $MER_{N,\alpha}$ for N=2 through 12 by step 2 and α=1 (up, left), α=2 (up, right), α=3 (down) of Example 3.

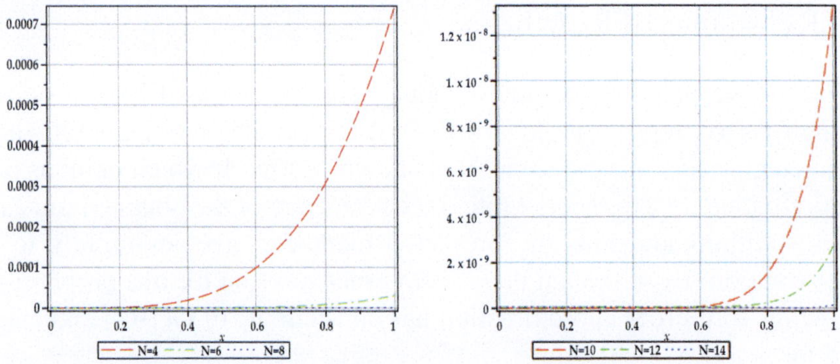

Figure 3: The absolute residual error functions $|ER_N(x)|$ for N=4,6,8 (left) and 10, 12, 14 (right) of Example 5.

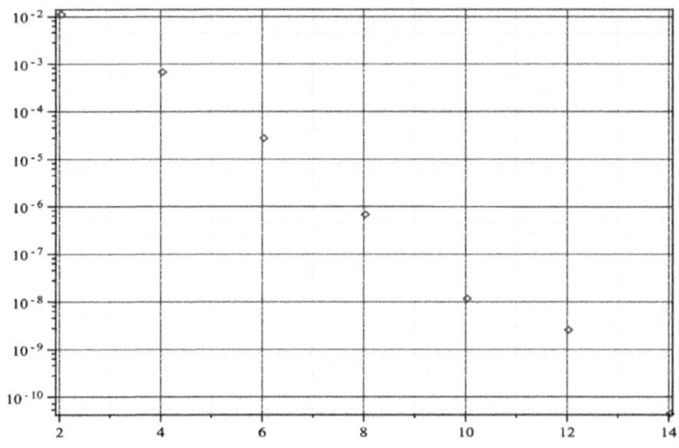

Figure 4: The logarithmic plot for the maximal error remainder parameters MER_N for N=2 through 14 by step 2 of Example 5.

CONCLUSION

In this work, a reliable approach based on the IDTM is presented to handle the numerical solutions of a class of nonlinear SBVPs arising in various physical models. This scheme takes the form of a truncated series with easily computable coefficients via the Adomian polynomials of those nonlinearities in the given problem. With the proposed algorithm, there is no need of discretization of the variables, linearization or small perturbation. Numerical results show that the proposed method works well for the SBVPs (1–3) with a satisfying low error. Besides, it is obvious that evaluation of more components of the approximate solution will reasonably improve the accuracy of truncated series solution by using the proposed method. Comparisons of the results reveal that the present method is very effective and accurate. Moreover, we are convinced that the IDTM can be extended to solve the other type of functional equations involving nonlinear terms more easily as the Adomian polynomials are applicable for any analytic nonlinearity and can be generated quickly with the aid of the algorithm proposed by Duan.

It is necessary to point out that algebraic Eq. (23) is a nonlinear one, and we shall inevitably encounter the bad roots while solving it. The criterion to separate the good root from a swarm of bad ones is convergence because it represents the value of unknown function at the origin and will not change for the different N.

AUTHORS' CONTRIBUTIONS

All authors contributed equally in this article. They read and approved the final manuscript.

ACKNOWLEGEMENTS

This work was supported by the Scientific Research Fund of Zhejiang Provincial Education Department of China (No.Y201430940) and K.C. Wong Magna Fund in Ningbo University.

REFERENCES

1. Abbaoui K, Cherruault Y, Seng V (1995) Practical formulae for the calculus of multivariable Adomian polynomials. Math Comput Modell 22(1):89–93
2. Abdelwahid F (2003) A mathematical model of Adomian polynomials. Appl Math Comput 141(2–3):447–453
3. Adomian G (1990) A review of the decomposition method and some recent results for nonlinear equations. Math Comput Modell 13(7):17–43
4. Adomian G (1994) Solving frontier problems of physics: the decomposition method. Kluwer Academic, Dordrecht
5. Adomian G, Rach R (1983) Inversion of nonlinear stochastic operators. J Math Anal Appl 91(1):39–46
6. Ames WF (1968) Nonlinear ordinary differential equations in tranport process. Academic press, New York
7. Anderson N, Arthurs AM (1985) Analytical bounding functions in a spherical cell with Michaelis–Menten oxygen uptake kinetics. Bull Math Biol 47(1):145–153
8. Azreg-AÏnou M (2009) A developed new algorithm for evaluating Adomian polynomials. CMES-Comput Model Eng 42(1):1–18
9. Babolian E, Eftekhari A, Saadatmandi A (2015) A Sinc-Galerkin technique for the numerical solution of a class of singular boundary value problems. Comp Appl Math 34(1):45–63
10. Çağlar H, Çağlar N, Özer M (2009) B-spline solution of non-linear singular boundary value problems arising in physiology. Chaos Soliton Fract 39(3):1232–1237
11. Chandrasekhar S (1939) An introduction to the study of stellar strcture. Dover, New York
12. Chang SH (2012) Electroosmotic flow in a dissimilarly charged slit microchannel containing salt-free solution. Eur J Mech B Fluid 34:85–90
13. Chang SH (2014) Taylor series method for solving a class of nonlinear singular boundary value problems arising in applied science. Appl Math Comput 235(25):110–117
14. Chang SH, Chang IL (2008) A new algorithm for calculating one-dimensional differential transform of nonlinear functions. Appl Math

Comput 195(2):799–808

15. Chawla MM, Subramanian R, Sathi HL (1988) A fourth order method for a singular two-point boundary value problem. BIT 28(1):88–97
16. Duan JS (2010) Recurrence triangle for Adomian polynomials. Appl Math Comput 216(4):1235–1241
17. Duan JS (2010) An efficient algorithm for the multivariable Adomian polynomials. Appl Math Comput 217(6):2456–2467
18. Duan JS (2011) Convenient analytic recurrence algorithms for the Adomian polynomials. Appl Math Comput 217(13):6337–6348
19. Duggan RC, Goodman AM (1986) Pointwise bounds for a nonlinear heat conduction model of the human head. Bull Math Biol 48(2):229–236
20. Elsaid A (2012) Fractional differential transform method combined with the Adomian polynomials. Appl Math Comput 218(12):6899–6911
21. Fatoorehchi H, Abolghasemi H (2013) Improving the differential transform method: a novel technique to obtain the differential transforms of nonlinearities by the Adomian polynomials. Appl Math Modell 37(8):6008–6017
22. Flesch U (1975) The distribution of heat sources in the human head: a theoretical consideration. J Theor Biol 54(2):285–287
23. Gökdoğan A, Merdan M, Yildirim A (2012) The modified algorithm for the differential transform method to solution of Genesio systems. Commum Nonlinear Sci 17(1):45–51
24. Gray BF (1980) The distribution of heat sources in the human head–theoretical considerations. J Theor Biol 82(3):473–476
25. Hiltmann P, Lory P (1983) On oxygen diffusion in a spherical cell with Michaelis–Menten oxygen uptake kinetics. Bull Math Biol 45(5):661–664
26. Khuri SA, Sayfy A (2010) A novel approach for the solution of a class of singular boundary value problems arising in physiology. Math Comput Modell 52(3–4):626–636
27. Kumar M, Singh N (2010) Modified Adomian decomposition method and computer implementation for solving singular boundary value problems arising in various physical problems. Comput Chem Eng 34(11):1750–1760

28. Lin SH (1976) Oxygen diffusion in a spherical cell with nonlinear oxygen uptake kinetics. J Theor Biol 60(2):449–457
29. McElwain DLS (1978) A re-examination of oxygen diffusion in a spherical cell with Michaelis–Menten oxygen uptake kinetics. J Theor Biol 71(2):255–263
30. Momani S, Ertürk VS (2008) Solutions of non-linear oscillators by the modified differential transform method. Comput Math Appl 55(4):833–842
31. Odibat ZM, Bertelle C, Aziz-Alaoui MA, Duchamp GHE (2010) A multi-step differential transform method and application to non-chaotic or chaotic systems. Comput Math Appl 59(4):1462–1472
32. Pandey RK (1997) A finite difference method for a class of singular two point boundary value problems arising in physiology. Int J Comput Math 65(1–2):131–140
33. Pukhov GE (1980) Differential transforms of functions and equations. Naukova Dumka, Kiev
34. Pukhov GE (1982) Differential transforms and circuit theory. Int J Circ Theor Appl 10:265–276
35. Pukhov GE (1986) Differential transformations and mathematical modeling of physical processes. Naukova Dumka, Kiev
36. Rach R (1984) A convenient computational form for the Adomian polynomials. J Math Anal Appl 102(2):415–419
37. Rach R (2008) A new definition of the Adomian polynomials. Kybernetes 37(7):910–955
38. Rashidinia J, Mohammadi R, Jalilian R (2007) The numerical solution of non-linear singular boundary value problems arising in physiology. Appl Math Comput 185(1):360–367
39. Ravi Kanth ASV, Aruna K (2010) He's variational iteration method for treating nonlinear singular boundary problems. Comput Math Appl 60(3):821–829
40. Ravi Kanth ASV, Bhattacharya V (2006) Cubic spline for a class of non-linear singular boundary value problems arising in physiology. Appl Math Comput 174(1):768–774
41. Russell RD, Shampine LF (1975) Numerical methods for singular boundary value problems. SIAM J Numer Anal 12(1):13–36
42. Singh R, Kumar J (2014) An efficient numerical technique for the

solution of nonlinear singular boundary value problems. Comput Phys Commun 185(4):1282–1289

43. Turkyilmazoglu M (2013) Effective computation of exact and analytic approximate solutions to singular nonlinear equations of Lane-Emden-Fowler type. Appl Math Modell 37(14–15):7539–7548
44. Wazwaz AM (2000) A new algorithm for calculating Adomian polynomials for nonlinear operators. Appl Math Comput 111(1):33–51
45. Wazwaz AM (2011) The variational iteration method for solving nonlinear singular boundary value problems arising in various physical models. Commun Nonlinear Sci Numer Simulat 16(10):3881–3886
46. Wazwaz AM, Rach R, Duan JS (2013) Adomian decomposition method for solving the Volterra integral form of the Lane-Emden equations with initial and boundary conditions. Appl Math Comput 219(10):5004–5019
47. Xie LJ, Zhou CL, Xu S (2016) A new algorithm based on differential transform method for solving multi-point boundary value problems. Int J Comput Math 93(6):981–994
48. Zhou JK (1986) Differential transformation and its applications for electrical circuits. Huazhong University Press, Wuhan (in Chinese)

INDEX

D

E

F

G

H

I

J

K

L

M

N